International Energy Agency

WORKSHOPS ON ENERGY SUPPLY AND DEMAND

ORGANISATION FOR ECONOMIC CO-OPERATION AND DEVELOPMENT
1978

The International Energy Agency (IEA) is an autonomous body which was established in November 1974 within the framework of the Organisation for Economic Co-operation and Development (OECD) to implement an International Energy Program.

It carries out a comprehensive programme of energy co-operation among nineteen* of the OECD's twenty four Member countries. The basic aims of IEA are:

 i) cooperation among IEA Participating countries to reduce excessive dependence on oil through energy conservation, development of alternative energy sources and energy research and development,

 ii) an information system on the international oil market as well as consultation with oil companies,

 iii) cooperation with oil producing and other oil consuming countries with a view to developing a stable international energy trade as well as the rational management and use of world energy resources in the interest of all countries,

 iv) a plan to prepare participating countries against the risk of a major disruption of oil supplies and to share available oil in the event of an emergency.

* IEA Member Countries: Austria, Belgium, Canada, Denmark, Germany, Greece, Ireland, Italy, Japan, Luxembourg, Netherlands, New Zealand, Norway, Spain, Sweden, Switzerland, Turkey, United Kingdom, United States.

The Organisation for Economic Co-operation and Development (OECD) was set up under a Convention signed in Paris on 14th December, 1960, which provides that the OECD shall promote policies designed:

— to achieve the highest sustainable economic growth and employment and a rising standard of living in Member countries, while maintaining financial stability, and thus to contribute to the development of the world economy;

— to contribute to sound economic expansion in Member as well as non-member countries in the process of economic development;

— to contribute to the expansion of world trade on a multilateral, non-discriminatory basis in accordance with international obligations.

The Members of OECD are Australia, Austria, Belgium, Canada, Denmark, Finland, France, the Federal Republic of Germany, Greece, Iceland, Ireland, Italy, Japan, Luxembourg, the Netherlands, New Zealand, Norway, Portugal, Spain, Sweden, Switzerland, Turkey, the United Kingdom and the United States.

* *

CONTENTS

INTRODUCTION

This book offers twenty papers first presented at two Workshops, one on energy supply in November 1976 and the other on demand in December 1977. Both seminars were held in Paris under the auspices of the International Energy Agency. These meetings were organised with the purpose of bringing together recognised industry, government and academic researchers to exchange their latest work and impressions based on analytical or empirical research in energy supply and demand. The variety, scope and quality of the papers give evidence of the dynamism of the present research effort that is being directed at understanding the transition of the international energy market to a new equilibrium.

The reader will note the diversity of topics addressed under the broad subject of energy supply and demand. This arises out of the desire of the organisers of the Workshops to pit academic researchers against industry planners, as well as to afford an opportunity to researchers from the major regions of the industrial countries that are members of IEA. All this had to be achieved while holding the size of the group to twenty or so in order to allow an easy exchange. Arranging these papers in a suitable way for publication proved more difficult than talking them out in the Workshops; the juxtaposition of theoretical and empirical approaches and global and country-specific viewpoints went far better for the participants than it may for readers.

It was decided at last to order the papers on the basis of the approach chosen by the authors. Papers that are mainly concerned with the elaboration of methods or analytical techniques are presented first, followed by the papers that are largely empirical in approach. Where practical, the papers are further ordered by geographical areas. Wherever it was important to do so, the papers were updated by the authors in late 1978 to take account of changes that occurred since the presentation at the Workshops. Further gaps occur because not all authors found it possible to put their papers in final form.

The papers are gathered into three sections of equal length: seven supply papers - all empirical in approach, six theoretically-oriented

demand papers and seven empirical ones. It should be no surprise that
the papers on energy supply - generally the province of investigators
studying physical properties of specific producing regions - are all
categorised as empirical while those treating energy demand, where
behavioural relations are better understood and are the frequent subject
of study by economists, have a theoretical bent.

Three papers treat North American energy supply. Mr. Ball discusses
supply decision-making in the oil industry in the United States, with
specific reference to the synfuels. Professor Gordon tackles the complex
supply questions surrounding United States coal supply. Mr. Mattinson
reviews the oil supply potential of the Canadian frontier regions.

Professor Robinson in his "Financial Analysis of Oil Production"
looks into the economics of North Sea oil as related to the United
Kingdom's energy balance and balance of payments. Professor Odell is
also concerned with the North Sea but in the larger perspective of
Western Europe's energy independence. Mr. Querol's contribution investigates the supply prospects offered by the oil industry in Spain. The
Japanese energy situation is placed in a global energy context by
Mr. Ikuta. Unfortunately, several important papers presented at the
supply Workshop on the behaviour of the nuclear power industry in the
United States, and supply issues of Middle Eastern oil, were not made
available for publication herein.

In the first set of papers on energy demand there is a greater
application of economic theory. Professor Waverman elaborates his model
to explain in detail how industries adjust over time in response to
higher energy prices and policy changes. Professor Sweeney presents a
model which he applied to analyse changes in policies affecting gasoline
consumption. Professor Berndt investigates the effects of higher energy
prices on employment, investment and economic growth using pooled, cross-
country data within the OECD regions. More general analysis is found
in Professor Eden's "Methodology for Studying World Energy Demand". A
more empirical flavour is found in Dr. Bernardini's sector by sector
examination in the rapid increase in the ratio of energy to the gross
national product in Italy since World War II. More methodological, but
with important implications about energy efficiency, is Mr. Frisch's
proposal on the valuation of electricity in final energy consumption in
energy balances.

Professor Griffin's results were based upon a model of demand for
total energy and for individual fuels in end-use sectors in 18 OECD
countries, employing a trans-log function - an approach found in several
papers presented at the Workshops. Projections are made to the year 2000

using five different scenarios. The same subject is developed differently for the EEC by Mr. Waeterloos with projection to 1990. Other multi-national studies are Mr. Choe's demand estimates for non-OPEC developing countries and Mr. Da Silva's analysis of energy consumption in Latin America. Mr. Choe's work regresses per capita commercial energy demand against per capita GDP, the real price of energy and per capita commercial energy demand in the preceding year. The study is based upon time-series-cross-section analysis of data for 35 developing countries over the period 1965 to 1975. In Mr. Da Silva's paper, four simultaneous equations determines oil demand, non-oil imports, investment and gross national product. Simulations to 1990 for each Latin American country were made.

The remaining three demand papers are concerned with particular countries. Mr. Bain presents some projections of Canadian energy demand, discusses the methods used and investigates the possibilities for energy savings. Mr. Kraft discusses, in two scenarios, the development to 1990 of energy consumption in Western Germany. Finally, Mr. Schmitt, also dealing with the Federal Republic of Germany presents projections made jointly by three research institutes in Germany, with the stress on nuclear power and coal.

The Workshops allowed the Secretariat staff and government experts to sample the richness and variety of today's research on energy supply and demand by offering a sample of different kinds of research techniques being employed, and to hear contrasting viewpoints and judgements on the results obtained and their policy significance. It was also pleasing to hear how many of the participants appreciated the opportunity to meet not only a multinational group but one composed of researchers with widely divergent work environments. It is hoped that readers of these papers can share some of that experience.

ENERGY SUPPLY

AUTHORS AND PARTICIPANTS IN WORKSHOP ON ENERGY SUPPLY

Mr. B.C. Ball, Jr.
  retired Vice-President of Gulf Oil Corporation;
  now President of Ball & Associates of Cambridge, Mass
  & Prof. Adjunct of Management & Engineering at M.I.T. Mass. USA

Mr. T. Espedal
  Statoil
  Norway

Prof. R. L. Gordon
  Dept. of Mineral Economics
  Pennsylvania State University
  USA

Mr. T. Ikuta, President
  Institute of Energy Economics
  Tokyo, Japan

Prof. H. Jacoby
  M.I.T.
  Department of Economics
  Cambridge, Mass. USA

Prof. P. Joskow
  M.I.T.
  Dept. of Economics
  Camb. Mass. USA

Dr. M. Liebrucks
  Economic Research Institute for Germany
  Berlin

Mr. M. Lovegrove
  Wood McKenzie & Co.
  Edinburgh, Scotland

Mr. C. Mattinson
  Consultant
  Shell Canada Resources Ltd.
  Alberta Canada

Prof. P. Odell
  Erasmus University
  Rotterdam, Netherlands

Prof. E. Penrose
  INSEAD
  Fontainebleau, France

Mr. R. Querol
  Hispanoil
  Spain

Prof. C. Robinson
  Economics Dept.
  University of Surrey
  England

Mr. A. Scanlon
  BP
  London, England

Prof. H. Schneider
  Direkton des Energiewirtschaftlichen
  Cologne University
  Germany

Mr. D. Sherman
  Mobil Oil Corp.
  New York, USA

Mr. W. Timms
  International Bank for Reconstruction & Development
  Washington, D.C.

Mr. D. Wade
  Shell
  London, England

SYNFUEL SUPPLY ISSUES FROM THE COMMERCIAL PERSPECTIVE: NEW KINDS OF DECISIONS

by

Ben C. Ball, Jr.

One of the reasons for the complexity of the "Energy Crisis" is that the contexts for decision making by vitally affected groups (e.g., producers, consumers, public, industry, governments, etc.) impinge with little overlap. Resulting is an adversarial procedure comparable in magnitude (if not significance) to issues of world prosperity and world peace. One of the least understood facets is the role of the private sector's decisions, how they are made, and how the decision-making procedure itself is adapting to maintain authenticity and viability.

Unprecedented in history is a public demand that the petroleum industry with investments in the hundreds of billion of dollars largely replace itself in less than a generation with synfuels. Many industries in the past have been replaced, but by "natural selection". Today, we are talking instead of what amounts to "biological engineering" in the industrial sector on a massive scale, and in a basic and crucial area.

In addition to that, in the Western World and especially in the United States, we are talking in terms of an approach never before undertaken: solution of a long-term economic problem through government guidance of the free marketplace. The United States Government has traditionally played the role of "policeman", to keep the marketplace "fair", and of "fireman", to deal with crises. We have yet to understand how new and untried we are at having the Government determine and carry out such a thing as an "energy policy". No matter how desirable it may be, we must realise that we are without the tools for implementation, the analytical skills for good planning, or even the conceptual framework within which to operate.

But, we shall turn again to these points, for this paper will deal with the new demands placed upon the decision making processes of the energy supplying corporations by the changing commercial dynamics of the

energy sector, by the new kinds of policy issues which result, and by
the new dimensions of uncertainty which must be dealt with. To put this
all into context, it is worth our while to deal with the dynamics which
brought us to where we are.

HISTORICAL AND ENVIRONMENTAL PERSPECTIVE

The current energy situation is a logical outcome of trends in the
world energy market over the past half century. It does not reflect any
sudden change in the relationship between man and his natural environment - although the trends may be building pressures for more substantial
adjustments than in the past. The most fundamental and rapid changes
appear to have been political - representing basic changes in property
rights and the incentives affecting the various actors in the petroleum
marketplace. The combined effect of the economic development and political changes may have substantial impact on companies in the energy
business.

The first eighty years

The early history of the petroleum industry was marked by successive
gluts and shortages of oil, wide price fluctuations, and precarious
existences for many in the industry. "Energy crises" - fear that the
United States or the world was running out of oil - have occurred in
every generation: in the 1850s the fear was that decline in whale oil
production would plunge people into darkness; in the later 1800s, as
Pennsylvania oil ran low, it was feared that petroleum was running out;
again in the early 1920s there was a real fear that the United States
was running out of oil. Each time, new discoveries were made, and the
shortage soon became a glut. The thrust of Rockefeller's Standard Oil
Trust was to gain wide enough control of supplies at all stages to smooth
out the swings and, particularly, to prevent the periodic drastic price
declines. Despite the breakup of the Trust, much of the later history
of the petroleum industry, both domestically and abroad, has been marked
by attempts to control surplus production and maintain stable prices.

Production outside the United States began in Mexico in 1910, in
Venezuela in the 1920s. Oil was found in the Middle East in the late
1920s, but significant production did not occur until the late 1940s.
As late as 1938, the United States accounted for 68 per cent of world
production, with Russia and Venezuela the next two largest producers.

From the late 1920s onwards there has been a persistent surplus production relative to world demand, punctuated by a few crises or cutoffs.

Internationally, the industry was highly concentrated, with few
antitrust prohibitions, encouraging attempts at cartel action to control
production and prices. In the late 1920s, 50 per cent of sales outside
the United States were controlled by three companies: Jersey Standard
(23 per cent), Shell (16 per cent), and Anglo-Persian (11.5 per cent);
another 6.5 per cent was controlled by the Russians, and the rest mostly
by large United States corporations. In the Middle East, the "Red Line
Agreement" was signed in October, 1927, to curb competition - although
several companies were left out and subsequently discovered vast pools
of oil elsewhere in the Middle East, leading to the demise of the agreement in 1948. In 1928, the "Pool Association" was formed among the
larger international companies with the intent of specifying market
shares, reducing capital investment, and co-ordinating world prices
(through "Gulf plus" pricing based on the price of crude from the Gulf
of Mexico). The attempt was too ambitious, as some companies did not
choose to participate, and it only worked well in a few isolated markets
to stabilize prices. Although the OPEC cartel currently appears to
possess far greater market power than the oil companies of pre-World
War II, prognoses for their long-term success should be made against the
background of repeated but not very successful attempts to organise
world markets and restrict output.

In the United States, the instability of the industry was addressed
during the Great Depression. State controls were imposed on production
and the Connolly "Hot Oil" Act of 1934 assured their general success.
Although not all states adopted and strictly enforced prorationing, the
major producing states did. The result has been a higher and more
stable United States price than that of world markets.

The post-World War II world petroleum situation

The post-war period has seen world production, consumption, and
reserves grow at an extremely rapid rate - considerably faster than
within the United States. As shown below, world production multiplied
by a factor of five from 1950 to 1974, compared to less than a doubling
within the United States. World reserves increased by a factor of eight
in the same period, compared to a United States increase of only one-
and-a-half times. The rapid production increase was due not only to
increased energy consumption per se, but also to a relative increase in
consumption of oil and natural gas: from 1960 to 1972 they increased
from 48 per cent to 66 per cent of total world energy consumption.

CRUDE OIL PRODUCTION
Thousand of Bbls.

	1950	1960	1970	1974
United States	1,973,574	2,574,933	3,517,450	3,199,328
Total World	3,803,027	7,674,460	16,718,708	20,518,139

ESTIMATED PROVED RESERVES OF CRUDE OIL
Thousand of Bbls.

	1950	1960	1970	1974
United States	24,649,489	31,719,347	29,631,862	35,299,839
Total World	76,452,739	290,035,071	530,524,067	626,706,479

Source: API Data Book

Despite the rapid increase in consumption, and the depletable nature of world energy resources, the period has been characterised by almost continuous excess productive capacity and an increasing ratio of reserves to production. The current energy crisis notwithstanding, these under-lying economic facts have not changed significantly in the past couple of years. The current excess productive capacity will exert a pressure downwards on prices - how successfully that pressure can be resisted in the long run remains to be seen.

Events leading to the current energy crisis represent not a basic shift in the relationship between man and nature, but rather a shift of ownership and power among men. Outside of the United States, the major consuming nations have never been the major producing ones. The post-war period has seen a shift, slow at first but very rapid in the last five years, of control of production from companies of the consumer nations to the producing nations themselves. As early as 1948, Venezuela introduced an income tax law to take 50 per cent of the profits from operations in that country. In 1950, Saudi Arabia introduced the prac-tice of using posted prices as a basis for taxation. In 1951, Iran nationalised oil production - with disasterous results for Iran. By 1957 the Italian company, ENI, was willing to work on a real partnership basis with the National Iranian Oil Company. OPEC was formed by Iran, Iraq, Kuwait, Saudi Arabia, and Venezuela in 1960 to protest a reduction in posted prices (resulting from world oversupply and the effect of

United States import quotas). OPEC power grew slowly: first there were discussions of royalties and market allowances; then it was agreed to eliminate all discounts on gravity by 1975; in June, 1967, the United States and Britain were embargoed because of their support for Israel; in June, 1968, OPEC adopted a resolution that producer governments have the right to participate in ownership and to change terms of existing concessions or contracts under a doctrine of "changing circumstances." In 1970, Libya pushed for higher posted prices and finally forced the operating companies to cut production until an agreement was reached. The other OPEC nations followed suit, and new agreements were reached in 1971, only to be altered in 1973 and the years since. The first participation agreement went into effect in 1973. By early 1976 most producing countries were approaching 100 per cent participation, or nationalisation.

The precise history of price increases following the 1973 embargo need hardly be recounted here. In essence, by significantly cutting production, the OPEC nations have managed to increase prices by a factor of five and at the same time to considerably reduce the amount of profits attributable to the oil companies.

The growing power of the producing nations is mirrored by fundamental changes in the positions of the major oil companies. In 1950, the world market outside of North America and several regions producing primarily for their own consumption was supplied largely by eight large companies - Exxon, BP, Shell, Gulf, Texaco, SoCal, Mobil, and CFP. By 1969, their share had been reduced to just over 80 per cent, still substantial, but allowing for large expansion of new firms and national oil companies due to the rapid growth of consumption. The trend is increasing for national oil companies to take over control in the producing countries, leading to less concentration (apart from cartel actions) in the world market and a change in role for the major oil companies).

RESOURCE AND MARKET TRENDS

United State reserves outside of Alaska peaked in 1962 and have declined since then, but only significantly since 1970. The addition of Alaskan reserves in 1971 provided a 30 per cent jump in reserves in that year, from which point total United States reserves have declined 10 per cent in four years. United States production peaked in 1970, has declined slightly since then, but will increase when the Alaskan oil is produced.

Imports have risen steadily since World War II, both absolutely and as a percentage of United States production, but have spurted since 1970 as shown below.

SHARE OF IMPORTS IN UNITED STATES DOMESTIC PETROLEUM DEMAND
Thousand of Bbls.

	1950	1960	1970	1974(Prel.)
United States Imports	310,261	664,111	1,248,062	2,222,179
Domestic Demand for Ref. Prod.	2,392,974	3,585,820	5,364,473	6,060,471
Imports as % of Demand	13.0	18.5	23.3	36.6

Source: API Data Book

Exploration activity, both in terms of wells drilled and real dollars spent, has declined slowly since the mid-50s to about half its former level. On the one hand, this reflects the availability until recently of lower cost sources overseas which reduced the incentive to explore, and increasing efficiency of exploration activities so that current reserve to production ratios (11.4 average for 1970-74) are not much below those of the 1950s (12.7 average for 1950-54). This would seem to bode well for increased exploration in the next decade. On the other hand, costs of exploration and production are rising as oil must be sought in ever more hostile environments, much of the country has already been explored and depleted, and activities are ever more tied to government leasing timetables, decreasing the initiative to explore and the ability to spread risks by the oil companies. In any case, it is clear that United States petroleum resources are substantially less plentiful and higher cost than in the world generally.

Domestic production of oil and gas may expand considerably to re-place imports so long as the foreign price stays high. But given the world surplus of oil, the risk remains that domestic producers of high cost petroleum can be undercut by foreign supplies. Domestic producers would then have to rely on United States government protection - a commit-ment to high cost energy.

INDUSTRY STRUCTURE AND PERFORMANCE

The domestic energy industry is enormous by almost any measure. It is also one of the least concentrated industries in the United States. There are more than 10,000 firms involved in some aspect of petroleum

exploration and production; over 3,700 producing natural gas; over 130
in refining; several hundred in transportation; and several hundred
thousand in marketing (including service stations). There are over
3,800 coal producers and 260 companies owning or producing uranium.
These activities are dominated by several dozen very large, fully integrated companies - usually referred to as "majors". However, in 1972 no
single company produced more than 7 per cent of the energy consumed in
the United States, 11 per cent of the crude oil, or 12 per cent of coal.

Large amounts of capital are required, but there are no insur-
mountable barriers to entry and growth. A number of relatively recent
entrants (e.g. Amerada Hess and Commonwealth) now rank among the top
20 oil companies. At least 13 companies have entered the refining
business since 1960.

Concentration levels of various energy activities are given in
Table I. In petroleum, the greatest concentration is in the ownership
of reserves and interestingly enough, the least is in production. Moreover, the trend towards concentration has been greatest in production.
Thus, in the mid-50s the major oil companies were buying crude from small
producers to keep their refineries running. A number of small producers
have been purchased since then. It appears that domestically the major
oil companies were expanding upstream (and diversifying into chemicals,
other energy sources, etc.) at the same time they were integrating downstream overseas. Only in the late 60s did United States crude self-
sufficiency reach 100 per cent for Gulf, Exxon and Texaco, but it has
declined since then.

Independents have increased their share of the accelerated exploration and development activity in the United States. Since 1969, independents have drilled 9 out of 10 new field wildcat wells and made 75 per
cent of the new discoveries, although some were joint efforts with
majors. In 1973, independents' portion of exploratory success was up
8.9 percentage points from 1972 while shares of both majors and drilling
funds fell. The independents' success continued in 1974 as they drilled
85.2 per cent of the exploratory wells and completed 78 per cent of
the discoveries. While independents make about 75 per cent of the discoveries; they find about one-half of the oil and gas. Independents
have been increasingly active offshore.

Long-run profitability of the major petroleum companies has been
about the same as for other industrial United States companies. During
the period 1951 to 1971 the average return on stockholder's equity for
the eight major petroleum companies was below that of Moody's 125
industrials for 16 of the 21 years. Profits increased substantially

Table I

CONCENTRATION LEVELS OF FIRMS IN
DOMESTIC UNITED STATES ENERGY ACTIVITIES

% of Proved Domestic Crude Reserves

	1970
Top 4	37.17
Top 8	63.88
Top 15	85.67
Top 20	93.55

% of United States Crude Oil Production

	1955	1960	1970	Increase 1955-70
Top 4	18.8	26.5	30.5	62.2
Top 8	31.1	43.8	50.1	61.1
Top 15	41.7	NA	65.1	56.1
Top 20	46.2	63.0	68.2	48.0

% of United States Coal Production

	1955	1965	1970	1972	Increase 1955-72
Top 4	16.5	26.6	30.2	30.4	84.2
Top 8	24.0	36.4	40.7	40.4	68.3
Top 20	37.5	49.5	55.9	55.1	46.9

% of United States Uranium Mining and Milling

	1967	1971
Top 4	57.0	54.4
Top 8	78.7	78.5
Top 16	100.0	99.8

in 1973 and 1974, but have dropped back to historic levels since. These figures tend to confirm the implications of the concentration levels for the industry that the majors do not possess substantial market power.

The industry has experienced surplus capacity in production throughout most of the post-war period, in large part due to governmental activities. On the one hand, production was restricted through prorationing and other regulatory activities in most producing states. On the other hand, the depletion allowance, expensing, and other incentives stimulated the development of capacity. As the 1950s progressed, the availability of cheaper foreign crude began to push domestic prices down and to threaten domestic production. Voluntary import controls were instituted in 1957, followed by mandatory quotas in 1959.

The excess capacity was slowly eliminated by the late 60s and imports grew. Domestic production approached 100 per cent of maximum efficient recovery under state regulations. By 1973 the import quotas had to be lifted and imports have spurted. At about the same time, however, the depletion allowance was eliminated. Thus, many of the conditions causing a domestic surplus have changed, and the surplus position is not likely to recur unless international prices fall substantially or large new domestic reserves are found and produced.

As the narrative of events of the last several decades suggests, the federal and state governments have been deeply involved in the petroleum industry for a long time. The nature of that involvement appears to have changed in a fundamental way. The extent of that change and its permanence will have a great effect on the entire energy industry.

From the 1930s until very recently, most governmental impacts have been supportive of at least some aspects of the industry. Prorationing and import restrictions protected domestic producers and supported domestic prices. Depletion allowances and other tax provisions encouraged the industry. Most of these measures have been eliminated, however. In their place have come price controls, allocations and entitlements, threats of divestiture or excess profits taxes, and other indications of hostility. The effect has been to raise costs, impose uncertainties, and in many ways complicate an already difficult energy situation. The FEA programme has locked the refinery and marketing part of the industry into its May, 1973 position, which was already affected by price controls, and prevented adjustments to changing circumstances.

The government's basic goals do not seem to have radically changed from the past. It still is concerned that the nation have the energy resources it needs. Project Independence is based on the same argument that was used in 1959 when import quotas were imposed.

CHANGING COMMERCIAL DYNAMICS

Thus, during the past decade or two, the profit dynamics in the "traditional" petroleum business have made dramatic shifts:

Within the United States, the cost of petroleum lay largely in the risk caused by low wildcat success ratios. Although lease acquisition and development costs were low (compared to today), so were success ratios. Economic success was then based either on isolated discoveries for small companies or, for the large companies, the ability to spread the risk over a large number of wildcats and to finance the dry holes with the successes.

However, now, in the United States the geology is comparatively well known, so that traditional "wildcatting" has been largely replaced with paying huge sums for lease bonuses on tracts where success is relatively more certain. In addition, in place of shallow, relatively inexpensive development wells, we now have expensive deep offshore wells. Therefore, the cost of United States petroleum today lies largely not in the risk of dry wildcats, but rather in concrete capital outlay of lease bonuses and expensive wells. This shift is an example of what I mean by a "changing commercial dynamic". The rules of the game and the keys to success are simply different.

Outside the United States, cost formerly was also largely the risk of failure. However, here two shifts have taken place simultaneously. One, as in the United States, the capital cost of wells has skyrocketed as we moved from the deserts of Kuwait to the North Sea. In addition, to this, there seems little chance in the future of covering the costs of dry holes in one country with a successful discovery in another. At best, a newly producing nation is willing to allow the operator a "fair return" on that production in that nation. This leaves the question still un-answered: Who is to pay for the losses on the dry holes in the other countries?

Now, with synfuels under consideration, these shifts seem destined to continue, that is, the shift from exploratory risk to the outlay of front-end capital for hardware, along with a shift from profits derived from development of supply and markets to profits which must be derived from processing.

These shifts are not in themselves "bad", but it would be disasterous to fail to recognise their significance on the success keys and conse-quently the new decision-making processes required, the shift in impor-tance of the factors considered, and the trauma possible before the new industry structure and culture have been hammered out.

Another important shift is the role of technology. The two techno-logical changes which had the greatest impact on the petroleum industry came from outside the industry. As electricity replaced kerosene for light, the development of the oil burner opened a whole new market. The second change was the automobile, which offered petroleum its other huge market. Technological change within the petroleum industry itself has been largely evolutionary, as would be expected in a very capital intensive industry; and one that is growing from a large base with large demand inertia, so that shifts in demand or technology can largely be accomodated through additions to capacity. (The fixed cost alone on an improved technology can often be higher than the variable cost of the technology it is designed to obsolete.)

With the prospect of synfuels on the horizon, we are now talking in terms of a rapid, massive infusion of a new technology, driven not by evolutionary changes in product demand or by profitability, but by traumatic changes in supply, and now under the "guidance" of government.

CHANGING CONTEXT FOR DECISION

Not only are the internal dynamics of the industry undergoing major shifts, but also the real world environment to which current and future decision must relate is also shifting. The present industry is composed largely of the results of selection through the process of the survival of the fittest. Those companies with too little luck, not enough skill, and/or too many mistakes are no longer with us: their investments are on no books, their costs are in no statistics. However, it is inherent that we have learned from their failures; through the trial and error process, we have learned the better way. (A dry hole teaches us where not to drill.)

In the past, individual failures were mostly small, though the total of them all may be large. Now, a crucial difference arises in synfuels: individual failures in a trial and error process will each be enormous by any standards. The failure of a single 50 MB/D synfuel plant would be equivalent in size to the bankruptcy of the 100th largest United States corporation. This is not to say that learning by trial and error with these kind of stakes is impossible, but it is certainly unprecedented in the private sector. It would at best be naive to believe that new frameworks are not required. We'll bet on five to one odds when we roll a die for a quarter, but Russian roulette is a whole other game, not because the odds are different, but only because the stakes are higher.

The private sector has grown largely through mastering the art and science not simply of matching risk to return, but also of covering risks, and/or reducing them. Statistics have a different meaning when applied to many events, as compared to few events. The decisions surrounding synfuels over the next decade or so amount to "you-bet-your-company" type decisions - and here we're talking about the largest companies in the world.

The time frame is also new. For a synfuel investment to be attractive, the industry must be viable for a quarter century or more. This is
also true in other areas, but presents no exotic problems in well defined,
more mature industries, where competitive forces are perceived to be well
known. However, in a totally undefined industry such as synfuels, the
familiar tools for evaluating investments over a long time span (e.g. DCF)
become little more than academic exercises. We are capable of spreading
risks in the short term or in a familiar environment, but we have no
tested tools for deciding today the viability of a huge new industry
over a long time frame.

Our culture, and especially the free enterprise sector, has become
accustomed to including as a significant criteria the dimension of growth.
Even though the entire energy sector may continue to grow through all of
the next century, there are few with any hope that any one of the currently available technologies will grow for more than a couple of decades,
before they decline due to, probably, the emergence of a "backstop"
technology (e.g., breeder, fusion). It is not at all clear what basic
attraction there is to rush into an industry whose death certificate is
already being filled out as rapidly as possible in the finest R & D
laboratories in the world.

(Note here that we first talk of a few decades as a "long" time
in terms of analytical tools, and a "short" time in terms of the life of
the industry. It is exactly this contradiction that illuminates the
basic dilemma. By the time we will have learned how to deal with the
issues, the industry is likely to be on the decline because of technological progress).

What might appear to be a fine point here must not be missed, for
it is in fact pivotal. First, a little background. We in the United
States long ago gave up "chartered company" in the classic sense of
creating a company with a mission. Rather, incorporation in the United
States implies the right to existence as long as profits are satisfactory
in the eyes of the investors and as long as the generalised rules of the
game are obeyed. If a company decides to be a shoe company or an energy
company, that is their own decision, not one that is forced upon them,
and is a means to the end of profits, etc., and is not the end in itself.
By the laws of classic economics, if there is a "shortage" of shoes or
energy, the shoe and energy companies would not be obligated to fill the
gap, but rather to maximise profit after considering all factors. Therefore we are speaking in schizophrenic terms when we say that energy

companies are obligated to furnish plentiful, low priced energy; that
may well be the results of the system we now have in operation, but it
would be a result, not an objective.

A perhaps even more subtle but more crucial point should also not be
missed. This needs especially to be pointed out to our colleagues from
countries like Japan, where there is a long history of successful co-
operation between the private sector and the government. By contrast,
the United States government was designed and intended not so much to
co-operate with the private sector as to regulate it, and to deal with
crises on a short-term basis. We have neither the practical tools, the
conceptual basis, nor the cultural framework for the government to guide
the private sector towards predetermined goals over a long time span.
Despite our cry for a "national energy policy", one must wonder just
what it could be, and just how the government would go about effecting it.
There is no precedence for the government dealing with a specific economic
issue through the private sector over a long time frame.

NEW DIMENSIONS OF THE UNKNOWN

We speak often of "risk and uncertainty"; usually we know at least
the parameters about which we are uncertain and in which we face risks.
However, in the area of synfuels, literally a new dimension is added in
that we are uncertain about which parameters contain the significant
risks. For example future price is always an uncertainty. However, here
we are not even certain about which forces will determine prices.
Similarly, competition is always a risk, but here, where the structure of
the industry will not be known for decades, we do not know what the competition will be.

Once an industry is developed, the competition is homogeneous and
the rules are relatively uniform. Success involves standing out in a
field where the competitors face similar decisions, similar problems,
and similar costs bases, and where changes in the rules affect the field
more or less similarly. However, in the area of synfuels, even this
degree of relative comfort is not available. Basic issues arise, such
as: Will the prime competition be OPEC, with a cost base different by
one or two orders of magnitude, and with significant non-economic
criteria? Will public policy set prices (and volumes?), with profitability determined arbitrarily? Even if, somehow, we could be told now
what the actual price will be in the future, removing all risk and uncertainty from that dimension, we would still be very limited in making

decisions for lack of knowledge and understanding of the dynamics of how that price got determined. A price determined by, say, the incremental cost of conventional petroleum is qualitatively very different from the same price determined by, say, an international cartel or by public policy.

In simple terms, a business cannot be defined without also defining the structure and dynamics of the industry of which it is a part. Uncertainty about one increases the uncertainty about the other by an order of magnitude.

An industry is more than the sum of the businesses which make it up. The form of the businesses determine the dynamics of the industry and the structure of the industry determines the commercial opportunities that lie within it. We are essentially void of the analytical and intellectual tools which would be necessary for conceptually synthesizing such a framework before the fact. The skills of, say, financial analysis are slightly helpful, but essentially ignore the institutional and structural issues; they probably are reliable on the average over the long run, but reveal little about the particulars or mechanisms.

One area relevant here is that of mystique. An emerging industry usually has as a primary barrier to entry and foundation for profits its transparency and mystique. It takes a while for imitators to have something to imitate. It is only after maturity is approached and visibility is gained that classic economic analyses become really useful. It is not at all clear how the deliberate and open building of a new industry through public policy, under the bright lights of public scrutiny with enforced transfer of knowledge and with the expectation of frequent accountability will effect the attraction of firms and businesses to the industry. It is quite possible that the "spoiler effect" may apply, at least directionally: great public attention may well deter commercial entry.

The focus of these issues may well be termed an "identity crisis" for the firm, whether the firm be old or new. Aside from any complications arising from legislation proposed to impose arbitrary constraints: What does it mean to be an "energy company"? Are we investors, explorers, operators, or sellers of services? How is diversification of risk accomplished, especially when a synfuel plant costs more than all but a few companies in the world. How do you sell services? How much do you charge per hour for the services of the man who discovers the next large field of inexpensive oil?

All that has been said up to this point (the shift from evolutionary change to discontinuous change, the new commercial dynamics, the enlarged context and added criteria for decisions, and the added dimension of unknowns) indicate the need not for marginal improvements in traditional management skills, but rather for a new approach to the decision-making process. In this sense, it is entirely possible that the most significant breakthrough to occur over the rest of this century in dealing with the energy issue will be in the development of a new and adequate decision-making process, which can be used authentically by both the public and private sectors, and which will permit dealing with real issues in a significant way.

Management sciences have made unbelievable progress over the last couple of decades, giving us such useful tools as econometrics and operations research. It is in the area of policy analysis and synthesis that reliable tools are yet to be developed. And, to continue the "tool" analogy, we lack the "toolbox" - the integrative framework which pulls together all our skills and permit the building of a single course of action which is comprehensive and serious, and which weighs each factor appropriately.

Today, as I perceive it, the most promising attempts to develop such a technique are being conducted under the category of "strategic planning", although the name is not important here. What is important is that management styles are emerging which are new and radically different, and are far more than the simple addition of new skills to the old style. The new style deals not with decisions at the margins, but with decisions at the core. Not with increased efficiency, but with increased effectiveness. (Efficiency deals with how well you do something; effectiveness deals instead with what you ought to be doing at all.) Not with maintaining momentum, but with deciding the speed and direction that is suitable. Not with what rewards make a risk desirable, but with how risks can be reduced in fact.

Strategic planning as a management style includes the newly crucial and dynamic relationships between technology, finance, government, the public, the stakeholders, and competition (both homogeneous and heterogeneous). It is not simple, and certainly no panacea, but conceptual frameworks are being built within which national decisions can be made in a world of increasing options and increasing ambiguity.

Even less well developed, however, is progress beyond the stage of problem-solving essentially by adversarial proceedings or even through compromise. It is becoming almost common wisdom that our futures are so interlinked, the stakes are so large, and the time horizons of current commitments are so great that some kind of real "co-operation" between, say, industry, government, academia, the consumers, labour and the public is essential. However, we don't have the mental picture, the conceptual outline, or even the vocabulary for creative thinking in this dimension. Old, obsolete modes keep popping into mind (e.g., "socialism"), which block our imagination. Crisis (e.g., war) creates "co-operation" (and this isn't the right word) ad hoc, but is without sustaining power. If there is an energy issue at all, it will be with us 50 or a 100 years, and the word "crisis" therefore doesn't apply. Current experiments at institutionalising the cohesive forces which will be necessary in the future are today only very embryonic.

The greatest danger is to believe that the new kinds of decisions facing us today are not really new, but are merely different, or that they differ only in degree or magnitude. My argument here is that they are truly new, that we have some ideas about the ways in which they are new. There is a lot of promise in the current struggle to develop the management styles for handling them adequately.

COAL SUPPLY PROSPECTS IN THE UNITED STATES -
AN APPRAISAL OF CURRENT KNOWLEDGE

by

R.L. Gordon(1)

1) The present discussion draws upon work done under research grants from
the Electric Power Research Institute, the U.S. Bureau of Mines and the
National Science Foundation. Some work was also undertaken as a consultant to the Federal Energy Administration. Material was also taken
from articles from Science and Regulation magazines. As usual, reading
my views should make clear I speak for myself and not for any of my
sponsors.

The supply of coal, as with other energy resources, is quite poorly
understood. The differences between knowledge about coal and about other
fuels is most pronounced in the nature of data presentation. Coal resource data are, as shown below, collected on a physical availability
basis, with little effort to determine the costs of exploiting different
portions of the resources. Oil and gas reserve data relate only to economically viable already discovered resources.

In this paper, the discussion begins with reviews of coal resource
data, proposes a conceptual basis for coal supply analysis, proceeds to
discuss the U.S. Bureau of Mines (USBM) Process Evaluation Group (PEG)
coal costing studies, turns to examination of various efforts to estimate
coal supply curves and then notes the political problems affecting coal
production. Conclusions follow. The supply analysis begins with an overview of pre-1976 work and then turns to particularly extensive review of
the work done for the Federal Energy Administration. The paper defines
coal supply in the narrow sense of production economics. This is by no
means the only possible choice and indeed is more restrictive a concept
than employed in most of my prior work. A broad concept of the supply
of usable energy from coal is necessary for a full analysis. This requires review of transportation, transformation and final use economics.
A full examination of all these issues would be an unmanageable task for

a single paper (and besides reviews of these issues appear in recent and forthcoming research I have conducted or directed).(1)

The bulk of this paper is concerned with reviewing analytic studies on coal supplies. However, given the interest in securing insights into the prospects of coal, I supplement the review of these studies with conjectures on supply prospects.

At the outset, the position being taken about coal should be made as clear as possible. It is argued here that until at least 1985, any United States or European desires to reduce oil imports radically are most easily met by expanding United States coal output. Thereafter, there are far too many uncertainties for anyone to be confident about what constitutes the most likely development. Therefore, the ebullience often expressed about the prospects for coal may or may not be valid but the available evidence is unsatisfactory to resolve the argument. In any case, many of the concerns expressed here about coal prospects are identical to those of industry sources. We both feel public policy is impeding expansion of coal output. (Naturally, the industry is more certain than I that the impediments are ill-advised.)

A NOTE ON COAL RESOURCE DATA

As should be well-known, available data on mineral resource endowment were not designed to permit appraisal of long-run supply prospects and obviously, therefore, should not be used for such evaluations. This principle unfortunately is frequently violated. A prime problem is the tendency to compare proved oil and gas reserves to coal resource data. Proved reserves include only discovered resources that are known to be economically recoverable at current prices. Just what coal resource data cover is quite unclear but it certainly is not anything comparable to the proved reserve concept. However, efforts have been made to generate figures for the United States that are more closely akin to proved reserves and this effort may be reviewed prior to viewing economic studies of coal supply.(2)

---

1) See, e.g. Historical Trends in Coal Utilization and Supply, an August 1976 report to the U.S. Bureau of Mines. A report on synthetic fuels by Edward and Louise Julian, my graduate assistants on a National Science Foundation sponsored project on western coal, reviews synthetic fuel prospects. My 1978 book, Coal in the U.S. Energy Market synthesizes and updates much of this work.

2) This section is an analytic summary of Chapter V of my USBM report which may be consulted for details and references.

For many years the only available coal resource data were those of the U.S. Geological Survey (USGS). It produced two basic aggregates. One covered both reasonably well measured (identified) resource and those less well measured (hypothetical). The other and more widely used figures related only to identified reserves. More recently, the USBM has developed a third concept - the demonstrated reserve base. This is an effort to identify the resources most likely to be exploited.

This reserve base was constructed by developing rules of thumb for inclusion. Part of the adjustment was based on the certainty about the existence of the resources. The USGS subdivides identified resources into three categories - measured, indicated and inferred - each successively representing a lesser degree of certainty about the estimates. The first step in developing the reserve base was to exclude inferred resources. Then seams of less than a minimum thickness and below a maximum depth (both of which differed among states and coal ranks) were excluded. Finally, a maximum was set for the ratio of seam depth to seam thickness for strip mineable coals. Earlier work by the Bureau and the National Petroleum Council (NPC) suggests that the reserve base may overstate the amount of coal recoverable at costs roughly comparable to those presently prevailing. The NPC work on underground resources indicated that an even stricter minimum seam thickness criterion should be applied than that used in the USBM work (e.g. only bituminous coal seams thicker than 42 inches should be included instead of the seams thicker than 28 inches placed in the reserve base) and that only half the coal be considered recoverable.

The earlier USBM work on strippable resources involved starting with figures that apparently were conceptually similar to those used to define the demonstrated reserve base, applying a recovery factor, and finally making further deletions on the basis of information that such factors as resource location under cities and other things that could not be disturbed, difficult mining conditions and poor coal quality made economic recovery unlikely.

The demonstrated reserve base data on underground mineable coal are presented so that it is possible to apply to them the NPC adjustment process. However, the dependence of the USBM adjustment method on unavailable data on conditions at specific sites precludes accurate adjustment of the new reserve base figures to an "economically recoverable" reserve basis. A crude adjustment is provided here by assuming that currently economically strippable reserves are the same fraction of the demonstrated reserve base as was true in the earlier USBM work. Table 1 provides a summary of my efforts to apply these data adjustments to the latest available USBM figures.

The most that should be concluded is that adjustment of resource figures by rough indicators of what are known economic resources implies less than a tenth of identified resources meet the stricter cirterions. It may be noted parenthetically that since the revised total strippable reserve base is larger than the earlier estimate, so are the strippable reserves I estimated. The reserve base increase, however, is the net effect of reduction of figures for bituminous in the East and increases of estimates for sub-bituminous and lignite in the West. The underground reserve figure, however, is lower than that calculated by the NPC.

Nevertheless, these reduced figures are quite large relative to current or projected coal consumption so do not by themselves indicate problems of maintaining coal output. On the other hand, even neglecting consumption growth and losses in transformation of coal to synthetic oil and gas, the reserve would only cover total United States energy consumption at 1975 levels for about 50 years. Beyond this lie the basic uncertainties about any figures of this type. Plausible arguments can be made supporting the alternative propositions that by using too liberal cutoffs and neglecting cost trends the figures are too optimistic or that by ignoring technical advance and discoveries the figures are too pessimistic. This we cannot resolve; we can simply look at the state of the art of supply analysis.

The whole process of trying to project supply prospects from adjusting physical availability data seems an unsatisfactory compromise. Nothing short of a thorough reworking of data gathering procedures oriented toward the needs of supply analysis would be a satisfactory approach.

CONCEPTS IN COAL SUPPLY ANALYSIS

Three basic components of coal supply analysis can be distinguished — the development of relationships between the physical conditions of mining and costs, the use of data on physical conditions and the cost relationships to estimate the cost function for the industry and the appraisal of short-run divergences between prices and costs as derived in the first two steps.

Costs generally depend on the physical conditions of mining and the cost and performance of the labour and capital resources used. To complicate matters further, regulation, technical progress and price changes produce shifts over time in the costs of mining under any specified mining

Table 1

COMPARATIVE ESTIMATES OF UNITED STATES COAL RESOURCES
(billion net tons)

USGS Resource Data for 1st January, 1974

Estimated total identified and hypothetical resources	0-6,000 feet	3,968
Hypothetical resources	3,000-6,000 feet	388
Estimated and hypothetical resources	0-3,000 feet	3,581
Hypothetical resources	0-3,000 feet	1,850
Identified resources	0-3,000 feet	1,731
of which Bituminous		747
Sub-bituminous		486
Anthracite		478
Lignite		20

USBM Demonstrated Reserve Base as of 1st January, 1976

Bituminous	228
Sub-bituminous	168
Anthracite	7
Lignite	34
Total	438

Reserve Base Adjusted for Recoverability Factors

Bituminous	78
Sub-bituminous	57
Anthracite	2
Lignite	15
Total	151

Source: R.L. Gordon, Coal in the U.S. Energy Market, Lexington: Heath Lexington Books, 1978, p. 67-74.

conditions.(1) Therefore, an ideal costing formula must take account of these changes and show not only the current cost function but how costs are most likely to shift over time.

The problem in analysing current costs then becomes determining whether a feasible method of costing exists. Given the wide variation in physical conditions, supply estimation based solely on cost-engineered studies would be impractical. Every deposit would have to be separately costed. A statistical approach (perhaps, however, applied to analysis of a large number of engineering cost studies) must be taken. Such an

1) A further distinction can be made between cost impacts due to general changes in the price level in the economy as a whole and changes that alter the relative relationship between coal and the rest of the economy. It is conventional in cost analyses to isolate and stress the latter cost impacts by conducting the analysis in "real" or "constant dollar" terms - i.e. by ignoring as well as possible the impact of general price level changes.

approach is designed to distinguish explicitly significant influences from trivial details that essentially can be ignored. In particular, we should use regression analysis to determine which physical conditions significantly affect costs, the magnitude of these impacts and the extent to which an adequate cost model can be built using a manageable number of variables.

These principles suggest how one would determine both the feasibility of statistical cost analyses and how far short of the ideal existing costing techniques have been. As shown below, the adequacy of existing techniques has already been subjected to severely critical statistical analysis. However, the feasibility of improvement remains unclear.

Analysis of time shifts in input prices, technology and regulation by definition cannot be adequately conducted using current data. To make matters worse, using the standard first approximation method for predicting time shifts - assuming that historic trends will continue - appears untenable for analysis of the coal industry.

The long-term trend up until 1969 was towards rising output per worker and constant or falling costs in constant dollars; the reverse has been true subsequently. The extent to which this change in conditions can be slowed or even reversed remains controversial. Therefore, the analyst is faced with the need to warn about possible cost changes but cannot accurately forecast what they might be. The best that can be done is to test the sensitivity of future consumption to different assumptions about cost trends. This rarely has been done.

Use of the resulting costing model developed to analyze data on coal resources is another conceptually simple but practically difficult task. Ideally, one secures data on the characteristics of all coal deposits that one wishes to evaluate and uses the data and the cost model to develop a cost function. It is shown below that such a procedure is difficult to implement.

Finally, the recent history of the coal industry makes clear that, to analyze supply at any actual moment, one must go beyond the cost analysis just outlined. Implicit in such an analysis is the assumption that coal mines will be opened as rapidly as demand expansion makes them necessary and no faster. In the convenient terminology of economic theory, most costing models assume the maintenance of long-run equilibrium. The usual assumption in cost studies is that firms that can recover with interest their initial investment and other costs will operate. This, by definition, abstracts from the widespread tendency for unexpected surges or slumps in demand cause prices to exceed or fall short of that required by mines actually operating - a problem of great importance to some analysts.

Existing studies stress predominantly the determination of current cost relationships. Only a few have attempted to examine time shifts in the cost. Generally, fairly simple models that can be employed with existing resource data are developed, although one major effort has been made to appraise the use made of these resource data. Similarly, very little has been done to treat disequilibrium.

THE U.S. BUREAU OF MINES MODEL MINE STUDIES

The PEG reports, largely written by Sidney Katell and one or more collaborators, have since the late 1960s provided a series of cost engineering studies relating to coal mines. Each case study involves first equipping and manning the mine and then pricing the labour, capital, supplies and other expenses. Given accurate cost estimates, it is possible by use of standard techniques of financial analysis to determine the per ton selling price which, if received continuously over the life of the mine, will recover costs including the required return on investment.

The PEG studies involve a standard format for providing tables summarising the data on the model mines. Thus, tables list equipment needs and their costs, manpower and its cost, the depreciation schedules for the equipment, power and water costs, a tabulation of accounting costs, a calculation of development expenditures, a capital cost summary and a table calculating the present value of investment. A final table then calculates the selling price that will recover with interest the investment in the mine.

Most of these tables are quite straightforward and need no discussion here. Instead attention is concentrated on the accounting-cost and required-price calculations. Tables 2 and 3 show, respectively, samples of the standard cost and required price calculation tables of a PEG costing analysis. The first four items in Table 2 - on labour costs - are taken directly from the labour cost tables provided with the case studies. Similarly, the power, water and depreciation figures are transferred from other tables. The remaining figures are computed using the rules of thumb shown on the table.

The calculations shown in Table 3 are designed to derive required selling prices for the case in which output, cash costs and prices are constant. Under these assumptions, the net before tax cash flow - the difference from revenues and cash operating costs is a constant. This reduces the income requirement problem to one of the simplest in financial analysis - calculating the constant annual payment required over a fixed

Table 2

ESTIMATED ANNUAL PRODUCTION COST, 1.03-MM-TPY MINE

	Annual cost	Cost per ton
Direct cost:		
Production:		
Labour	$2,108,900	$2.05
Supervision	507,000	.49
	2,615,900	2.54
Maintenance:		
Labour	324,700	.32
Supervision	75,000	.07
	399,700	.39
Operating supplies:		
Mining machine parts	648,600	.63
Lubrication and hydraulic oil	257,400	.25
Rock bolts and timber	319,200	.31
Rock dust	133,900	.13
Ventilation	195,600	.19
Bits	123,600	.12
Cables	61,800	.06
Miscellaneous	154,400	.15
	1,894,500	1.84
Power	314,400	.30
Water	1,000	
Payroll overhead (40 per cent of payroll)	1,206,200	1.17
Union welfare(1)	1,099,300	1.07
Indirect cost:		
15 per cent labour, supervision, and supplies	736,500	.72
Fixed cost:		
Taxes and insurance, 2 per cent of mine cost	471,100	.46
Depreciation	1,065,000	1.03
	1,536,100	1.49
Total	9,803,600	9.52

1) Effective 6th December, 1974, under the Bituminous Wage Agreement of 1974.

Source: Katell, 1975b, p. 11.

period to repay with interest a given investment. This always could be determined by use of standard financial tables and the electronic calculators designed for handling financial problems generally are designed to implement the calculation. Under the assumptions used in the 1975 and 1976 PEG studies, about 16 per cent of the investment must be received annually. PEG chose to use instead the reciprocal of this factor - this present value of a repayment of $1 per year over a specified life ($6.259).

Table 3

DISCOUNTED CASH FLOW, 1.03-MM-TPY MINE

---

15 per cent - 20 years

R = $28,537,300 ÷ 6.259(1) = $4,559,400
   less depreciation   1,065,000
 Depletion + net profit   = 3,494,400

Depletion = 10 per cent of sales
Federal income tax = net profit
Depletion + net profit = cash flow - depreciation

Sales = 1/0.55 (1/2 operating cost + cash flow - depreciation)
     = 1/0.55 (4,901,800 + 3,494,400)

Sales	$15,265,800
Operating cost	9,803,600
Gross profit	5,462,200
Depletion	1,526,600
Taxable income	3,935,600
Federal income tax	1,967,800
Net profit	1,967,800

Annual cash flow = net profit + depreciation + depletion
             = $1,967,800 + $1,065,000 + $1,526,600
             = $4,559,400

Selling price per ton = $15,265,800 ÷ 1,029,600 = $14.83

---

1) Uniform series present worth factor.

Source: Katell, 1975b, p. 14.

The main problem with the calculation is in handling the complications introduced by tax laws. The tax laws affecting minerals divide before tax cash flow into three elements - depreciation, depletion and "profits". Specific rules govern how much of the cash flow may be considered depreciation and depletion; taxable profits are the difference between the cash flow and allowable depreciation and depletion. To simplify the analysis, PEG sets depreciation conservatively by use of "straight line" depreciation over the actual life of the asset. This means that depreciation of an asset costing $X and lasting n years would have an annual depreciation of $X/n. For example, depreciation on an asset costing $1 million and lasting 10 years would be $100,000 per year.

Actually, the tax laws and their administration allow more rapid write-offs but PEG ignores this because it would produce annual variation in cash flow.

Depletion allowances are set by law as the lesser of half the cash flow less depreciation and a per cent of gross sales income specified for each mineral. Until 1975, PEG assumed that the 50 per cent of cash flow less depreciation rule always applies. Thus only half the cash flow less depreciation is taxed. The 1975 and 1976 studies assumed that the full ten-per-cent-of-sales-income depletion allowance could be earned. As a further simplification, the 48 per cent corporate tax rate was rounded to 50 per cent. Thus, under the earlier assumptions, 75 per cent of cash flow less depreciation is retained by the firm – the 50 per cent that is depletion plus the half of the remaining 50 per cent that is left after taxation. Three dollars are retained for every four received; thus to get $1 of after tax cash flow less depreciation, $4/3 in before tax income are required.

With the new methodology, PEG first calculates the required net cash flow. Then depreciation is deducted to determine the required contribution from depletion and net profits. It can immediately be calculated that gross profits are double required net profits and the required gross cash flow is the sum of depletion, depreciation and gross profits. Thus, Table 3 shows that the required $4.6 million in net cash flow is recovered by a gross flow of $6.5 million consisting of $1.1 million of depreciation included in operating income, $1.5 million of depletion allowances and $4.0 million in taxable profits.

While these estimates have proved valuable to analysts of coal supply, many problems exist with the studies. The basic problem is one of establishing the empirical relevance of the studies. The concept of a typical mine is inherently problematic and serious questions arise about the implementation of the analysis.

The fundamental problem with the Bureau effort is one endemic to energy research – the lack of adequate support from funding agencies. The Group's work was not and did not pretend to be anything more than the efforts of a small number of knowledgeable people to provide indicators of coal-mining costs. The effort is not extensive enough to allow analysts to be confident that the Bureau of Mines studies (or the calculation of further cases by TRW Systems for the Project Independence Task Force Report on Coal) are based on an adequately extensive collection of cost data.

Many additional problems arise from the execution of the cost analysis. On a a priori basis, the most obvious problems are the failure to specify precisely what is meant by typical conditions and the stress on mines for gasification plants. The absence of details on what constitutes typical conditions means that no satisfactory way exists to adjust for the impact of deviations of conditions from whatever was assumed. The stress on costing mines for gasification plants means that every case study has related to mines producing from one to five million tons per year and many cases fail to include costs of unit train or other loading facilities and cleaning costs are ignored. The main problem with the stress on large mines is that a substantial portion of output comes from smaller mines than those for which the USBM makes cost estimates.(1)

In any case, the underground mining analysis of the Bureau does implicitly provide a two-variable - mine size and seam thickness - model of costs. A number of analysts have expressed tacit disagreement with the model by adopting radically different assumptions. However, the fullest explicit evaluation is that of Martin Zimmerman (1975) and he feels that several defects exist in the USBM analysis. He contends that the decrease in average costs that the Bureau shows to occur with increases in mine size arises from the particular costing technique used and this technique does not accurately reflect actual conditions. He also presents evidence that the costing model understates the effect of seam thickness on costs and excludes significant additional influences.

Zimmerman shows by statistical regressions between costs and the number of operating sections that the number of sections is the main source of differences from one USBM case to another in costs. The Bureau's costing method essentially consists of assuming a constant direct cost per section and overhead expenditures that do not increase significantly with mine size. This leads inexorably to lower cost for larger mines since the overhead is spread over a higher output.

Given the critical role of output per mining section, Zimmerman concentrates on developing, as is possible by analysis of available data, a statistical appraisal of the behaviour of output per section. A comparatively simple test is employed - that of determining how well seam thickness and determination of how accurate an estimate of output per section

---

1) Due to the peculiarities of data tabulation, no precise measure of the defect is readily calculable. Neither of the two available sources of systematic information provides information on the role of million ton per year mines. The Bureau of Mines' data on distribution of output among mines of different sizes have as the largest category mines producing 500,000 tons per year or more; in 1974, these mines accounted for 54 per cent of output. McGraw-Hill has begun listing all known mines producing more than a million tons per year and for 1975 mines producing more than a million tons accounted for by about 40 per cent of output.

can be made on the basis of seam thickness. He finds that the average impact of seam thickness is considerably greater than assumed by the Process Engineering Group and, more critically, a highly inaccurate forecast is provided by a model based only on seam thickness.(1)

The relationship between seam thickness and output per section is combined with the relationship between number of sections and costs to provide an alternative means of estimating the effect of mine size on costs. The first equation is transformed to show the number of sections as a function of output and seam thickness and then the transformed equation is substituted into the cost equation.

Thus, the output per shift equation has the form $q = 0.93\ Th^{1.42}S^{-0.33}$, where q is output per section during a single work shift, Th is the seam thickness and S is the number of sections in the mine. To transform this equation into the desired form, Zimmerman uses the definition of the relationships among number of sections, section productivity and annual output. In particular, the annual output of a section is (q)(SPD)(DPY), where SPD denotes the number of shifts worked each day and DPY stands for the number of days worked per year. Then the mine's total output (Q) equals (S)(q)(SPD)(DPY) or S = (Q)/(q)(SPD)(DPY). By substituting $0.93\ Th^{1.42}S^{-0.33}$ for q and solving, we get the equation showing number

---

1) The approach is to develop a regression between the logarithms of seam thickness, shift number and output per shift. Regression coefficient size measures the impact of thickness. Zimmerman's results imply that output per shift increased significantly more rapidly than seam thickness; a doubling of thickness raises output per shift to a level 2.7 times that previously prevailing; the Process Evaluation Group's data indicate that such a doubling would only increase output per shift 15 per cent. The inaccuracy argument is justified by the large size of the variance of ε – his error term. The multiplicative nature of his equation makes a simple explanation of his analysis difficult but we can roughly explain what is involved. In statistical analyses of the type conducted by Zimmerman, an estimate of the relationship between OPS and Th is made by writing the equation in the form of log OPS=a+b log Th + log ε where a and b are values determined in the analysis. The estimation technique assumes that, on the average, log ε =0 or ε =1. However, this can be an average of substantial overestimates offset by equally substantial underestimates. A measure known as the variance which is computed by computing the deviations, squaring them so that they all become positive numbers, and summing and averaging them is used to provide the basis of a measure of deviation. The square root of the variance is known as the standard deviation and roughly can be interpreted as a measure of what constitutes a reasonable error. In particular, with a standard deviation of σ , a value of ε of 1 $\pm$ σ would be considered reasonable. Given Zimmerman's logarithmic approach, the best way to interpret the results is to transform the equation back into a function of the original variables – i.e. OPS = K Th^{b} ε . In this form, the error term becomes an indicator of the likely proportional deviation of actual OPS from that predicted by the measured variables. The ratio ranges from about .38 to 2.66.

of sections as a function of Th and Q. The cost equations have the form
TC = a + bS where TC is total cost, S again is section number, and a and
b are numbers obtained in the regression analysis. By division by Q, these
total cost equations become average cost equations. The equation relating
S to Q and Th then can be used to transform the average cost equations into functions of Th and Q. By assuming a given seam thickness, the average
cost equations become functions of output alone and Zimmerman differenti-
ates each equation with respect to Q, solves and therefore finds the Q
that minimises average costs at a given seam thickness. These average-
cost minimising outputs prove to be much lower than suggested by the USBM
analyses. For example, a deep mine in a six-foot seam minimises average
costs with an output of slightly more than 600,000 tons. Thus, the a
priori reservations about the USBM approach to costing are reinforced by
Zimmerman's statistical tests. This result reinforces the basic impression that the Bureau was only undertaking exploratory studies and the work
by and large did not pretend to suffice for the use, to which others have
put it, of providing the data to characterise the cost distribution for
the whole of the United States.

THE ASSIGNMENT OF RESERVES

A major problem in implementing costing analyses, particularly for
eastern underground mining, is the limited data availability. Unless the
analyst is willing to accept the argument that the remaining reserves are
similar to those now in use, it is difficult to decide what assumptions
should be used. Here again, Zimmerman provides useful insights. He attempts to estimate a plausible pattern for the cost distribution of remaining coal resources in the eastern United States. He proceeds by arguing that both seam thickness and the estimated deviations of costs from
those predicted by the assumption that thickness is the main influence on
costs are distributed statistically, following a lognormal distribution.
In the lognormal distribution, the logarithms of the variables are distributed according to the normal (bell shaped) distribution that is so important in mathematical statistics. The assumption of lognormality is frequently made in mineral resource evaluation and Zimmerman feels that the
data on the distribution for existing mines support its use in the coal
case.
Without going into the details of his calculation, we can simply note
that his assumptions, together with his prior provision of an equation relating costs to seam thickness, shift number and other (unmeasured) factors,
permit him to develop a cost distribution for reserves. His evaluation of

the estimates leads him to argue that the availability of reserves comparable in costs to those currently being exploited is far smaller than generally argued. In the most favourable case considered, the Illinois basin could maintain production rates for 21.5 years before costs rose 5 per cent above present levels, 43.7 years before costs rose 11 per cent, and 89.9 years before costs rose 22 per cent. The respective figures for low sulphur Appalachian coal are 2.9, 5.5 and 14.1 years and for other Appalachian coals 8.9, 18.3 and 38.0 years. Thus, the future prospects for eastern coal depend critically upon cost trends in other fuels. Growth in demand will depend upon the comparative rates of technical progress and depletion of rival fuels.

Only a few examples are available of efforts to develop a supply function by combining the costing equations with data on the characteristics of extant mines or unexploited reserves. The problem of differences in the quality of available resources is ignored in many studies and in others, most notoriously the Coal Task Force Report for the 1974 Project Independence study, do not provide explicit information about how the availability of coal in different cost classes was determined.

Four examples of cost allocations can be identified. The North Central Power Study contained estimates made of the cost ranges for exploiting the most promising coal deposits in the region. A Battelle study contained an extensive compilation of the characteristics of the coal seams in the United States. By the use of the costing formulas noted below and the information on reserves, a supply function could be generated.

However, the most explicit supply estimates are those in the two Charles River Associates (CRA) studies. In its earlier study of Appalachia, stress was placed upon determing the cost distribution for existing coal mines. Information on the cost determinants was pieced together by use of data in the State mining reports for Pennsylvania, Ohio and West Virginia. Pennsylvania reports all three cost determinants noted by CRA (size, depth and seam thickness) but explicit thickness figures were not available for the other two States. These were estimated on the basis of assigning representative depths to each of the reported mine types - drift, slope or shaft. The information was then used to determine the distribution of underground output by cost ranges in the three States. Distributions for other States were developed by assuming that the West Virginia or Pennsylvania pattern would prevail. Strip mining cost distributions were assumed to be identical to those underground.

The later study of low sulphur coal in southern West Virginia involved assigning reserves to cost categories. The costing model described below was combined with data on the characteristics of the coal in the region.

A privately compiled inventory of coal seams was used as the basis for seam thickness and most mine depth data estimates. The proportions of inside production workers in the labour force and gassy mines in the total were set at their 1973 levels. The allocation of reserves to different size mines was determined on the basis of the size distribution that will prevail among the mines whose opening has been announced.

To develop supply curves, Charles River Associates proceeded to develop three basic scenarios of a resource development; each scenario is subdivided into two cases - one in which low sulphur coal reserves are developed first and another in which lower cost reserves are developed first regardless of their sulphur content. The basic scenarios each involve a series of assumptions about labour availability for the industry as a whole, the needs of existing mines as affected by productivity advance and depletion and, therefore, what is left over for new mines. The principal cost change distinguished in the model is the net effect of a 5.6 per cent average annual increase in output per man-day, a 2.4 per cent rise in real wages in the private sector of the economy and a ratio of 1.60 to 1.45 of coal-mining wages to wages in the private sector. The first two scenarios show the respective effects of annual entry into the coal-mining labour force equal to 6 and 2 per cent of the present labour force and thus models of supply limited by labour availability. The third case is a no growth model in which labour is no problem.

ALTERNATIVE COAL-COSTING RELATIONS

Even among those using USBM cost studies as a basis for supply analysis, there has been a widespread tendency to modify or extend the cost models to improve their utility. One of the earliest efforts in this area was made in the CRA study of Appalachian coal. CRA employed early unpublished USBM studies that treated mines at different depths, as well as at different sizes and seam thicknesses. However, the number of cases was so few that CRA had the bare minimum of information needed to deduce the impact of each variable on costs. For example, the cost per foot of shafts was derived by dividing the total cost for shafts in a single case study by the depth of the mine. The effect of seam size was calculated by comparing the costs of two mines with identical outputs produced from seams of different thicknesses.

CRA developed a more sophisticated approach to cost analysis in its 1975 study of southern West Virginia low sulphur coal. Here CRA concentrated on varying costs from the level realised by a particular model underground mine – one producing 1.03 million tons per year from a four-foot seam. The costs and required revenues were derived by adjusting, largely for price-level changes, data in a USBM report.

By appraisal of how the different elements of costs are affected by variation in output per man-day and different kinds of cost shifts over time, CRA produced an equation that indicated that the required selling price P at any future time for any new underground mine in southern West Virginia must equal $(75.64/OMD)LI + (13.44/OMD) \times GI + 0.801WI$, where OMD is output per man-day of the mine being costed, LI is an index of labour costs, GI is the wholesale price index and WI is an index of payments per ton to the union welfare fund. The equation thus decomposes the determinants of costs into first the labour costs themselves, the other OMD related costs such as supplies and those costs that do not vary with OMD and for each component defines an appropriate index of cost changes over time.

This equation was derived by simple inspection of the cost data. Regression analysis was then used to estimate OMD as a function of seam thickness, mine output, whether or not the mine was gassy, the proportion of workers engaged in underground activities and whether or not the mine was producing at least 25 per cent more in 1973 than in 1969. The resulting analysis indicated that this simple model could only explain 23 per cent of the variation in OMD. Moreover, neither seam thickness nor the proportion of workers underground proved to be a significant influence on OMD.

A similar method of adjusting specific model mine data, decomposing the costs and developing formulas to show how the components varied with departures from the conditions assumed for the model mines was developed by Battelle Pacific Northwest Laboratories in a study for the Energy Research and Development Administration. However, the workers at Battelle derive the basic relationships on the basis of unexplained engineering criteria rather than statistical analyses.

Thus, the analyses for underground mining adjusts costs for differences in depth, seam thickness and the dip of the mine. The costing formula is $C = [(6/t)(A) + (B) \times 100/(100-DIP)] + D_e$, where C is the cost, t is the thickness of the seam, A consists of all costs of the base mine affected by seam thickness and dip, B are those base mine costs affected by dip alone, DIP is the dip in degrees and D_e is a measure of cost increase associated with greater depth. The 6/t factor implies a strict inverse proportionality between costs and seam thickness. The thickness-dependent costs of a mine in a three-foot seam, for example, would be double those of the model mine in a six-foot seam. A similar inverse proportionality is assumed between DIP and cost. The effect of depth is derived by provision of numerical estimates of the rise of depth-dependent costs with depth. For example, it is calculated that a mine, 1,000 feet deeper than the standard mine would have $9.114 in additional costs.

A NOTE ON TIME SHIFTS IN COSTS

Curiously, it has been the most avowedly simplified approaches to cost analysis that have devoted the most attention to time shifts in costs. For example, my 1975 book on coal concentrated on evaluating what consti- tuted the marginal new eastern mine to supply steam coal and suggesting plausible ranges for the changes in its labour costs. Such an appraisal involved indicating possible developments in wages and output per man-day worked. Unpublished work by Milton Searl adopted the same basic philo- sophy. However, he costed more types of mines, had the USBM data to draw upon and his assumptions about wages and output per worker are quite dif- ferent from mine. Still another study in the same spirit is that of Asbury and Costello dealing with western coal. They argue that western reserves are so ample that in the long run one can open as many mines of the most efficient size (five million tons per year or more) as the market can ab- sorb but that short-run constraints on equipment supply may force the re- course to million ton mines. Thus, the short-run supply curve consists of the output from those five million ton per year mines that can be equipped and additional output coming from million ton per year mines. The analysis provides cost projections based on constancy of non-labour costs, a stable output per man-day, rising wages and steady increases in the per ton payment to worker pension funds.

Whatever may be wrong with these approaches, they at least provide a critical warning of the peril of not at least testing for the impacts of price changes. I was able to show that rising labour costs alone could prove a severe threat to the competitive position of coal even if no de- pletion effects occurred. Thus, the neglect of such cost changes may have produced over-optimism by coal forecasters, such as those preparing the 1974 Project Independence report.

FEDERAL ENERGY ADMINISTRATION COAL ANALYSES

The U.S. Federal Energy Administration (FEA) has attempted to deve- lop a more complex refined analysis of coal supply. The work took place in two phases. An earlier simpler version of the analysis was provided for use within FEA's Project Independence Evaluation System (PIES) model; a more complex model for independent use was then developed, both by ICF, Inc.

The initial ICF study was a model of careful presentation of the analysis and frank recognition of the drawback of the analysis. Particular care was taken to disaggregate regionally coal supply in a fashion that treated all important producing areas as distinct entities.

For the purpose of the report, three broad types of coal - metallurgical and low and high sulphur - were distinguished. Metallurgical coal could be produced only in Appalachia, the Central West and Rockies. In Appalachia, the Midwest, Northern Great Plains and Southwest, both high and low sulphur steam coal could be produced. The steam coal in the Central West, Gulf and Northwest was all high sulphur; that in the Rockies and Alaska, low sulphur.

However, FEA was particularly concerned with the prior lack of an explicit, verifiable presentation of the derivation of the supply figures and stressed provision of such information. Thus, the analysis began by listing five basic "assumptions" (actually often consisting of several related points) and proceeded to go through an eleven step supply analysis procedure.

The first basic assumption was that within a region the lowest cost resources of a given type are developed first and that in the long run prices of that type equal the costs of the highest cost mine operating. This simply amounts to the assumption that the coal industry is competitive and economically efficient and thus the view accords reasonably well with reality.

As the second assumption, it is argued that little change in mining technology is likely by 1990 and that current real dollar costs of labour supplies and equipment will prevail throughout the period. Given the absence of evidence that major changes in mining technologies are sufficiently close to perfection that they can be widely employed by 1990, the technological prediction seems reasonable. However, this lack of technical progress combined with pressures for sharply rising real wages suggests that real costs will rise over the period considered. Thus, the analysis perpetrates the widespread failure to consider possible rises in costs.

Assumption three is one explicitly indicated to be questionable but the only one possible using available data. Cost calculations are made presuming that only the overburden ratio (the cubic yards of cover that must be removed to mine one ton of coal) and mine size affect strip mining costs; only seam thickness, depth and mine size influence deep mining costs.

The fourth presumption is that "step functions are a good approximation". This assumption essentially amounts to arguing that an engineering costing approach is the best way to forecast supply and the approach should be implemented by defining a large but finite number (102) of possible mine types and limiting options to those types. Enough mines of each type would be opened in each region so that the supply curve would consist of a series of flat portions (see Figure 1). However, the closely related fifth assumption was that many more steps than were shown in the 1974 study should be employed and the size distribution should be derived by an explicit cost minimisation analysis. Figure 1, developed by superimposing material from two graphs in the report, shows that, as a result of both allowing more options and not aggregating figures, the 1976 analysis involved many more steps in its supply curves than did the 1974 Bureau of Mines study.

Figure 1

COMPARISON BETWEEN USBM AND ICF COAL SUPPLY CURVES
FOR A REPRESENTATIVE REGION

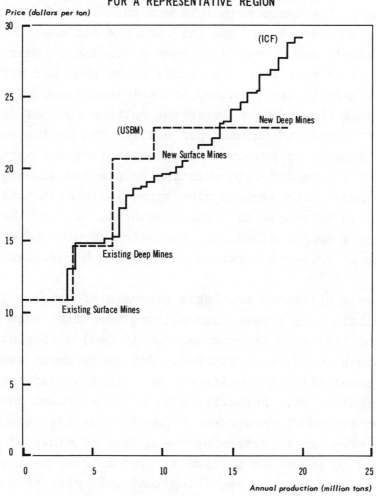

The actual methodology involved moving systematically by explicitly stated procedures to convert data on coal resources into coal supply estimates. The first four steps were largely mechanical manipulations of the data. In particular, the USBM demonstrated reserve base data discussed above provided the basic figures for the ICF analysis.

The basic procedure followed was to rearrange the USBM data so that they related to the supply areas, subdivide the reserves among the three quality types and then eliminate reserves that were considered negligible. The exclusions were reserves unlikely to be exploited - predominantly (87 per cent) in Alaska. The allocation by coal type was possible because the USBM provided analyses of the coal in different seams and a distribution by sulphur content. The identification of the metallurgical coal reserves was made by use of a definition of coking coals in still another USBM publication (Mutschler, 1975); the publication also had estimates of the metallurgical coal reserves by States that were used as a check on the ICF estimates of such reserves. The sulphur content distribution data then allowed allocating the remaining coal between the high and low sulphur categories. The basic principle was that low sulphur coal was that which could be burned without special controls and meet the Environmental Protection Agency's New Source Performance Standard. This requires that the coal contain no more than 0.6 pounds of sulphur per million Btu. Subbituminous and lignite are required to meet that standard without cleaning but it is assumed cleaning can lower the sulphur content of bituminous coals with 0.72 pounds of sulphur per million Btu to the required level.

The remaining steps involved first estimating the portion of the reserve base that is committed to existing or planned mines, allocating the remaining reserves among various mine types (with distinctions made about size, depth, seam thickness and mining method), costing these types and thus generating a supply schedule. The steps adopted to effect this analysis need not be reviewed here. A long series of approximations had to be employed.

The analysis allocated available reserves of each coal type in each region to specific mine types. Since there were five seam thicknesses, four depths and five size categories, 100 (5x5x4) different types of underground mines could be considered. Similarly seven overburden ratios and six size possibilities considered imply that 42 different types of strip mines might exist. Actually, only 66 underground and 36 strip categories contain enough reserves to justify consideration.

The next step was to determine the number of mines of each type that could be opened to produce a particular type of coal in a given region. This calculation was made by requiring that each mine have reserves sufficient to support 20 years of operation given the same recovery factors assumed for old mines.

The analysis was completed by developing an estimate of the cost of each type of mine and then proceeding to generate a supply curve. The costing procedure was to start with base case strip and underground mines of specific characteristics and vary costs on the basis of estimates of the impact of the variables included on costs. The base underground mine was a slope mine 700 feet beneath the surface using continuous mining machines, possessing unit train loading facilities and lacking cleaning facilities. The size, loading facility and cleaning plant assumptions were the same for the base strip mine. A 10/1 overburden ratio was assumed.

The base case costs were obtained by updating USBM and TRW model mine estimates. A further change was that it was assumed that the cost of capital was 8 per cent rather than the 12 to 15 per cent used in USBM studies. It was argued that the higher rate includes compensation for inflation and, therefore, is inappropriate for constant dollar analysis. Otherwise, the costing techniques were quite similar to those of the USBM and TRW.

A series of rules were developed to vary costs with differences from base mine conditions. Four types of changes were specified: (1) variation in initial investment, (2) variation in deferred investment, (3) changes in output per man-day and (4) changes in expenditures on supplies and power. In each case, further calculations were needed to translate the effect into a measure of the change in required price. Thus, the changes in capital costs must be translated into the annual receipt that, given existing tax laws, will ensure that the required rate of return is secured.

Throughout the analysis, ICF was forced to use extremely rough assumptions both to indicate the effect of physical conditions on costs and the distribution of resources among categories of physical resources. The nature of this methodology is sketched below in the review of the National Coal Model.(1)

A flat figure of $4.00 a ton was added to metallurgical coal costs to cover cleaning; a $2.00 figure was used for cleaning bituminous steam coal. Lignite and sub-bituminous were assumed to be used uncleaned.

The first nine steps, therefore, indicated the numbers of new mines of any type that can be opened in any region. Step 10 then provided costs for each type. The final step was simply to combine the data to generate supply curves.

---

1) For a fuller review see ICF, 1976a and 1976c and Gordon, 1976c.

This involved arraying mine types in increasing order of costs and indicating how much output of a given grade of coal can be produced from mines of each type. Table 4 shows a sample calculation for the lowest cost resources of low sulphur coal in the Western Northern Great Plains. Table 5 provides selected data on each supply curve. The first entry shows the smallest size mine considered in each case; everywhere but in the Gulf and Eastern Northern Great Plains, some mines in the 100,000 ton per year size class are costed. The second column shows the largest mine feasible in each case.

To give some idea of the range of output coming from mines of different types, the next two columns show how much output is accounted for by the new mine types, whatever they might be, making the smallest and largest groupwide contribution to supply. Due to lack of space, the table does not identify the specific category of mine making the largest or smallest contribution.

Table 4

ICF SUPPLY CURVE FOR WESTERN NORTHERN GREAT PLAINS LOW SULPHUR COAL - 1990
(Cheapest 1.5 billion tons of annual output only)

Mine Type		Marginal Cost (Dollars/Ton)	Output of Group (Million Tons)	Cumulative Output (Million Tons)
Existing Strip		$3.80	2.0	2.0
Planned Strip		4.14	97.3	99.3
Existing Underground		4.38	0.1	99.4
Planned Underground		4.43	1.8	101.2
New Strip 5/1	4(1)	$4.48	72.0	173.2
5/1	3	4.75	78.0	251.2
5/1	2	5.16	78.0	329.2
10/1	4	5.48	92.0	421.2
10/1	3	5.82	96.0	517.2
5/1	1	5.90	76.0	593.2
10/1	2	6.30	96.0	689.2
15/1	4	6.54	92.0	781.2
15/1	3	6.95	96.0	877.2
10/1	1	7.16	94.0	971.2
15/1	2	7.50	96.0	1,067.2
20/1	4	7.65	92.0	1,159.2
20/1	3	8.13	96.0	1,255.2
15/1	1	8.46	94.0	1,349.2
20/1	2	8.78	96.0	1,445.2
20/1	1	9.84	94.0	1,539.2

1) The first figure is the overburden ratio; the second, annual output in millions of tons for the individual mines.

Source: ICF, 1976a, p. 79.

50

Table 5

SELECTED CHARACTERISTICS OF ICF 1990 SUPPLY CURVES

(Million tons per year except as noted)

Area	Coal Type(1)	Smallest New Mine	Largest New Mine	Production Of Smallest Output Group Costed	Production Of Largest Output Group Costed	Strip Mine Production Costed	Total Production Costed	Marginal Cost Of Total Production Costed	Marginal Cost Of Production One-Half Maximum
Northern Appalachian	M	0.1	1.0	0.3	2.5	5.6	50.5	$ 73.90	$ 22.07
	L	0.1	0.5	0.1	1.2	5.2	19.6	71.90	18.42
	H	0.1	2.0	1.0	68.0	230.8	1,954.5	71.90	19.54
Central Appalachian	M	0.1	2.0	1.0	34.6	92.1	381.4	73.90	23.18
	L	0.1	2.0	0.5	5.1	63.7	179.8	71.90	17.51
	H	0.1	2.0	0.9	19.0	131.7	467.0	71.90	20.52
Southern	M	n.a.	n.a.	n.a.	n.a.	0.2	11.0	14.70	n.a.
	L	n.a.	n.a.	n.a.	n.a.	5.0	7.4	18.81	n.a.
	H	0.1	1.0	0.2	2.4	2.7	41.3	29.10	20.52
Midwest	L	0.1	1.0	0.2	1.0	6.7	20.0	29.10	19.51
	H	0.1	3.0	24.0	81.0	685.4	2,806.5	29.10	18.57
Central West	M	0.1	1.0	0.2	2.0	3.5	4.7	31.10	26.69
	H	0.1	3.0	4.0	34.0	209.8	508.4	29.10	20.07
Gulf	H	1.0	4.0	3.0	10.0	128.8	128.8	9.84	6.54
Eastern Northern Great Plains	L	1.0	4.0	6.0	10.0	105.0	105.0	9.84	7.16
	H	1.0	4.0	68.0	70.0	832.1	832.1	9.84	7.16
Western Northern Great Plains	L	0.1	4.0	54.0	116.2	1,537.3	3,930.2	23.47	11.10
	H	0.1	4.0	8.0	54.0	805.0	1,165.1	23.47	7.16
Rockies	M	0.1	2.0	0.2	2.0	1.5	35.7	31.10	19.70
	L	0.1	2.0	0.5	4.0	28.4	142.2	29.10	17.75
Southwest	L	0.1	1.0	0.3	1.0	9.7	13.0	23.47	6.55
	H	0.1	4.0	0.3	8.0	90.8	105.2	23.47	7.16
Northwest	H	0.1	3.0	1.0	3.0	20.0	54.5	23.47	11.37
Alaska	L	0.1	3.0	1.0	4.5	15.7	96.4	35.20	21.28

Notes: See text for more details.

1) M is metallurgical coal; L, Low Sulphur; H, High Sulphur; n.a. means no new mines feasible due to reserve depletion.

Source: ICF, 1976a, p. 44-94.

The next column shows how much of the total supply costed is strip mined. The next column shows the total amount of production costed. The next to last column then shows the marginal cost of maximum production; the last column shows that marginal cost of output half the maximum level.

The discussion was concluded by noting the problems and raising questions. Thus, it noted that the impacts of such characteristics as roof or floor conditions, the gassiness of the mine, the dip of seams, the extent of water problems, the hardness and cleavage of the coal and the amount of prior mining have not been considered. Then the authors noted many other issues that they feel should be resolved.

Basically, as ICF is well aware, the analysis is a fairly straight-forward extension of the USBM work. As already noted, the main changes were providing an explicit allocation of reserves to mines and presenting the full details of the costing analysis.

However, no significant amounts of additional data were obtained and the costing model apparently adopted roughly the same assumptions about the effect of physical conditions on coal as was employed by the U.S. Bureau of Mines.

This discussion of the 1976 update may conclude by noting the price levels that the analysis indicated would prevail. Only limited data are available on this - namely prices for the reference case for 1985 are shown in the National Energy Outlook. These are summarised in Table 6.

THE NATIONAL COAL MODEL

In the fall of 1976, programming work was completed on an extension of the ICR coal supply analysis called the National Coal Model (NCM). This model is designed to operate independently, rather than as a compo-nent of the FEA's overall supply model. The NCM is a linear programming model and, thus, is capable only of optimising the supply of a present level of demand. However, to lessen the problems of such an approach, the analysis was broadened to include optimisation of electric utility supply. Therefore, the model is really a coal and electricity model but discussion here stresses the coal portion.

The model is designed to minimise the costs of supplying the sum of a preset amount of electricity demand and preset non-electric utility de-mands for coal. Thus, the model determines the optimal combination of fuels to meet electricity demand and, thus, the share of coal is deter-mined within the model. Other coal demands are set outside the model. Clearly, the analysis provides a great deal more than an analysis of coal

Table 6

FEA ESTIMATES OF COAL PRICES PREVAILING IN 1985 REFERENCE CASE
FOR $13/BARREL OIL

Region	Btu content per ton of coal (millions)	Price of High Sulphur Coal		Price of Low Sulphur Coal	
		Dollars per ton	Cents per million Btu	Dollars per ton	Cents per million Btu
Northern Appalachia	24	12.942	54	24.941	104
Central Appalachia	24	12.662	53	24.147	101
Southern Appalachia	24	14.452	60	25.977	108
Midwest	22	10.780	49	22.779	104
Central West	22	11.350	52	n.a.	n.a.
Gulf	14	6.610	47	n.a.	n.a.
Eastern Northern Great Plains	14	5.248	37	8.160	58
Western Northern Great Plains	19	4.080	21	5.844	31
Rockies	22	n.a.	n.a.	10.028	46
Southwest	19	4.680	25	8.901	47
Northwest	19	5.950	31	n.a.	n.a.
Alaska	19	n.a.	n.a.	7.995	42

Source: U.S. FEA, 1976b, p. F-30 - F-37.

supply. The model incorporates estimates of electricity demand by region,
estimates of efficiency in electricity distribution, the shape of load
curves (i.e., the extent to which demand varies over the course of the
day or the year), existing fuel use capabilities of power plants, plans
to expand or add stack gas scrubbers, pollution regulations, power plant
thermal efficiencies, prices of rival fuels, capital and non-fuel operating costs of different types of electric power plants, transmission
capabilities and costs, non-utility demands for coal and costs and capabilities for coal transportation.

The model is designed so that all the basic assumptions can easily
be varied but the initial write-up presents a data base for 1980 that
illustrates the degree of detail possible with the model. A major difference between this model and prior ICF work is that much greater detail is
possible and was provided in the data base. The number of coal supply
regions was increased from 12 to 30; the number of consumption centres
from 9 to 35; the number of coal types considered from 8 to 40. In particular, the original model considered four ranks of coal - metallurgical
and high, medium and low Btu subdivided between high and low sulphur.
Five different Btu levels and eight different sulphur levels were defined

for the NCM; metallurgical coal no longer is distinguished but metallurgical demands are met from the coals with Btu and sulphur contents that meet the specifications of coking coal. This process allows explicit modelling of the competition for such coal among different users. Where the prior model assumed fixed input costs and this was subject to the criticism of neglecting the deterioration of the relative position of eastern underground mining, the revised model is designed to consider changes in real costs as well as the effect of general inflation. The model expresses values in actual rather than in constant dollars (the earlier report used constant dollars).

The executive summary of the description of the model sums this up by noting that basically the model has the capability of evaluating the implications of a wide variety of public policy and market developments. The key characteristics are described as providing equilibrium solutions, a high degree of resolution, analysis of price sensitivity, flexibility, understandability and usability. The claims of an equilibrium solution, flexibility, understandability and usability are closely related. To a considerable degree, these characteristics are inherent to a linear programming model. However, special effort has been taken to go beyond the basic virtues of linear programming to provide the desired characteristics. Flexibility is re-enforced by special efforts to facilitate changes in assumptions. Each step of the analysis was spelled out with considerable care. A final report format was developed that was intended to make the results easier to follow.

It is price sensitivity and resolution that are of most interest here, however, and they are attained by special effort on the part of the programmers. Price sensitivity mainly denotes that this model follows the example of the earlier FEA work of assuming upward sloping supply curves. While this may seem an obvious approach, most linear programming models of the coal industry (see e.g., Schlottmann) have assumed that marginal costs are constant up to some arbitrarily set capacity figure. The high degree of resolution refers to the large number of supply and demand areas and coal types treated.

The 30 supply areas treat 22 states with "substantial" reserves each as a single region but West Virginia, Kentucky, Colorado and Montana are each divided into two regions. The five Btu content classes are 26 million Btu/ton or more, 23-25.99 million Btu/ton, 20-22.99 million Btu/ton, 15-19.99 million Btu/ton and less than 15 million Btu/ton. The eight sulphur level groups range from a low of 0.40 pounds or less per million Btu to more than 2.50 pounds per million Btu. (The pound per million Btu basis was adopted because U.S. Environmental Protection Agency New Source

Performance Standards set controls on the basis of pounds of sulphur oxide per million Btu.) New features in the model include an improved treatment of coal washing, the ability to make runs involving use of surge capacity (ICF's concept of extra capacity available only in times of peak demand) and handle other short-run conditions.

The supply analysis begins with a modification of the methodology of the earlier study. Most of the changes are simply those necessitated by the greater disaggregation but a few other changes were made.

Thus, the discussion begins with an allocation of the USBM demonstrated reserve base for each State to different coal types. Then an elaborate effort is made to determine the capacity of existing mines to produce coal of each type in each State in 1980. This involves first estimating the 1975 pattern and then projecting closings. The 1975 allocations were based on available data on distribution of coal. Estimates on closings are based on the premise that some fraction of the large (200,000 plus tons per year) mines operating 30-35 years before would be retired - 3 per cent of underground and 5 per cent of the strip mine capacity.

The prices required to keep old mines operating were estimated by simple rules. Since, in most cases, demand is sufficient to necessitate use of new as well as existing mines, refined analysis of existing mines was considered less critical. The required price was assumed to be that prevailing in 1973. This year was selected because it was a year without marked supply shortages so prices did not include any premiums for such shortages. It was further assumed that coal sold on the spot market for 10 per cent more than coal sold under contracts because of higher risks. The required price for spot and contract price is calculated by applying the 10 per cent premium factor and estimates of the relative share of spot and contract coal to the given 1973 price. For example, in Pennsylvania the average was $10.30 and the split was 70 per cent contract, 30 per cent spot. The contract price (X) then is found from the equation

$$0.7 X + 0.3 (1.1) X = 10.30$$
$$.7 X + .33 X = 10.30$$
$$1.03 X = 10.30$$
$$X = \$10.00.$$

These calculated prices were converted, by use of state averages, into costs per million Btu and the price of coals of a specific Btu content was calculated by multiplying the estimated cost per million Btu by the Btu content and then converted to 1975 dollars by use of the implicit deflator for the gross national product. Spot supplies were further divided so that one portion was available at a lower price midway between the higher spot price calculated as just described and the corresponding contract price. The first increment of surge capacity was assumed to become available at prices 150 per cent of the higher spot price and the rest of surge capacity opened at prices double the higher spot price. Specific escalation factors (not shown) for labour and supplies were used to calculate the rise in the required price to reflect rising costs.

The next step was to allocate reserves to new mines. The first step was, as before, to delete reserves dedicated to existing mines. A new second step was added of estimating how much of the strippable demonstrated reserve base was in locations such as under cities, highways or parks that precluded extraction.

The next steps were modifications of those undertaken in the earlier study. First, strippable reserves were allocated by "overburden ratios" (the ratio of cover thickness to coal seam thickness). The minimum ratio in the area generally was set at 15 per cent above the 1970 average. For example, in northern Appalachia a 14.8 to 1 ratio prevailed in 1970 so new strip mines in the region were assumed to have overburden ratios of at least 17 to 1. It was further assumed that reserves with ratios greater than 45 to 1 were uneconomic.

The model further assumes that economic reserves as a whole can fall in the overburden ratio categories - 5/1, 10/1, 15/1, 20/1, 25/1, 30/1 and 45/1 (the last category is triple the size of the others). Reserves are assumed uniformly distributed among ratios between the minimum feasible and 45/1. Then strippable reserves were assigned to mine sizes. A maximum feasible size was set for each state and for most states the reserves were considered to be split equally among the feasible size categories. However, in the West, only 10 per cent of the reserves were allocated to the smallest size classes and the remaining 90 per cent was distributed equally among the three largest size classes (see Table 7). Analogous procedures (see Tables 8 and 9) were used to assign deep reserves to seam depth and mine size categories.

Having allocated the reserves in this fashion, supply curves were developed by a modification of the costing rules used in the earlier study. A special feature of the 1980 analysis was that construction lead times are such that all the large mines that will operate in 1980 probably have already been announced and data on such announcements were incorporated into the analysis.

The basic costing analysis apparently is that used in the earlier study by which first USBM model mine data were revised to relate to two base mines - a million ton per year slope mine working a six-foot seam

Table 7

DISTRIBUTIONS OF STRIPPABLE RESERVES TO MINE SIZE

NCM Region	Mine Size (10^6 tons/year)				
	0.1	0.5	1.0	2.0	3.0
PA	.334	.333	.333	-	-
OH	.334	.333	.333	-	-
MD	.500	.500	-	-	-
NV	.334	.333	.333	-	-
SV	.334	.333	.333	-	-
VA	.334	.333	.333	-	-
EK	.334	.333	.333	-	-
TN	.334	.333	.333	-	-
AL	.334	.333	.333	-	-
IL	.200	.200	.200	.200	.200
IN	.200	.200	.200	.200	.200
WK	.200	.200	.200	.200	.200
IA	.334	.333	.333	-	-
MO	.334	.333	.333	-	-
KS	.50	.50	-	-	-
AR	.50	.50	-	-	-
OK	.50	.50	-	-	-
TX	.050	.050	.300	.300	.300
ND	.200	.200	.200	.200	.200
SD	.200	.200	.200	.200	.200
EM	.200	.200	.200	.200	.200
WM	.050	.050	.300	.300	.300
WY	.050	.050	.300	.300	.300
CN	.200	.200	.200	.200	.200
CS	.200	.200	.200	.200	.200
UT	.334	.333	.333	-	-
AZ	.050	.050	.300	.300	.300
NM	.050	.050	.300	.300	.300
WA	.050	.050	.300	.300	.300
AK	.050	.050	.300	.300	.300

700 feet deep using continuous mining, having unit train loading facilities but no cleaning plant and a million ton per year strip mine working a six-foot seam at a 10/1 overburden ratio - again with unit train loading facilities but no cleaning plant. Rules of thumb, also taken from the earlier study, were developed to deal with deviations from the assumed conditions. For example, each 100 foot reduction in depth reduced initial investment by $500,000 and another $5,000,000 was removed from costs if the mine were a drift rather than a slope mine.

However, several modifications were made in the analysis for the NCM. First, the analysis was modified so that it could handle the applicable depletion allowance - the lesser of 10 per cent of gross revenues or 50 per cent of accounting profits. When coal was cleaned, the losses in cleaning were taken into account in the costing. For example, in

Table 8

DISTRIBUTION FACTORS USED TO ASSIGN
RESERVES TO SEAM DEPTH CATEGORIES

PIES Region	Feet Below the Surface			
	Drift	400	700	1,000
Northern Appalachia*	.050	.250	.350	.350
Central Appalachia	.050	.250	.350	.350
Southern Appalachia	.050	.250	.350	.350
Midwest	X	.300	.350	.350
Central West	X	.300	.350	.350
Gulf	X	X	X	X
Eastern Northern Great Plains	X	X	X	X
Western Northern Great Plains	X	.333	.333	.334
Rockies	X	.333	.333	.334
Southwest	X	.333	.333	.334
Northwest	X	.333	.333	.334
Alaska	X	.333	.333	.334

*) Ohio was treated as an exception and not assigned any drift
   reserves. The 400-foot category production was increased to
   .300.

Table 9

DISTRIBUTION FACTORS USED TO ASSIGN
RESERVES TO MINE SIZE CATEGORIES

Seam Thickness Categories (in inches)	Mine Size Categories (10^6 tons per year)			
	0.1	0.5	1.0	2.0
72	.250	.250	.250	.250
60-72	.334	.333	.333	-
48-60	.334	.333	.333	-
36-48	.500	.500	-	-
26-36	.500	.500	-	-

Appalachia output was reduced 30 per cent (and costs per ton of clean
coal increased commensurately) when cleaning was undertaken. Account
was also taken of the fact that for tax purposes depletion is secured on
the value of cleaned coal. The state by state nature of the analysis
allowed for explicit consideration of state taxes on coal mining.

58

Finally and most critically, provision was made for rising costs of mining. The initial cases assumed 5 per cent per year cost increases but the model can handle different cost increases for different cost components. As the illustrations of the methodology suggest, much of this model is still built on undocumented assumptions. Nevertheless, the model has great potential for improving coal supply analyses. The flexibility of the model should be most helpful in allowing the ultimate calculation of numerous scenarios that test such questions as the impact of rising real labour costs and relationships between seam thickness and cost different from those used in this model (for example, the quite different relationship proposed by Zimmerman). Thus, the model seems a significant contribution to coal supply analysis. Results, as yet unpublished, of work with the model confirm that it does provide useful insights into the impact of public policy on coal supply pattern.

PUBLIC POLICY, LABOUR PROBLEMS AND THE FUTURE OF UNITED STATES COAL

Before turning to a synthesis of the cost studies, it is essential to pause to recognise explicitly the numerous public policy issues that complicate prognosis. The most publicised questions are environmental ones relating to the air pollution associated with coal use and the land disturbance arising from surface mining of coal. A second set of issues involves the various concerns about the structure of the present and future coal industry. A third problem area - that of labour relations - is not strictly speaking a public policy issue but, as much for convenience as because the difficulties have influenced public policy, the labour situation may be noted here.

Air pollution regulations have been the primary policy influence on coal use. Given the complexity of air pollution regulations, stress may be placed on discussion of the one example most relevant to appraisal of the prospects for coal. Several different air pollutants have been identified as the source of sufficient damage to justify controls. The three that are associated with, among other things, coal burning are particulates, sulphur oxides, and nitrogen oxides. The sulphur oxides constitute the critical issue.

Particulates are a catchall for a wide range of materials such as ashes and airborne wastes of grain milling. Neither coal nor fuel use as a whole constitutes the predominant source of particulates. More critically, particles produce the most visible and long-recognised pollution. Control regulations have long existed and control technologies are well established. Nitrogen oxides are primarily the result of combustion; the high temperatures involved lead to formation of the oxides in the air. However, automobiles are the main source of nitrogen oxides, and to date pressures to limit such emissions from stationary fuel-burning facilities have been moderate.

In contrast, it is the fuels used by large stationary boilers that are the predominant source of sulphur oxides. Such boilers frequently burn coal or residual fuel oil (so-called because its principal constituent is the residuum that remains at the bottom of the distilling towers used in petroleum refining). Historically, both the coal and oil used have contained significant amounts of sulphur that are transformed into sulphur oxides during combustion.

The policy for sulphur dioxide emission control is, not only the most important policy affecting coal, but probably the most complex. Two types of complexities may be distinguished. First, many different groups have influenced the policy adopted, both in form and substance. Second, the policies themselves have become increasingly convoluted.

The groups include the actual policy makers, the lobbyists for controls (mostly environmentalists), and the coal-using industries affected. The second of these groups needs little treatment, its success being by now familiar. In contrast, the policy makers within our intricate regulatory structure and the companies responding to that structure merit fuller review. The principal policy makers have been the Congress, the Environmental Protection Agency (EPA), and the state governments. (A fourth group - Indian tribes - also have been given some authority.) EPA has naturally pushed vigorously to ensure compliance with congressionally set air pollution control objectives - although at one point in the 1970s, it is true, EPA did suggest that the states stop trying to comply more rapidly than necessary with federal requirements. This tendency towards as much vigour as possible suggests EPA tends to give more weight to the views of its environmentalist supporters than to other points of view.

Firms in regulated industries have frequently sought relief in the courts and, although the enforcers have ultimately prevailed, the firms have at least won delays in enforcement. In Ohio, for example, the state pollution control agency was unwilling to impose restrictions as stringent as EPA believed were required. EPA therefore imposed federal restrictions in 1975. Several Ohio electric utilities took the matter to court, and the implementation of the restrictions is being delayed while the litigation between those utilities and EPA continues.

Similarly, a number of utilities won relief from the pressure applied by EPA to use stack gas scrubbers (discussed below) when some district courts found that the scrubbers were not technologically or economically feasible. On appeal, the Supreme Court decided in 1976 that, since the law was intended to force technological developments, feasibility was not germane. Again, we can wonder about the wisdom of this type of decision making.

Congressional action has involved most of the predictable political conflicts. Press accounts periodically report rumours that various legislators are vying to maintain or assert a dominant role over pollution control. Members of Congress from coal states continually try to find ways around air pollution regulations. A blatant example of local protectionism came in the provision of the 1977 Clear Air Amendments that allowed requiring the maximum possible use of local coal if serious damage to the local economy would be avoided. Either the U.S. or the state government could make the request, but federal concurrence in state requests is required. In early 1978, the states of Illinois and Ohio were trying to use this provision to prevent local electric utilities from shifting to western coals.

When policy makers impose goals that seem unreasonable to the industries affected and provide no credible incentives for compliance, it is improbable that the goals will be met. Managements recognise, at least subconsciously, that if their companies' products are essential, extensions will be granted to permit continued operation even if air pollution regulations are being violated. This seems a powerful incentive for dilatory behaviour and apparently is one of the forces that produces the complex policies of stringent goals hedged with numerous escape clauses.

To gain possible insight into the problem of air pollution, let us look first at sulphur oxide control policy. In present policy, maximum allowable average annual concentration of pollutants in the atmosphere is established by EPA, as is the deviation from the annual norm allowed on any one day (and even perhaps in any three - or, indeed, one-hour period). In addition, limits are imposed on the increases in pollution allowed in areas with air already significantly cleaner than required by the basic standards. This controlled increase policy, called "prevention of significant degradation", came into the enforcement process through a court interpretation of the statement in the Clear Air Act that air quality should be preserved. Lacking precise legal guidelines, EPA has difficulty developing workable rules to "prevent significant degradation" until the Congress, in the 1977 Clean Air Amendments, finally sought to lessen the difficulties by establishing definitions of non-degradation.

The new rules are based on concepts developed by EPA. Three levels of strictness were defined - Classes I, II, and III (Class I being the most stringent) - and specific limits were set on the allowable increase in pollution in areas assigned to each class. Some land had to be subject to the severest possible restrictions. Thus, all land previously set as Class I along with certain public lands (such as pre-existing national parks exceeding 6,000 acres in area) must be in Class I. Others such as national parks in areas greater than 10,000 acres faced the lesser limitation of being forced to be put into either Class I or Class II. No comparable provisions exist for automatically allowing lesser restrictions. All other lands are initially put into Class II, and the governor of the state (or states) involved may petition for up- or down-grading. The 1977 amendments also stipulated that areas not in compliance with air quality regulations be subjected to more stringent controls than are areas meeting the rules: new facilities would not be allowed unless the areas demonstrated that they had made vigorous efforts to comply or that the facility was vitally needed. In addition to these rules governing atmospheric concentrations, separate limits based on technical feasibility are imposed on invidivual facilities. Given the different bases of these two sets of rules, the compatibility is unclear.

Until 1977, the emission rules for sulphur oxides had required that new facilities (those whose planning was begun after the rules were promulgated in 1971) burning more than 250 million Btu/hour in fuel must limit their emissions to 1.2 pounds of sulphur oxides per million Btu burned. This standard could be met by using coal naturally low enough in sulphur content, by cleaning the coal to acceptable levels before combustion, or by using devices called stack gas scrubbers that capture the sulphur oxides after the fuel is burned thus preventing their discharge. In practice, precombustion cleaning has not been feasible, so that until 1977 the choice was between naturally low sulphur coal and scrubbers. In the 1977 amendments, the option of using naturally low sulphur coal was removed by introduction of the concept of the "best available control technology" (BACT) - a requirement that technically feasible sulphur removal technologies be employed to the maximum possible extent. EPA initially indicated an intention to issue precise rules in early 1978, but, as of June, these rules had not been provided. Statements have been made by EPA officials that the rules will combine minimum sulphur removal (probably around 90 per cent) and maximum emission (probably about 1.2 pounds of sulphur oxide per million Btu burned). Such rules will both force the use of control techniques and prevent the resort to coal so high in sulphur that, even with treatment, there would be more pollution than before.

It may be asked what benefit has been hoped for from the imposition of the BACT requirements and the resulting substitution of scrubbed high sulphur coal for natural low sulphur coal. BACT supporters seem to have been strongly motivated by hope that both requirements would prevent a shift from eastern to western coal - either because the shift would plunge Appalachia into still greater poverty than it now suffers or because it would lead to western strip mining. A better rationale is that BACT will produce a needed reduction in pollution. Given economic growth, one envisions that new emission sources will arise and with the old standards total pollution would increase, possibly to undesirable levels. BACT could prevent this rise. This defence would be more valid if BACT rules concentrated on emission rates rather than rates of sulphur removal.

Even the imperfect models now used for projecting coal industry production patterns show that BACT slows rather than eliminates the growth of western markets. Such displacement of western coal as is projected generally benefits coal producers in the Midwest more than those in Appalachia. The model builders believe that resource depletion has so seriously eroded the ability to expand Appalachian production that the efforts of local politicians to aid the region's coal industry will have little impact. The calculations clearly were designed to present what is readily calculable (the maximum benefits of BACT to eastern coals), and while they should be adjusted on the basis of more realistic assumptions, as it is, they still show little benefit to Appalachian coal. No models can adequately handle two critical possible offsets to BACT. First, BACT is a powerful incentive to shifts from coal to nuclear power. Conventional modeling approaches base the choice on least cost considerations and produce estimates of very high nuclear construction. The modelers doubt whether such levels are politically feasible. Similarly, no way exists to determine whether the production cost problems discussed above will cause eastern mining costs to rise enough to outweigh the effects produced by BACT.

THE STATE OF SCRUBBER TECHNOLOGY

BACT is a curious effort to further a controversial approach to sulphur oxide control. Evidence suggests that stack gas scrubbers may still not be as effective as their advocates contend. Let us review the available evidence. EPA receives from PEDCo Environmental, a research consultant, extensive bimonthly reports on scrubber utilisation, presenting detailed operating histories of scrubbers actually in place. Scrubber advocates tend to report the summary data and ignore the histories. The summaries show a growing number of "operational scrubbers (thirty-one as of January 1978), but the details make it clear the PEDCo uses a generous definition of "operational". Units are added to the operational category when test operations begin. No distinction is made among units that are operable, those that are out of commission, and those that fail to remove the required amount of sulphur.

Most of the successful units are engaged in mild scrubbing (50 to 60 per cent removal) of sulphur oxides from low sulphur coal. Those successes that have occurred in scrubbing high sulphur coal appear to have come at high cost. Extra units have been installed so that frequent cleaning can occur without the plants being put out of service. (The cleaning is necessary because the units quickly become clogged and corroded.) Long shakedowns are often required to make the scrubbers operational, and even with the shakedowns, frequent outages seem to occur. Finally, since scrubbers capture the sulphur oxides in some absorbent such as limestone, problems arise in disposing of the resultant sludge. In short, scrubbers create cost, reliability, and waste disposal problems.

How, then, did the devotion to scrubbers arise? The initial liking is easy to explain. Policy makers eagerly embraced the arguments of the equipment manufacturers and the coal industry (especially the high sulphur eastern coal industry) that scrubbers were a cheap solution to the problems of air quality, providing rapid results and low cost while maintaining existing regional production patterns. However, the advocates of scrubbers failed to anticipate the difficulties that arose.

It is often charged that electric utilities "did not try hard enough" to make scrubbers work. Accusations of this sort are too vague to permit reasoned evaluation, but it can be argued that public policy was poorly designed to stimulate vigorous scrubber development. The utility industry had good reason to expect that failure to perfect scrubber technology would simply lead to extended compliance deadlines, and the federal government failed to adopt effective incentives to encourage compliance (such as an emission tax) or to finance scrubber development.

The refusal to back off from scrubbers can be attributed to the interaction of three quite different forces - the usual reluctance of politicians to confess error, the pressures of coal-state legislators to protect local coal, of low sulphur western coal. These efforts having failed to stop a shift to western coal, the current justification for BACT, as I noted, is that it will slow this shift. Faith is now being placed on BACT which may prove equally ineffective.

In any case, the history of sulphur oxide control strategy to date has involved the setting of goals and the subsequent backing away from them - which happens when goals are irreconcilable with other goals.

Where we are heading cannot be told with certainty. However, the fact that several proposed new coal-burning power plants are having difficulties securing approval on environmental grounds is suggestive. A particularly dramatic example is provided by the problems of developing coal-fired plants to serve California.

Up to 1978, the emphasis for California's power developers has been on seeking sites outside the state where air pollution regulations are less stringent and then transmit the power into the state. Plants have been built in Nevada and New Mexico, and the last one that was planned - the Kaiparowits plant - was to have been built in Southern Utah. But in 1973, with environmental objections having grown quite substantial, Secretary of the Interior Morton indicated that he would not approve the plant. Extensive further efforts were made, and in early 1976, the Interior Department released a final environmental impact statement, with a decision expected shortly thereafter. Before this decision could be reached (or at least before it could be promulgated), the participating utilities cancelled the project on the grounds that regulatory uncertainties made it too risky. Another group of utilities has proposed an alternative plant in Southern Utah, but similar regulatory problems have arisen. Moreover, the proposed first coal-fired plant in Idaho has also been unsuccessful in finding an acceptable site. It will be interesting to see whether the suggestions that coal plants instead of nuclear units be built in California and Maine will ever come to fruition.

The Clean Air Amendments include numerous escape clauses that could delay attainment of the stated goals. It seems safe to conclude, however, that construction of coal-fired power plants is beginning to involve the lead times comparable to those already plaguing nuclear power. The question is whether these constraints will cause the inadequate capacity expansion so widely feared by the electric utility industry. What expansion rate they require and what the delays will be cannot be satisfactorily forecast. Moreover, matters are even more confusing for non-utility users of coal. Indeed, we have almost no idea about the economics of coal use by manufacturing plants under BACT. Thus, existing policies affecting coal use clearly conflict with alleged goals of greatly encouraging coal use.

Observers of pollution problems in coal fear that two further sources of pressures may emerge. Particulate emission standards may be tightened either because of fears that the small particles now not trapped are a serious health hazard or because of concerns over the trace elements such as uranium and lead contained in coal. Nitrogen oxide emission controls might also be tightened.

Underground mining faces at least two other regulatory problems beside the well publicised one of health and safety rules. First, the growing reach of environmental regulations in general has lengthened the lead times for mine construction. Numerous permits must be obtained from both the agencies regulating mining and the various environmental authorities, such as those charged with water pollution control. Second, the 1977 Surface Mining Act requires control of the surface effects of underground mines as well as of the effects of surface mines. Obviously, the main effects of the act are on surface (strip) mines.

Strip mining without reclamation can produce land disturbance that residents of and visitors to the area find highly objectionable and which precludes agricultural or recreational use of the land. The pits can be safety hazards, and rainwater washing through the mine can deposit silt and (in Appalachia) acidic material in waterways. In the West, surface mining activity may cause upheavals in the social framework of the affected area and put strains on local governments. The problems are a graphic example of what discussions of environmental issues refer to as the "site specific" characteristics of the situation, i.e., that every individual case involves different risks and must be judged on an individual basis.

In particular, quite distinct problems arise in Appalachia outside of Ohio, Ohio and the Middle West, and west of the Mississippi. The nature of strip mining and its damage is naturally quite different when the cover is hilly, as in most of Appalachia, rather than flat. Both mining conditions and restoration of the land are more difficult in such hilly land. However, a different complication occurs west of the Mississippi; the aridity of the land is such that revegetation can be extremely expensive.

Several studies have tried to undertake benefit/cost analysis of strip mine reclamation.(1) The universal condition is that easily measurable benefits of reclamation such as restoring the land to other uses fall far short of the estimated costs of a rigorous degree of reclamation. Thus, the policy of imposing such controls is implicitly based on high valuation of avoiding the aesthetic insults, changes in social structure, and strains on local government. (Certainly, many attacks on both Appalachian and western surface mining are presented entirely in terms of the upheavals caused for local communities.)

_____

1) See U.S. Congress, Senate Committee on Interior and Insular Affairs, The Issues Relating to Surface Mining, 92nd Congress, 1st Session, 1971 for a good collection of the early literature and, Energy and Environmental Analysis, Inc., Benefit/Cost Analysis of Laws and Regulations Affecting Coal, Washington; U.S. Government Printing Office, 1977 for a more recent appraisal.

In any case, individual states have imposed strip mine regulations of various degrees of severity.(1)   The U.S. Congress undertook discussions about surface mining regulations early in the 1970s. It was not until 1974 that a bill was passed, but it and a 1975 successor were successfully vetoed by President Ford. However, in 1977 a federal act was passed and signed by President Carter. The federal-state jurisdication problem was resolved by granting the states the right to enforce the rules so long as they were at least as strict as required by federal law. The state control could, moreover, be largely extended to federal lands although certain powers could not be delegated.

The Surface Mining Control and Reclamation Act, like most laws, is a complex collection of provisions. Several regulatory goals exist – supervising ongoing strip mining, restoring previously stripped land, and controlling the surface effects of underground mining. Reclamation of abandoned strip mines is to be financed by a tax on coal mining (underground as well as surface albeit at a lower rate - 15 vs. 35 cents per ton on surface mines). The law sets up rules for seeking and receiving a permit to mine. Applicants must demonstrate their ability to comply with the law. A bond must be posted to guarantee performance.

Section 515b sets forth 25 strip mine reclamation requirements, including the following:

- to "maximize" initial recovery so that a second disturbance of the surface to recover other seams is avoided,
- to restore land to a condition that allows a use at least as good as the use prior to mining, with extra effort required if the land is considered prime agricultural land,
- to restore land to the original contour, except where impractical,
- to stabilize areas so as to avoid erosion,
- to segregate and preserve the quality of topsoil or of a subsoil of better quality than the actual topsoil,
- to avoid disturbance of hydrologic balance with special emphasis upon alluvial valleys in the West (these being valleys where disruption of water flow would interfere with farming), and
- to revegetate the reclaimed land.

---

1) Energy and Environmental Analysis, op. cit., as noted, evaluates the impacts of the laws; the report analyzes extensive data on the laws presented in an earlier report, Energy and Environmental Analysis, Inc., Laws and Regulations Affecting Coal, Washington: U.S. Department of the Interior, 1976. Another summary of the laws appears in ICF, Inc., Energy and Economic Impacts of H.R. 13950 ("Surface Mining Control and Reclamation Act of 1976"), Washington: ICF, Inc., 1977.

State governments are authorised to prohibit strip mining on lands if it would produce specific hazards (such as increased chance of floods) or merely if it would be incompatible with land use plans. The Secretary of the Interior is required to prohibit new mines in national parks and other classes of federal lands, is authorised to prohibit new mines on the basis of state criteria, and must consider private requests for prohibitions. The law requires that, in cases where federal coal rights were retained when the surface was sold, the surface owner must give written permission for mining. Moreover, if substantial opposition to strip mining exists among surface owners, the secretary is authorised to prohibit the mining. To meet objections of those States that felt they had obviated federal legislation by passing their own stringent laws, the federal act allows transfer of administration to states that develop programmes acceptable to the federal authorities.

Opinions of the law obviously differ among observers. A study of the similar proposed 1976 strip mine bill made by ICF, Inc., for the Environmental Protection Agency noted that it would have somewhat greater effects than existing state regulations. ICF qualified this conclusion by warning that the law was sufficiently ambiguous to be interpreted so as to produce yet more severe effects. The coal industry has expected the worst and has sued the Interior Department for allegedly adopting more severe interpretations than the law required.

On its face, the current law does seem set up to disrupt coal production. The introduction of the alluvial valley provision, the special treatment for prime agricultural land, and the protection for surface owners seem particularly likely to produce implementation problems. Moreover, Secretary of the Interior Andrus has displayed a clear desire to shift his department from its traditional role as promoter to industrial development to a new role of protector of the environment with all that this implies - including more stringent interpretations of environmental laws.

Other problems affecting western coal come out of the complexities of land ownership patterns in the region. With the federal government the principal owner of coal rights, a substantial number of problems arise simply from the need to contend with federal policies on the leasing and exploitation of government-owned coal. In addition, much of this federal "ownership" consists of mineral rights on lands for which the right to use the surface has been sold. The Surface Mining Act contains, as mentioned, provisions requiring the consent of the surface owners before mining can proceed - creating a substantial likelihood of obstructionism.

Ownership of coal or coal rights is further fragmented by the long-standing "checkerboard" pattern of grants of land to railroads (so-called because the split between railroads and the federal government looks like a checkerboard when mapped), as well as by the fact that indian tribes, the states, and other private parties also own coal rights. But the most pressing problem here seems to be the difficulty in securing permission to exploit existing leases. Approval of exploitation plans is a major federal action requiring an environmental impact statement. Unfortunately, the Bureau of Land Management, which has primary responsibility for approvals, is inadequately staffed to complete the statements with any degree of rapidity.

A long-term problem arises from the combination of the moratorium on new leases that has prevailed since 1971 and the 1976 amendments to the leasing laws. The moratorium does allow leasing of limited tracts adjacent to previously leased land when these tracts are needed to permit optimal development of the property. However, the Interior Department has been quite dilatory in granting such leases.

The overall moratorium began because of concern that too few leases were being exploited. Delays in resuming leasing were the result of (among other things) a long gestation period for the environmental impact statement on the overall leasing programme Secretary Andrus's desire to re-appraise leasing policy, and a suit by environmental groups on the adequacy of the environmental impact statement. The 1976 amendments require greater planning by the Department of the Interior, increased reporting by firms, and the development of leases within a decade. A federal task force has charted the net effect of all the regulations affecting coal leasing and, as far as I can tell (winding my way through the four gigantic charts, each of which more than covers a large desk), once leasing resumes, it will take a decade to move from setting up a lease sale to producing coal.

SUMMARY AND CONCLUSIONS

In this section, the themes developed above are drawn together and their empirical significance is assessed. The paper stressed the deficiencies of coal supply data on the basis of an absolute standard of how close to appraising reality has the work come. Here this evaluation is recapitulated and a re-examination is taken using the more modest yardstick of whether the information is good enough to guide public policy decisions.

The basic thesis of this paper is that, as with most resource supply analyses, caol supply models have severe defects. Even in a simple world of constant technology and input, prices, stable public policy and persistent market equilibrium, the data on coal have serious defects. Clearly a coal supply model must provide a means to convert data on physical conditions in coal seams to estimates of the quantities of coal available at different costs. The prior discussion has shown that little agreement exists about what physical characteristics should be measured, let alone what their impact on cost will be. To make matters worse, almost all the coal supply studies reviewed rely on USBM cost studies that portray, with unknown accuracy, the costs of a representative mine whose characteristics are only loosely described. Even if these problems were solved, available data on mining conditions would not permit application of the costing theory to resource data. The analytic problems are exacerbated because of uncertain trends of eastern underground mining costs and of public policy.

We may ask now what problems these difficulties do and do not create for policy analysis. The only proposition that can be clearly stated is that the data do suggest that if one is willing to delay imposition of severe environmental controls on coal production and use, it clearly represents an attractive way to limit oil imports in the next decade. FEA estimates that real costs of eastern coal will be around 50 cents per million Btu by 1985 appear highly unrealistic. The analysis neglects the high probability of rising real costs. However, given $2.00 per million Btu oil, substantial rises in United States coal prices could occur and still leave a substantial cost advantage. The more difficult questions arise if, as is likely to occur, environmental regulations tighten. Given the rising costs of coal mining in the East and the costs of using eastern coal in an environmentally acceptable fashion, at least a westward shift of coal mining seems likely. Indeed, data collected on industry expansion plans (ICF, 1976b) in the West indicate that of 546 million tons of expansion expected for 1975 to 1985, only 212 million tons (39 per cent) will be in the East. Expansion in the Northern Great Plains alone is set at 244 million tons. As we move into the middle 80s, nuclear competition becomes a formidable threat. Here the uncertainties are great not only about coal but about all the elements affecting the decision.

As we move toward the 1990s, some envision the emergence of a major synthetic fuels industry. Here, coal supply data have proven a low priority item for the decisionmaker. The latest available data indicate that, even with 50 cent coal, synthetic fuels look unattractive. Decisions in this realm depend initially on good oil market forecasts. Only if the situation ultimately changes drastically, will coal supply forecasts be a major input into evaluation of synthetic fuel developments.

(This represents a radical change in my perceptions; when synfuel looked competitive at existing coal prices, I saw a critical need for coal supply forecasts to evaluate the effects on synfuel prices of expanding coal output.) The most important immediate use for coal supply data are to help decision makers in the next decade to determine the proper distribution of output between the East and West. In the longer run, the critical issues relate to providing good coal supply input data for the discussion of the coal-nuclear power issue. Looking beyond a narrow United States perspective, the critical question clearly is the position of eastern coals.

Many scenarios are conceivable here and I will mention only a few:

1. An ample supply - moderate demand case where the United States coal becomes an attractive source of boiler fuel and metallurgical coal.
2. A tight supply case in which the United States coal is an unlikely source of energy for Europe and Japan.
3. A case in which supplies are moderately adequate but United States demands are weak so that considerable net supplies are available for Europe and Japan.

Clearly, it is exceedingly important to Europe and Japan just what actually occurs but the state of the art precludes resolving this issue.

My own conjecture is that rising underground mining costs will ultimately lessen the role of eastern coal in the steam coal market. Ultimately here means a decade or more from 1978 due to the long lead times involved. Whether western coal or nuclear will be the key beneficiary of this development remains unclear. What is clear is that United States coal still looks attractive as a source of metallurgical coal for other countries but its role in providing steam coal for exports seems limited.

REFERENCES

Asbury, J.G. and Costello, K.W., 1974, Price and Availability of Western Coal in the Midwestern Electric Utility Market, 1974-1982, Argonne: Argonne National Laboratory.

Averitt, Paul, 1969, Coal Resources of the United States, 1st January, 1967, Geological Survey Bulletin 1275, Washington: U.S. Government Printing Office.

Averitt, Paul, 1975, Coal Resources of the United States, 1st January, 1974, Geological Survey Bulletin 1412, Washington: U.S. Government Printing Office.

Battelle Pacific Northwest Laboratories (Anderson, Donald L., Foley, Thomas J. and Reardon, William Q.), The Regional Analysis of the U.S. Electric Power Industry, v. 4B, Coal Supply Functions, Springfield, Va.: National Technical Information Service.

Charles River Associates, Inc., 1973, The Economic Impact of Public Policy on the Appalachian Coal Industry and the Regional Economy, 3v, Springfield, Va.: National Technical Information Service.

Charles River Associates, Inc., 1975, Analysis of the Supply Potential for Southern West Virginia Low Sulphur Coal, Cambridge, Mass: Charles River Associations.

Gordon, Richard L., 1975a, U.S. Coal and the Electric Power Industry, Baltimore: The Johns Hopkins University Press for Resources for the Future, Inc.

Gordon, Richard L., 1975b, Economic Analysis of Coal Supply: An Assessment of Existing Studies, EPRI 335, Palo Alto: Electric Power Research Institute and Springfield, Va.: National Technical Information Service.

Gordon, Richard L., 1976a, Economic Analysis of Coal Supply: An Assessment of Existing Studies, EPRI 335-1, Palo Alto: Electric Power Research Institute and Springfield, Va.: National Technical Information Service.

Gordon, Richard L., 1976b, Coal and Canada-U.S. Energy Relations, Montreal-Washington: Canadian-American Committee.

Gordon, Richard L., 1976c, Historical Trends in Coal Utilization and Supply, a report to the U.S. Bureau of Mines, University Park: The Pennsylvania State University.

Gordon, Richard L., 1978, Coal in the U.S. Energy Market, Lexington Heath Lexington Books.

ICF, Inc., 1976a, PIES Coal Supply Curve Methodology, Washington: U.S. Federal Energy Administration.

ICF, Inc., 1976b, Coal Mine Expansion Study, report to the Office of Coal, U.S. Federal Energy Administration, Washington: ICF.

ICF, Inc., 1976c, The National Coal Model Description and Documentation, Washington: U.S. Federal Energy Administration.

Katell, Sidney and Hemingway, E.L., 1974, Basic Estimated Capital Investment and Operating Costs for Coal Strip Mines, Information Circular 8661, Washington: U.S. Government Printing Office.

Katell, Sidney, Hemingway, E.L. and Berkshire, L.H., 1975a, Basic Estimated Capital Investment and Operating Costs for Underground Bituminous Coal Mines, mines with annual production of 1.06 to 4.99 million tons from a 72 inch coalbed, Information Circular 8682, Washington: U.S. Government Printing Office.

Katell, Sidney, Hemingway, E.L. and Berkshire, L.H., 1975b, Basic Estimated Capital and Operating Costs for Underground Bituminous Coal Mines, mines with annual production from 1.03 to 3.09 million tons from a 48 inch coalbed, Information Circular 8689, Washington: U.S. Government Printing Office.

Katell, Sidney, Hemingway, E.L. and Berkshire, L.H., 1976a, Basic Estimated Capital Investment and Operating Costs for Coal Strip Mines, Information Circular 8703, Washington: U.S. Government Printing Office.

Katell, Sidney, Hemingway, E.L. and Berkshire, L.H., 1976b, Basic Estimated Capital Investment and Operating Costs for Underground Bituminous Coal Mines, mines with annual production of 1.06 to 4.99 million tons from a 72 inch coalbed, Information Circular 8682A, Washington: U.S. Government Printing Office.

Mutschler, Paul H., 1975, Impact of Changing Technology on the Demand for Metallurgical Coal and Coke Produced in the United States to 1985, Information Circular 8677, Washington: U.S. Bureau of Mines.

National Petroleum Council, 1973, U.S. Energy Outlook Coal Availability, Washington: National Petroleum Council.

Schlottmann, Alan M., Environmental Regulation and the Allocation of Coal: A Regional Analysis, Knoxville: Appalachian Resources Project, the University of Tennessee.

Train, Russell E., 1975, "Clearing the Way for Coal", National Coal Association, First Symposium on Coal Management Techniques, Vol. 2, Washington: National Coal Association.

U.S. Bureau of Mines, 1971, Strippable Reserves of Bituminous Coal and Lignite in the United States, Information Circular 8531, Washington: U.S. Government Printing Office.

U.S. Bureau of Mines, 1972, Cost Analyses of Model Mines for Strip Mining of Coal in the United States, Information Circular 8535, Washington: U.S. Government Printing Office.

U.S. Bureau of Mines (Thompson, Robert D. and York, Harold F.), 1975a, The Reserve Base of U.S. Coals by Sulfur Content (In Two Parts) - 1. The Eastern States, Information Circular 8680, Washington: U.S. Bureau of Mines.

U.S. Bureau of Mines (Hamilton, Patrick A. et al.), 1975b, The Reserve Base of U.S. Coals by Sulfur Content (In Two Parts) - 2. The Western States, Information Circular 8693, Washington: U.S. Bureau of Mines.

U.S. Bureau of Mines, 1975c, The Reserve Base of Bituminous Coal and Anthracite for Underground Mining in the Eastern United States, Information Circular 8655, Washington: U.S. Bureau of Mines.

U.S. Bureau of Mines, 1975d, The Reserve Base of Coal for Underground Mining in the Western United States, Information Circular 8678, Washington: U.S. Bureau of Mines.

U.S. Bureau of Mines, 1976, Coal - Bituminous and Lignite in 1974, Washington: U.S. Bureau of Mines.

U.S. Bureau of Reclamation, 1971, The North Central Power Study, Billings, Montana: U.S. Bureau of Reclamation.

U.S. Federal Energy Administration, Interagency Task Force on Coal, 1974, Project Independence Blueprint Final Task Force Report, Coal, Washington: U.S. Government Printing Office.

U.S. Federal Energy Administration, Interagency Task Force on Coal, 1976, National Energy Outlook, Washington: U.S. Government Printing Office.

Zimmerman, Martin B., 1975, Long-Run Mineral Supply: The Case of Coal in the United States, unpublished Ph.D. thesis in economics, The Massachusetts Institute of Technology.

CANADIAN FRONTIER REGIONS OIL SUPPLY POTENTIAL

by

C.R. Mattinson

INTRODUCTION

My assignment in these proceedings is to say something about oil
supply potential in what is called the Canadian frontier regions. The
frontier regions have little hydrocarbon production history and only a
relatively short history of exploration. Manipulation of statistical data,
therefore, does not offer a very promising approach to the problem of predicting future results in the frontier regions - we have to fall back on
the basic factors of geological indications, technology, and the economic
and legal climate.

What I propose to do, therefore, is to provide you first with a
general overview of Canada as a petroleum producing region. This will be
followed by some thoughts as to where the major prospects for future supply
sources in the country are seen to be. I would then like to consider each
of the major frontier regions in terms of such factors as physical geology,
the results of work done to date, the peculiar operating problems of the
region, costs, and transportation modes for hydrocarbons, supposing any
are found.

Next we can give some consideration to the influence of the economic
and legislative climates on the production potential. On the basis of
these kinds of considerations, I will go out on a limb to present you
with a speculative forecast of total Canadian production potential to the
year 1995. Finally, we can compare that forecast of Canadian supply with
a forecast of indigenous Canadian demand for a look at Canada's longer
term prospects to be either a net importer or exporter of oil.

I should mention that the observations I am going to make while being
in part a result of work done by myself and others at Shell Canada Resources are in some part also a synthesis of the views of many people in the
Canadian oil industry as I have observed these views through such means
as the printed word, participation in inquiries before governmental

Figure 1

CANADA

MAJOR REGIONS WITH HYDROCARBON POTENTIAL

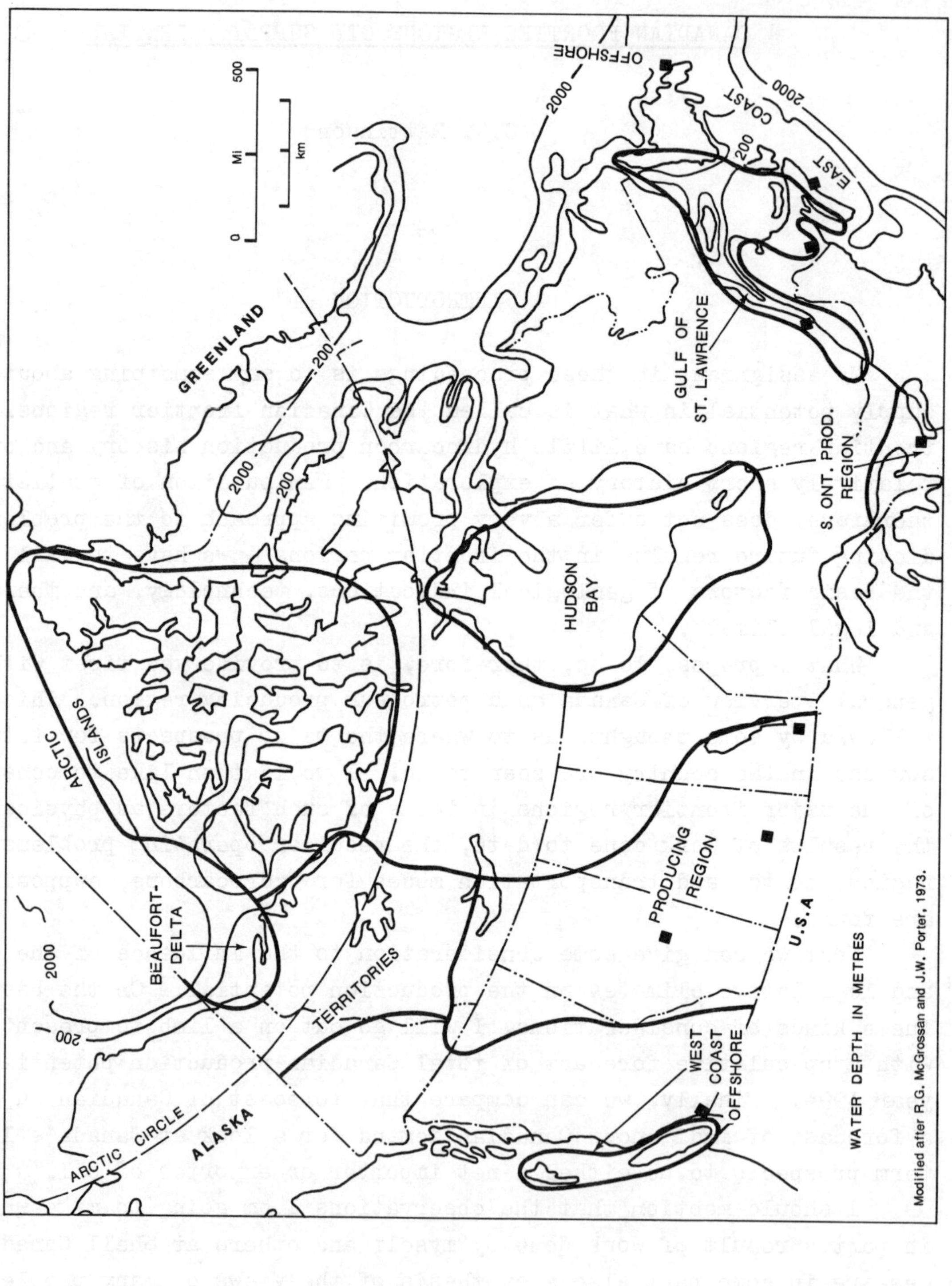

OFFSHORE

2000

EAST COAST

200

2000

GULF OF
ST. LAWRENCE

S. ONT. PROD. REGION

GREENLAND

200

2000

200

HUDSON BAY

ARCTIC ISLANDS

2000

BEAUFORT-DELTA

200

TERRITORIES

PRODUCING REGION

U.S.A

ARCTIC CIRCLE

ALASKA

WEST COAST OFFSHORE

WATER DEPTH IN METRES

500

Mi

km

0

Modified after R.G. McCrossan and J.W. Porter, 1973.

76

bodies and discussions with individuals in the business. For this reason
the views expressed in this discussion are not necessarily in all respects
those of Shell Canada Resources.

DEFINITIONS

Before proceeding, two definitions are in order. The producing
regions, as I use the term, comprise those areas of the country where
exploration has been going on more or less continuously for some decades.
All but two of the producing fields in Canada are in these regions.
Geographically the producing regions occupy the southern most part of
Ontario, parts of the four western provinces, and a small portion of Yukon
territory and MacKenzie district. All other parts of Canada that potentially may contain oil and gas deposits are called the frontier regions.
The major frontier regions are shown on Figure 1.

CANADIAN OVERVIEW

General

Let us begin with an overview of Canada against a world backdrop.
Canada encompasses some 6.4 per cent of the world's land mass and contains
some 0.5 per cent of its population. The nation is responsible for
3.1 per cent of world oil consumption and provides about the same share
of world production. Known reserves of oil in Canada amount to about
1.3 per cent of the world total.

Figure 2

CANADA'S SHARE OF WORLD:

Land area	6.4%
Population	0.5%
Oil consumption	3.1%
Oil production	3.1%
Oil reserves	1.3%

Geology

The most fundamental factors bearing on the oil potential of any
region are the amount and characteristics of the underlying rocks. The

normal habitat of oil is sedimentary rocks and, in particular, those sedimentary rocks deposited in ocean basins after organic life became abundant in the world's seas. Figure 3 shows schematically the four great time eras recognised by geologists.

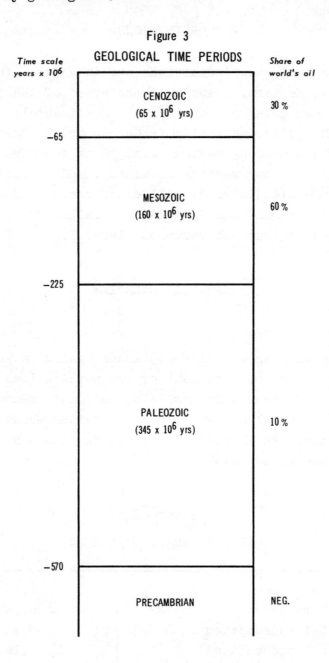

Figure 3

GEOLOGICAL TIME PERIODS

It was during the latter three of these eras that organic detritus, buried in sea bottoms under increasing depths of sediments, underwent the natural chemical reactions that converted it to oil.

Now oil deposits once formed are quite fragile and are subject to dissipation and permanent loss from a variety of geological accidents.

One might speculate from this that, other things being equal, younger rocks with a shorter and less complex geological history, would tend to have more oil than older rocks. The fact is that other things are not equal because the depth of burial of the sediments, the availability of organic material, and various other factors, have a major influence on whether, how much and how fast, hydrocarbons are generated. Nevertheless, younger rocks do indeed contain most of the world's oil. The two most recent major rock sequences, although spanning only 40 per cent of oil generating time, contain about 90 per cent of the world's oil. Statistically, the prospects for a country to become a major oil producer would appear to be enhanced if a high proportion of its rocks are mesozoic and cenozoic in age.

Figure 4 shows the distribution of favourable rocks - that is, unmetamorphosed sedimentary rocks - in Canada and adjacent offshore regions. The thickness is illustrated by a colour code. Observe that there are very large areas with no favourable rocks at all, and these areas are in consequence accorded no oil potential.

Figure 5 shows the distribution of paleozoic rocks in Canada. This group of rocks has the most widespread distribution in Canada of any of the three major systems.

Figure 6 shows the distribution of mesozoic rocks. Generally, they occupy the same locale as the paleozoic rocks but with some fairly significant differences. For instance, the Atlantic provinces, southern Ontario, parts of the northern territories and much of the southern arctic islands have no mesozoic rocks, although underlain by paleozoic rocks. We do, however, have a broad area of mesozoic rock in one area where no paleozoic rocks are shown, namely, the eastern offshore fringe from the high arctic to Newfoundland.

Finally, the cenozoic rocks.(1) Rocks of this age are very limited onshore in Canada. They find their chief expression in the Atlantic offshore region, along the northern continental margin, and as a narrow fringe along the west coast offshore region.

Harking back to what was said about younger rocks being the most likely habitat of oil, we have to conclude that nature has not been particularly kind to Canada in this respect. The areal extent of the highly prospective younger rocks not only is comparatively restricted but the younger rocks have been placed in those parts of the country that are the most difficult in terms of accessibility and operating conditions.

_____

1) Shown as tertiary (i.e., cenozoic less pleistocene and post-pleistocene deposits) on slide.

Figure 4
CANADA
DISTRIBUTION OF UNMETAMORPHOSED SEDIMENTARY ROCKS

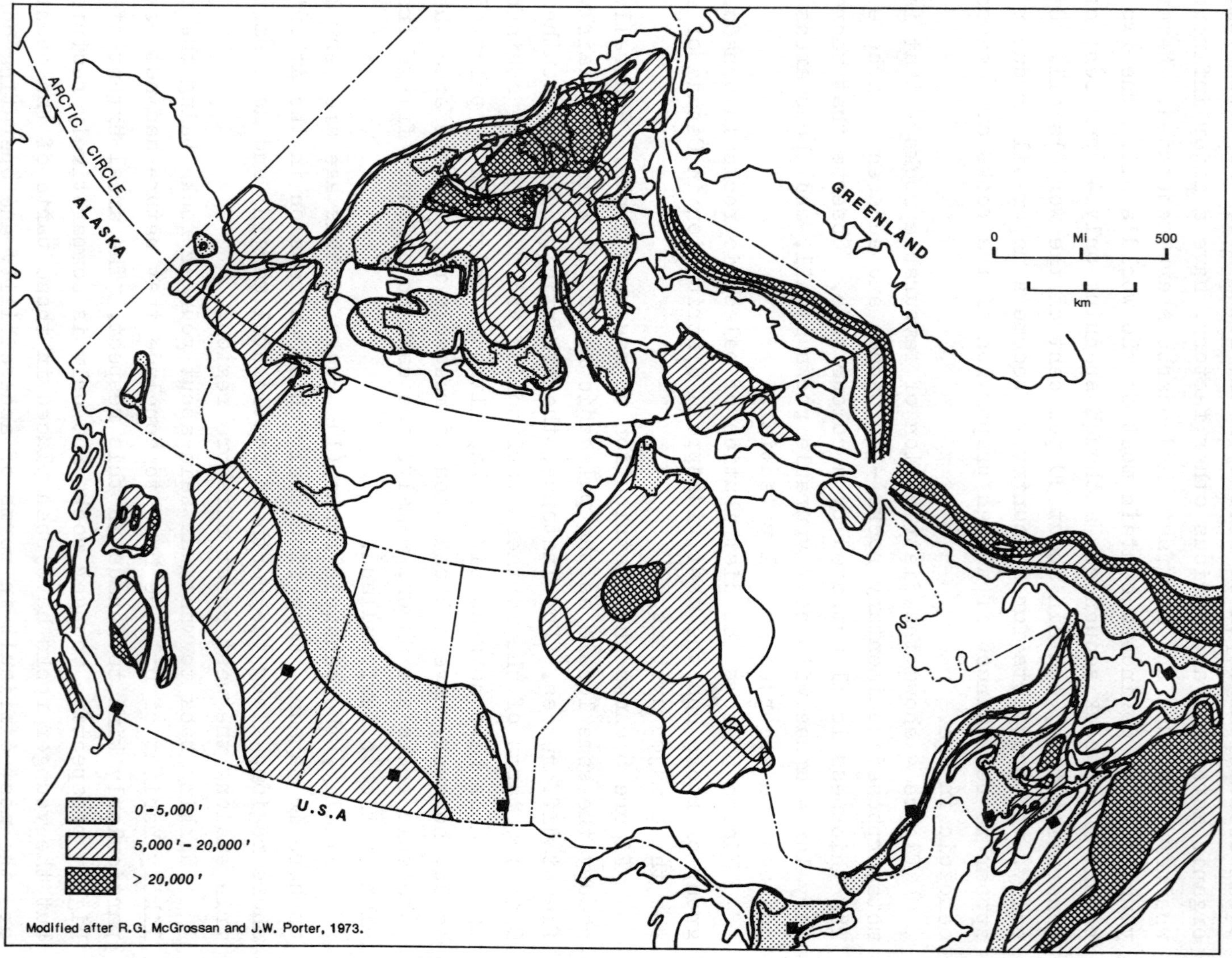

GREENLAND

ARCTIC CIRCLE

ALASKA

U.S.A

0 – 5,000 '

5,000' – 20,000 '

> 20,000 '

Modified after R.G. McCrossan and J.W. Porter, 1973.

500

Mi

km

0

Figure 5

CANADA

DISTRIBUTION OF PALAEOZOIC SEDIMENTARY ROCKS

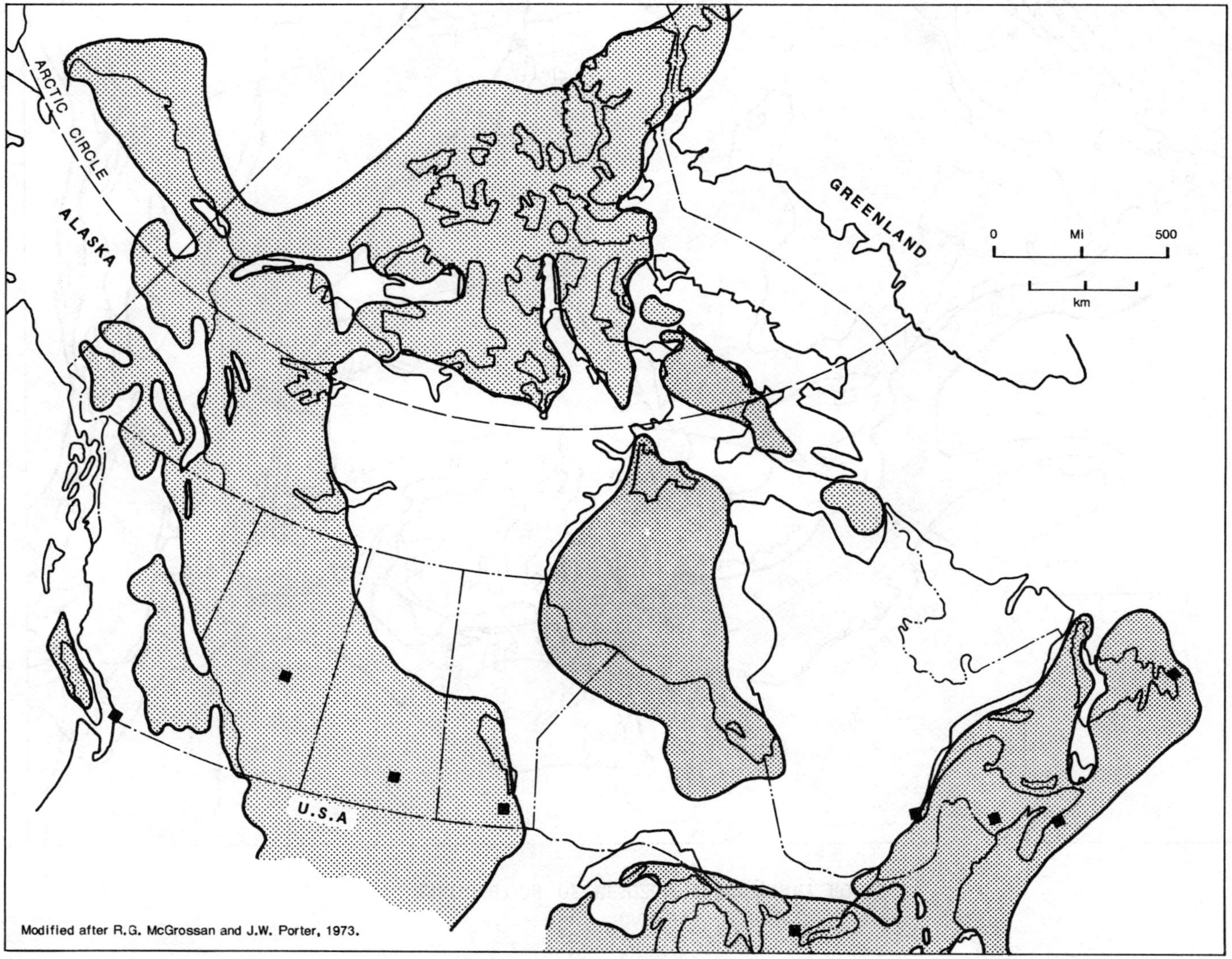
ARCTIC CIRCLE
ALASKA
GREENLAND
0
Mi
500
km
U.S.A
Modified after R.G. McGrossan and J.W. Porter, 1973.

Figure 6

CANADA

DISTRIBUTION OF MESOZOIC SEDIMENTARY ROCKS

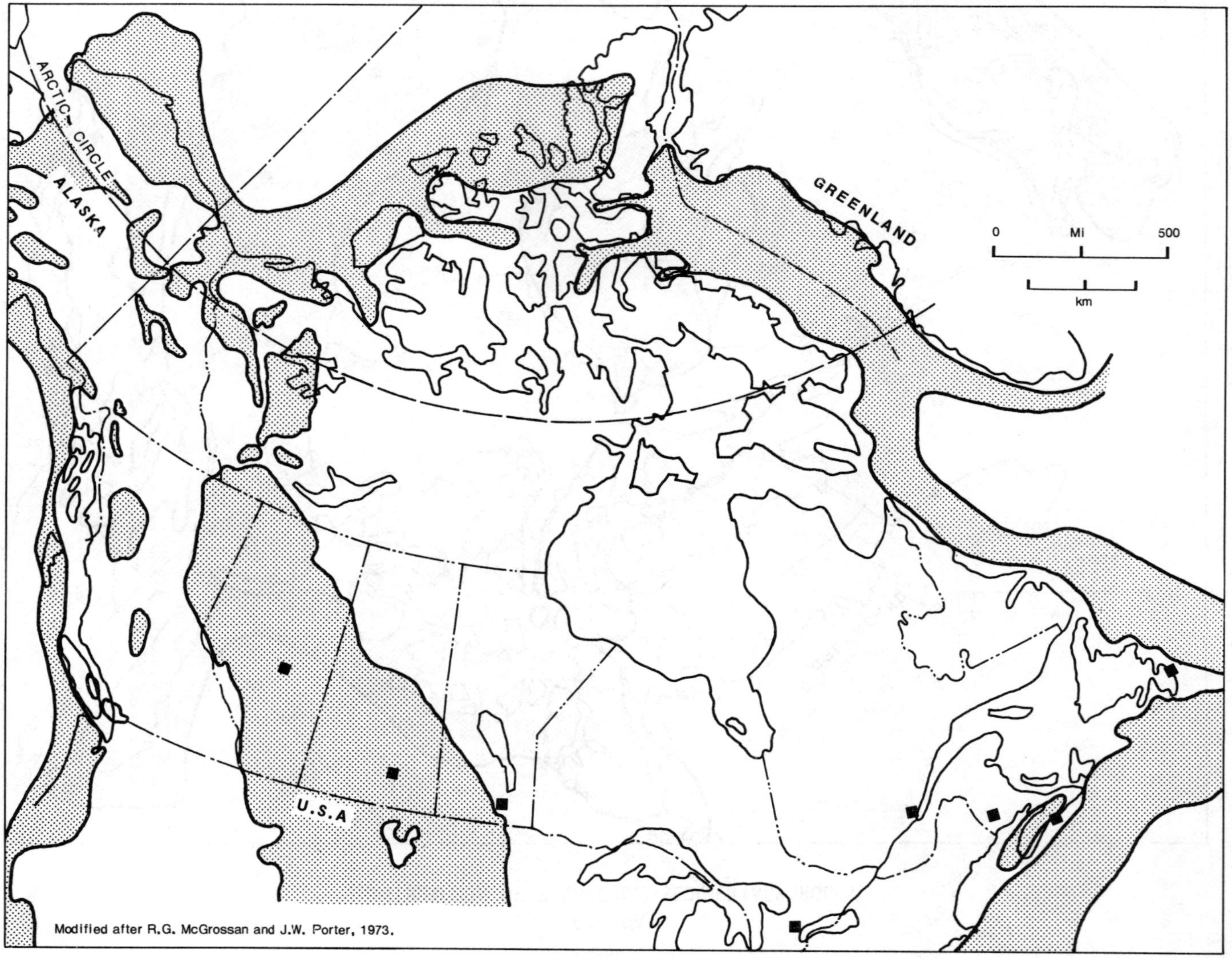
ARCTIC CIRCLE
ALASKA
GREENLAND
0
Mi
500
km
U.S.A
Modified after R.G. McGrossan and J.W. Porter, 1973.

Figure 7

CANADA

DISTRIBUTION OF TERTIARY SEDIMENTARY ROCKS

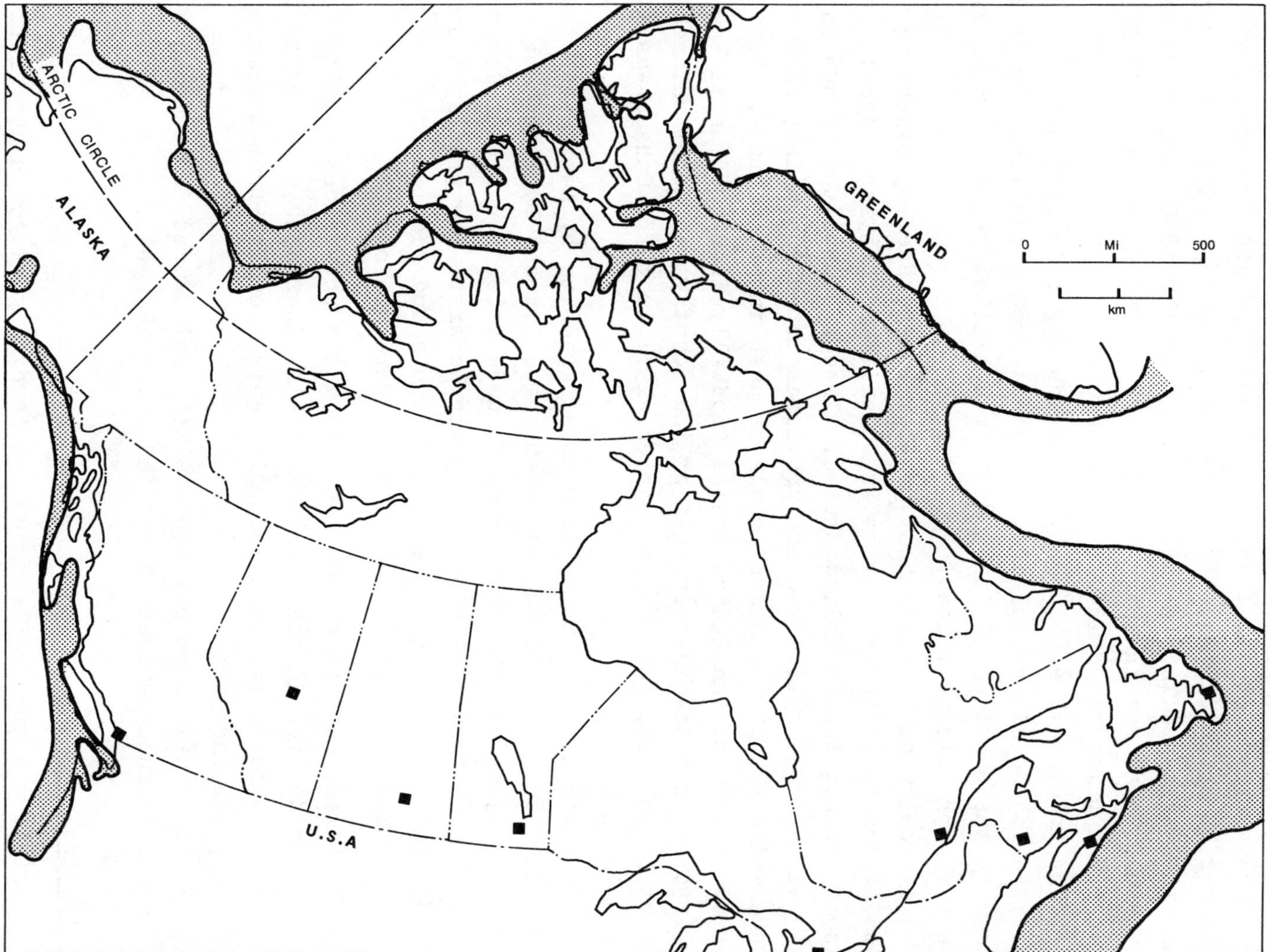

ARCTIC CIRCLE
ALASKA
GREENLAND
0
Mi
500
km
U.S.A

Modified after R.G. McGrossan and J.W. Porter, 1973.

Historical oil activity

The Canadian oil industry began almost 120 years ago with the development of a shallow field in southern Ontario. The area never became a major producer, but it still has a small oil and gas production and modest discoveries are still made from time to time. Gas discoveries were made in water wells drilled in western Canada in the 1880s and there was a considerable local gas industry by the 1920s. Only one large oil field had been found prior to world war II, however.

The modern oil era in western Canada began in 1947 when a major find was made in the Alberta plains. By about 1950 production capability had risen to the point where it exceeded the limited refinery requirements of the three prairie provinces.

The solution to the market problem was pursued by extending pipeline connections to refinery centres in Ontario and British Columbia and to export markets in the United States. In due course the situation stabilize with Canada's five western provinces depending mainly on domestic crude while the five eastern provinces used imported crude. During this period exports to the United States gradually increased until the volume of imports into the eastern half of the country was about balanced by exports. In fact, for a brief period, 1971 to 1975, Canada was a net exporter of oil.

The extreme crude price increases agreed to by members of the OPEC group late in 1973, and the sales embargos imposed by some members of the group at around the same time, focused attention by the Government and public on the oil question as never before. The principal effects on the oil industry, of the subsequent actions and reactions by the federal and the various provincial governments were twofold:

1. Taxes, royalties and other quasi tax provisions applicable to production income were drastically increased with a view to capture of presumed economic rents stemming from domestic crude price increases.
2. Restrictions on oil exports were introduced in accordance with a formula devised by the federal government.

Statistical summary

Let me now try to flesh out the foregoing skeleton description of the Canadian oil industry by means of a few statistics.

The first of this series of slides shows the volumes of oil discovered each year in Canada since intensive exploratory activity was initiated in 1947. It also shows annual current dollar exploration

Figure 8

CANADA

OIL RESERVES FOUND AND EXPLORATION EXPENDITURES

1947-1975

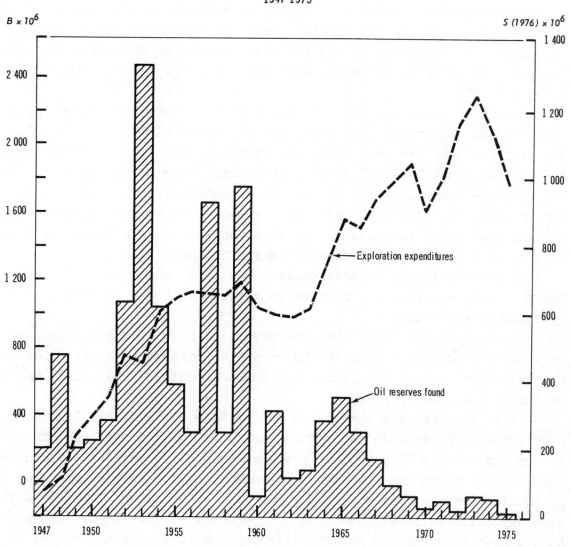

expenditures since that time. Notice the erratic character of the discovery line. The highs represent periodic discovery of a new exploration play and the rapid finding thereafter of the major fields in the play.

The last major new play was initiated in February 1965, now almost 12 years ago.(1)

Figure 9 shows Canadian requirements for oil against production. The sharp drop in the production line in 1975 and 1976 is in response to export constraints imposed by the federal government. This loss of export market was partly compensated for by deliveries to Montreal refineries following extension of oil pipeline service to Montreal. Nevertheless, the industry is producing well below its present capacity of about 2,100 MB/D.(*)

Figure 10 shows the trend in the cost of finding oil in the producing regions of Canada. The costs are in constant 1976 dollars. For the exercise, gas has been converted to oil on the basis of 10,000 cubic feet of gas equal 1 barrel of oil, this being a rough approximation of the relative economic value of the two commodities. This graph dramatizes the Canadian producing industry's great problem - how to reserve the finding cost trend in spite of the fact that the only big new oil potential is in remote, high-cost parts of the country.

As a final figure in this series, I want to show one that contrasts the producing and frontier regions in respect of a few aspects of their hydrocarbon potential.

These statistics show that the "hunting ground" in the frontier regions is much larger than in the producing regions - the frontier regions having almost five times the areal extent and over six times the volume of sedimentary rock.

It is obvious from the figure for seismic surveys, drilling and expenditures that the frontier regions are still at a rather youthful stage in the exploration life cycle. Somewhat ominously, however, it can be seen that although the frontier regions account for only about 8 per cent of total seismic survey work and 2 per cent of drilling, frontier exploration expenditures constitute about 25 per cent of the total. The much higher cost structure of the frontier regions is beginning to make its influence felt. To some degree the higher costs of frontier work have been offset by better results. Statistically, the all-time unit cost of finding oil equivalent in the frontier regions has been only about

1) Subsequent to delivery of this paper a new oil play was initiated in the West Pembina area of Alberta. Indications are that this play contains at least a few tens of millions of barrels; its full significance remains to be assessed.

*) About 1,900 MB/D as of mid-1978.

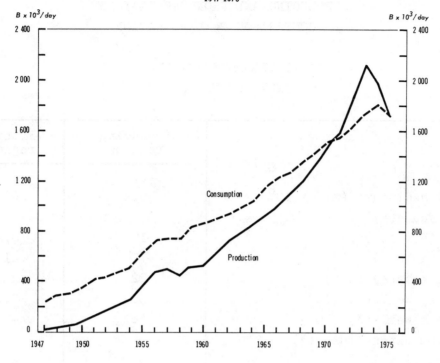

Figure 9
CANADA
OIL PRODUCTION AND CONSUMPTION
1947-1975

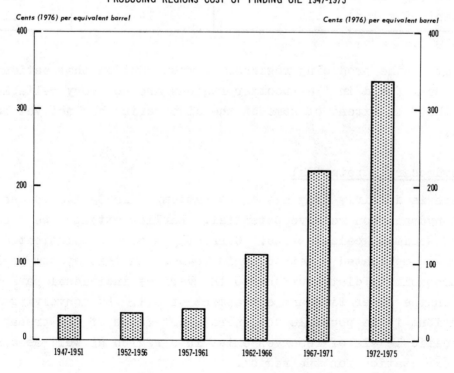

Figure 10
CANADA
PRODUCING REGIONS COST OF FINDING OIL 1947-1975

Figure 11

CANADA
PRODUCING AND FRONTIER REGIONS
COMPARISON OF EXPLORATION
&
DEVELOPMENT STATUS
31ST DECEMBER, 1975

	Producing regions	Frontier regions
Area (mi^2 x 10^6)	0.5	2.4
Volume of sediments (mi^3 x 10^6)	0.6	4
Reserves Found(*)		
Liquids (B x 10^9)	17	0.2
Gas (FT^3 x 10^{12})	77	21
Seismic months(*) (10^3)	22	2
Exploratory wells(*) (10^3)	30	0.6
Development wells(*) (10^3)	46	-
Exploration expenditures(*)		
$ Current ($ x 10^9)	9	2
$ Constant - 1976 ($ x 10^9)	17	3
Development expenditures(*)		
$ Current ($ x 10^9)	7	0.1
$ Constant ($ x 10^9)	14	0.1

*) 1947-1975.

twice that in the producing regions. I must caution that estimates of what has been found in the frontier regions are not very reliable at present as the full extent of some of the discoveries has not yet been assessed.

Future hydrocarbon potential

There is a fairly long history of attempts in Canada to assess the national hydrocarbon reserve potential. Earlier attempts were based on volume of sediment calculations. Currently a more sophisticated - or at least more complicated - method is in vogue. In this system geologists assign probability distributions to the various individual parameters bearing on the generation and entrapment of oil. By convolving these distributions it is possible to generate estimates of the probability that specific levels of reserves exist in a region as well as a mean reserve expectation for the region.

Figure 12 shows some of the estimates made over the last twenty years by different observers.

Figure 12

CANADA

OIL RESOURCE POTENTIAL ESTIMATES

Date	Authority	B x 10^9
1976	Geol. Surv. Can.	40(*)
Latest	Shell	60
1973	Can. Soc. Pet. Geol.	90
1973	Geol. Surv. Can.	100
1973	Shell	55
1972	Geol. Surv. Can.	140
1969	Can. Pet. Assn.	120
1958	Can. Pet. Assn.	50

*) Approximate mean as estimated from probability curve.

The comparison is not entirely a fair one, because the parts of Canada covered are not identical and the methods used differ. The 1958 estimate by the Canadian petroleum association, for example, excludes the offshore regions and the arctic islands. The 1976 estimate by the Canadian geological survey includes the arctic islands and the offshore shelf, but excludes the offshore slope. Shell's estimate includes everything onshore plus the slope and the shelves.

The differences, however, are not sufficiently great to obscure some fundamentals. Clearly there has been some retrenchment since the peaks of enthusiasms that were reached in the late 1960s and early 1970s. This is presumably in reaction to the relatively modest exploration results of the last several years. It appears that the early hopes for easy finds of large prolific frontier fields are not to be realised. Oil discovery and development in the frontier regions apparently is going to be as difficult and protracted as it is anywhere, perhaps more so than in most regions. Nevertheless, if the producing regions' potential of about 20 billion barrels is subtracted from the more recent totals shown, the indicated frontier regions' potential is in the range of 20 to 70 billion barrels - not a large figure in global terms, but certainly significant in relation to Canadian requirements.

FRONTIER REGIONS

West coast offshore

I would now like to briefly go through the main frontier regions in turn, briefly reviewing their current status and long-term prospects as I see them. I will begin with the west coast offshore region.

As is evident from Figure 1, this is one of the smaller frontier regions, and in fact, the areas of rock favourable to oil accumulation are much smaller than the generalised outline on Figure 1.

This area was the subject of exploratory interest during most of the 1960s with Shell Canada being the main explorer. By 1969 some 60 months of seismic surveys had been conducted and 37 exploratory wells drilled. The main prospects were considered to be in the offshore areas and this was the site of the first marine offshore drilling in Canada. The offshore drilling, all in less than 500 feet of water, presented no unusual problems except the necessity of providing strengthened rigs to meet ocean swell and wave conditions peculiar to those waters. The offshore holes were drilled to an average depth of around 10 thousand feet at an average cost of something under $1.5 million each.

Unfortunately, no production was established, although minor oil and gas shows were encountered. There has been essentially no exploration in the region since 1969.

Territories

Let us now move to a consideration of the region marked territories on Figure 1, so named because it includes part of the Yukon territory and part of the northwest territories.

The region contains a 50-million barrel field found in 1921, which produces oil from lower paleozoic carbonates for use at a small on-site refinery. During and immediately after World War II a drilling programme was carried out in the vicinity of the field, but no additional oil was found and interest in the area pretty well ceased after 1947.

Exploration of the territories resumed in the early 1950s exploration methods in this region are not different in nature from those used further south in Canada. There are, however, some new challenges occasioned by the climate and the lack of infrastructure.

The country is essentially without roads and entirely without railroads. Settlements are small and infrequent and year-round airstrips are scarce. Winters are rigorous, with temperature extremes of minus 55° celsius and long periods where the temperature remains below minus

30° celsius. Daylight hours are limited or nonexistent from about
November to February.

Barges on the MacKenzie river during summer, limited use of snow
roads in winter, and aircraft at any season are the only means of moving
freight into the country. The ground is permanently frozen, in some
places to a depth of 1,000 feet or more. Surface thawing to a depth of
a few feet occurs during the short summer season, creating swamp-like
conditions which make it difficult to move heavy equipment. The winter
thus becomes the preferred time for moving heavy equipment like drilling
rigs. Light equipment and personnel can be moved at any time of the year
by float or ski-equipped aircraft or helicopter. The helicopter became
the workhorse for light operations such as seimsic surveying.

As a result of these difficulties in the way of movement, lack of
local accomodation and the long haul from the nearest major supply centre
at Edmonton, exploration costs are decidedly more expensive in this region
than in southern Canada. For example, seismic surveys in southern Alberta
at present cost about $175,000 per crew-month versus $500,000 to $800,000
per crew-month in the territories region. A drilling rig that costs
$7,000 per day to operate in southern Alberta might have a daily cost of
£25,000 in the territories. These figures suggest costs in the range of
three to five times the cost of doing comparable work in southern Canada.

The presumptive outlet for oil from this region is southern Canada
via pipeline transportation. Conditions in the region impose some peculiar pipelining problems. Over the several hundreds of miles of permanently frozen ground a line would have to be elevated to prevent the melting
and caving of the surface that could otherwise lead to line rupture.
Special insulation may be required to keep the oil mobile in winter.

To date the territories region, which has some 250 thousand square
miles of prospective area, has been explored to the extent of about
610 months of seismic surveys and 245 exploratory wells. Apart from the
55-year-old Norman Wells field no commercial discoveries have been made.
Since 1974, exploration interest has languished.

MacKenzie Delta-Beaufort Sea

This relatively small region is entirely distinct geologically from
the territories region to the south. Parts of the Delta-Beaufort region
have been the site of intermittent rock deposition from the beginning of
paleozoic time onward. As a result there is an enormous thickness of
sedimentary rock in the deepest part of the basin, amounting to as much
as 30,000 feet. In addition to a thick rock sequence the basin is well
endowed with other attributes favourable to oil generation and entrapment.

These include an abundance of organic shale which could serve as the oil source, many porous zones in both sandstone and carbonate sections, and a large number of seismically definable structural features of various types and geological age.

Although the geology appears eminently favourable the operational and logistic conditions are not. For onshore work the adverse logistic features are similar to those in the northern part of the territories, creating similar operational problems and a cost structure three or four times that in southern Canada.

A quantum jump in operational problems and costs occur when the locale is shifted from onshore to the Beaufort Sea. Much of the Beaufort Sea is covered by permanent pack ice averaging perhaps 10-15 feet thick. The southern limit of the pack ice is not static, but its average position is defined by a line lying somewhat offshore from the 600 foot contour. South of the permanent ice, the sea is of course frozen in winter, but a summer open water season of eight or ten weeks exists.

Exploration drilling in the Beaufort Sea is performed, out to about fifty or sixty feet of water, from artificial islands formed of material dredged from the bottom. In the deeper water, drilling can be carried on from drillships during the brief open water season. Consideration is also being given to the possibility of winter drilling from drillships locked in landfast ice. An intermediate zone exists where the water is too deep for islands and too shallow for drillships. An untried concept for handling this zone is to drill from hollow concrete cones that would be fabricated onshore and floated to position, then sunk.

To date seven tests have been drilled from islands and the first two deep water holes were drilled this summer.(1) The cost of drilling from islands is speculated to be between perhaps $8 million for an island and well in very shallow water to perhaps £30 million for an island and well in 60 feet of water, at about the depth limit of island building. Holes from drillships might cost £20 to £35 million each, depending on equipment utilisation and depth of the hole.

If finds are made in the Delta-Beaufort region, development considerations onshore will be much the same as in the territories region. That is to say, difficult and two to four times as costly as in southern Canada but presenting no problems outside the capabilities of today's technology. Offshore, however, development problems will make the difficulties of exploring appear insignificant by contrast. The principal difficulty revolves around the necessity of maintaining a permanent production facility

----

1) By mid-1978 seven deeper water exploratory tests had been drilled or were drilling.

in a region subject to shifting pack ice. For water of less than 50 to 60 feet in depth artificial islands will probably be used. Beyond 60 feet more exotic concepts are under consideration. Bottom-supported structures shaped to permit moving ice to ride up the sides and collapse are a possibility out to about 200 to 300 feet. At still greater depths the possibilities include going completely under water, with either wet or dry wellhead systems. Another concept is a floating monopod supporting a column on which a production platform would be perched. The monopod would ride below, and the production platform above, the ice. The column itself would be equipped with a mechanical chopping device to chop a path through moving pack ice that infringed on the area.

From the offshore production facility oil would move to shore in a submarine pipeline. This brings up the problem of bottom scouring by ice. True icebergs are uncommon in the Beaufort Sea, but thick ice islands, driven by wind and current are known to abrade the shallow sea bottom to depths in excess of the 6 to 12 feet at which marine pipelines are normally laid. Handling this difficulty will necessitate the development of new techniques and equipment.

The industry's drilling effort in the Beaufort-Delta region comprises about 118 tests to date. A number of oil and gas accumulations have been found and a speculative estimate of the reserves in these finds is 16-8 TCF of gas and 1/4 billion barrels of oil. The reserves in the three largest gas fields can be economically developed provided the gas can move with Alaskan gas down a common pipeline. The amount of oil found to date clearly is insufficient to support a pipeline. If a threshold volume of oil is to be found it will probably be in the offshore part of the basin where the bulk of the undiscovered hydrocarbon potential is believed to exist.

Arctic islands

This region covering about 650,000 square miles, is abundantly endowed with all the basics of a major petroleum province - thick rock sequences, source beds, and favourable structural conditions. Environmental and logistic conditions, however, are most unfavourable. The arctic region is essentially a frigid desert. There are only eight or ten small permanent settlements in the entire region and normal infrastructure, except for airstrips, is unknown. Winters are characterised by extreme cold and months of continuous dark. Summers are short and, in spite of twenty-four-hour sunshine, the average July temperature at resolute, near the centre of the area, is only about 4° celsius. Access to the arctic islands is by vessel during a brief shipping season and otherwise by freighter aircraft. Inter-island moves of heavy equipment are made by aircraft or, for short distances, by truck on winter roads over land or over the sea ice.

Explorationists onshore in the arctic islands encounter the same kinds of problems as in the onshore Beaufort-Delta region. Offshore, the sea is ice-covered most of the year, ruling out marine seismic but offering an ice platform from which seismic surveys can be conducted. During winter the land-fast sea ice in the inter-island channels is quite stable. This has permitted the drilling of offshore exploratory wells from ice platforms where the ice has been thickened and strengthened. To date the technique has been restricted to holes drilled within 15 miles of land, but ice movement surveys are being conducted to determine whether the ice might be sufficiently stable farther from shore. Exploration costs in the region are generally in line with those in the Beaufort-Delta onshore.

To date some 121 wells have been drilled in this region and about 340 crew-months of seismic shooting has been performed. The result has been the discovery of some eight gas fields with reserves totalling about 15×10^{12} cf and one oil field of possible commercial significance. Exploration is continuing in the arctic islands region with a view to establishing the approximately 30×10^{12} cf of gas required to support a pipeline to southern Canada. The principal operator in the region, Panarctic Oils, considers that a comparatively small volume of oil (300 million barrels) in the right location would justify an ice-strengthened tanker fleet and ancillary facilities to move it out.

Hudson Bay region

The Hudson Bay region covers a total of about 350 thousand square miles, of which slightly over 60 per cent is offshore. The potentially favourable rocks consist of paleozoic clastics and carbonates with a maximum thickness of less than 10,000 feet. The thickest part of the sedimentary rock sequence, thus presumably the best oil prospects, lies offshore. Only a few wells have been drilled in the region, four of which were offshore.

Water depths range up to, and locally in excess of, 600 feet. Offshore drilling is conducted from semisubmersibles or drillships during the brief (late July to September) summer drilling season. Field development might be carried out from a bottom-supported platform, ice conditions permitting, or from underwater production facilities. Oil or gas production could be moved out of the region via laterals to pipelines originating in the arctic islands or by independent lines to southern Canada.

There has been no drilling in this region since 1974. Some seismic surveying was done in 1975 but no plans for additional work have been announced.

East coast offshore

The east coast offshore region stretches from about 48° north latitude to 78° north latitude. The total areal extent of the region, including the continental slope, is about 700,000 square miles. This region encompasses a considerable variety of geological and operating conditions. For our purposes we can break it into two subregions - that south of 48° north latitude and that north of 48°.

The region south of 48° appears very attractive geologically with a thick sedimentary sequence and an abundance of seismically visible structural features. The shallow water part of this subregion presents no unusual exploration or development problems. Exploratory drilling can be carried on most of the year from semisubmersibles or drillships. Reserves if found will be developed from bottom-supported platforms with the oil or gas moving to shore by submarine pipeline or possibly by vessel. Costs for 10 to 15 thousand foot exploratory wells are in the $2 to $4 million range. It is speculated that oil development costs will range upward from about $1.50/b, in some instances perhaps to three or more times that figure.

Some 103 wells have been drilled in shallow water south of 48°. Numerous oil and gas shows were reported and at least three gas and two oil accumulations were found, all non-commercial. Drilling is currently at a comparatively low level. The continental slope lying east of the shallow water shelf is also considered to have good geological prospects although it is not yet tested by the drill.

The portion of the region lying north of latitude 48° is characterised by differnet geological conditions and a different set of operating problems. The favourable rock sequences north of 48° are thought to be mainly mesozoic and cenozoic. Structural features formed by a variety of geological processes have been identified from seismic surveys. Some of the structures are extremely large.

The near shore fringe of shallow water is much narrower than to the south and most of the area currently under exploration permit is in water depths of 1,000 to 6,000 feet. Ice conditions prevent year round operations here. The open water drilling season lasts from a few months in the vicinity of latitude 48° to only a few weeks around latitude 78°. Even during the open water season the area is infested with icebergs.

Because of the iceberg hazard drilling has to be conducted from dynamically positioned drillships or other craft so that if necessary, drilling can be temporarily suspended while the drilling vessel retires out of the way of an oncoming iceberg. For the same reason, production systems in this region will in all likelihood involve underwater completion methods and movement to shore of oil or gas by submarine pipeline.

The cost of an exploratory well in this subregion may vary in the range of a few million dollars for holes in shallower waters up to $10 to $15 million for holes in the 3,000 foot plus depth range. Underwater production systems and development drilling technology have not yet been used at the depths of water that characterise most of this subregion, thus development costs are not well defined.

Only eight tests have been drilled and three are currently drilling, all off the Labrador coast between latitudes 53° and 58° north, in water depths of less than 1,000 feet. Three wet gas discoveries have been made. It is anticipated that tests farther north and tests in water in excess of 3,000 feet will probably begin within two years.

St. Lawrence

The last region to be discussed comprises a collection of small geological basins, about the only common feature of which is that the sedimentary rocks are very largely of paleozoic age. The prospective area within the regional outline is about 125 thousand square miles. The region which contains numerous oil seeps and one small gas field, has been explored intermittently for a long time. During the last 25 years about 75 tests were drilled onshore and six offshore. No find of commercial significance was made.

Tar sands

This completes my review of the major frontier regions in Canada. There are some other smaller basins, for example, in the interior of British Columbia and the Yukon, but their petroleum prospects are insignificant in comparison with the regions discussed.

There is, however, one other potential oil source in Canada that is germane to this discussion as it dwarfs the potential of all the frontier regions combined. I refer, of course, to the Alberta tar sands. These sands which occur in the northern part of Alberta, are impregnated with a heavy tar-like bitumen that can be upgraded to a high quality, light gravity synthetic crude. The reserves of this material are estimated at over 650 x 10^9 barrels in place. After allowing for losses contingent on recovery and upgrading, the recoverable reserve is in the order of 160 billion barrels of synthetic crude. About 26×10^9 barrels of synthetic crude are believed to be recoverable from near-surface deposits accessible to mining; the remainder of the deposit will have to be recovered by some in-situ process.

One plant with a capacity of about 55,000 barrels per day is now in operation. A second plant is expected to come on stream in 1978 with initial capacity of around 125,000 barrels per day. There are at present no active plans for further plant construction.(1) The factors inhibiting a more active programme of tar sands development appear to be the enormous capital cost - estimated at about $2.4 billion in today's dollars for a 100,000 barrel per day facility and the absence of government policy for oil sands development. There is also uncertainty as to the place at which oil will sell when new facilities come on stream and some dissatisfaction with fiscal and royalty terms that might be imposed by the federal and provincial government.

Canadian supply and demand

Having considered the Canadian reserves on hand and discussed prospects for new supply sources, my final task will be to develop a potential supply picture matched to demand. To begin, let us look at the supply potential from sources regarded as fairly predictable.

Producing regions supply

Figure 13 shows Canadian requirements for oil over the next 20 years in all of Canada, and in the traditional market orbit for domestic oil. This demand forecast, incidentally, is based on a rather subdued projection of economic expansion in Canada; also it incorporates the effect of active energy conservation measures.

Production from known fields, from the two tar sands plants, from new discoveries in the producing regions and from the application of new supplemental recovery techniques to known fields is expected to accomodate demand in the traditional market orbit until about 1985. Subsequently, production capability will decline. By 1995 Canadian oil from these sources will satisfy only about 45 per cent of a total domestic requirement of 2,700,000 B/D. The supply forecast may be optimistic because in the latter year it includes a considerable volume of production from

---

1) Plans for a third mining plant with combined capacity of about 265MB/D and initial production beginning 1985/86 are well advanced as of mid-1978.

Figure 13

CANADA

CRUDE OIL DEMAND VS PRODUCTION 1975-1995

LOW SUPPLY SCENARIO

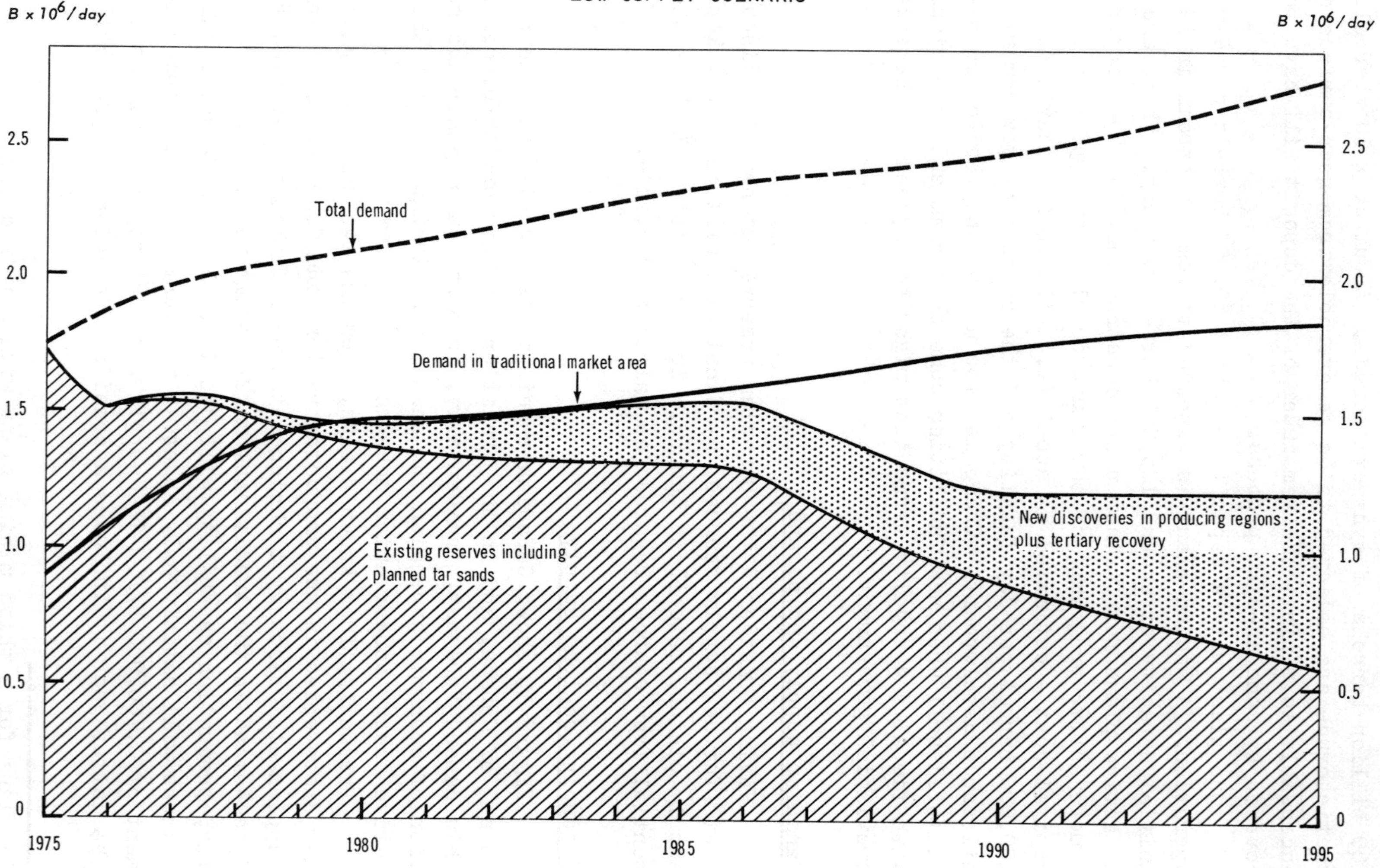
B x 10^6/day
B x 10^6/day
2.5
2.0
1.5
1.0
0.5
0
Total demand
Demand in traditional market area
Existing reserves including
planned tar sands
New discoveries in producing regions
plus tertiary recovery
1975
1980
1985
1990
1995

projected tertiary recovery processes that are not now economically or technically feasible.

Tar sands and frontier regions supply

Let us review the possibilities for improving the supply picture just shown.

In my view, prospects for bringing forth new frontier and tar sands oil, hinge on the extent to which the following factors are favourable:

1. Geological prospects;
2. Availability of markets;
3. Availability of investment capital;
4. Profit potential versus risk.

Three of the regions still have abundant untested geological potential, namely the Beaufort-Delta, the arctic islands and the east coast offshore. A ready market for any level of production capability that can reasonably be anticipated seems assured, considering the size of the projected Canadian supply shortfall and the proximity of the very large United States market. The extent to which the Canadian oil industry will generate capital that could be devoted to frontier exploration and development can be estimated by preparing a cash flow projection for all elements of industry revenues and expenditures for the period 1977-85.

One of my colleagues has recently made such a projection and it indicates that, after satisfying all other claims, about 24×10^9 of cash flow from operations may be available for frontier purposes. In addition, the recently formed national oil company might contribute 2×10^9 over the period. At the rate of inflation projected to apply to energy-related projects in Canada this translates to about 16×10^9 in 1976 constant dollars. Of this 16×10^9, about 6.5×10^9 could be required to develop gas reserves in the Beaufort-Delta region and oil and gas in the arctic islands or elsewhere. At least 1×10^9 will be required for seismic surveys, leaving say 8.5×10^9 in for drilling. The cost of a frontier well varies from $2 million to $35 million, depending on location and depth. Assuming a drilling distribution roughly in accordance with the estimated geological potential and present exploration status of the several regions, the available funds would provide for around 800 wildcats. This seems rather paltry considering that 30,000 wildcats have already been drilled in the much smaller area of the producing regions. Of course, frontier drilling is likely to be much more selective than in the producing regions and the average depth per well close to twice that in the producing regions.

Figure 14

CANADA
OIL INDUSTRY CASH FLOW 1977-85

		$ x 10^9
Total revenue		130
Operating cost	11	
Government take	69	80
Funds from operations		50
Producing regions		
Expl. & Dev.	12	
Other requirements(*)	15	27
Surplus		23
Assumed Petrocanada contribution		2
Cash available for frontier Expl. & Dev.		25
Cash available for frontier Expl. & Dev. in 1976 dollars		16

*) Includes dividends, increase in working capital, and equity invest-
ment in pipelines and tar sands.

The answers to our first three questions on geological prospects,
markets and capital availability are reasonably encouraging. It remains
to assess the frontier risk/reward ratios. An excellent opportunity to
sound the oil industry's opinion on the attractiveness of frontier investment was afforded by the hearing on future Canadian oil supply and demand,
held last month by the national energy board. My reading of submissions
and testimony at this hearing is that the industry is quite pessimistic
as to the likelihood of early translation of frontier potential to flowing frontier oil.

Much of the pessimism is related to a recent federal policy statement
on land tenure, royalty and other conditions governing exploration and
development of frontier lands. The industry contends that this package
of rules will greatly reduce the incentive to explore and develop reserves. The most discouraging aspect of the new policy is a progressive
incremental royalty system that will divert to government coffers a large
share of the profits from any unusually prolific and profitable fields.
The rules of the system are such, however, that in computing field profits only the cost of prior exploration activities in the immediate vicinity of a field will be deductible.

This treatment of exploration costs does not appear to give adequate recognition to natural phenomena. One of the striking characteristics of oil occurrence is that most of the oil of a region occurs in a few large fields. One recent estimate is that 79 per cent of the world's oil occurs in 2 per cent of the world's oilfields. The industry has alwasy depended on high profits from large, prolific fields to pay for the costs of the unsuccessful and marginally successful exploration that is inherent in the widespread programmes needed to turn up major fields. Under the incremental royalty system this opportunity will be constricted and the profitability of the industry's overall frontier venture may well be below acceptable levels.

Other aspects of frontier land tenure policy, that tend to create uncertainty and diminish industry incentive, include preferential treatment for the national oil company and proposed broad ministerial discretionary powers. Further problems include the costs and time of making studies and instituting special procedures to avoid damage to the fragile arctic environment, the existence of unsettled original peoples land claims, and the uncertainty created by long regulatory delays in processing and approving pipeline applications.

In view of the foregoing, I believe that the oil industry in Canada is unlikely to mount an aggressive, broad-based exploration programme in these high-cost, high-risk areas. More likely, some companies will withdraw from active participation in the frontier play, or will slow down and defer exploration in the hope that some uncertainties will be reduced with time. Many operators will attempt some form of high-grading, drilling only the larger structures that are readily observable from seismic or other surveys. Early stage exploration is legendary for the frequency with which the only reward of drilling an attractive looking prospect is a geological surprise. The efficiency of exploration improves once some threshold of geological knowledge in a region is acquired. The success that will attend a programme of attempted high-grading of leads in a geologically mysterious region is, therefore, questionable.

Because of uncertainty as to how the industry might respond to the new legislation and what if any changes might later be made, my company felt unable to attempt a forecast of frontier production at the recent national energy board hearing on Canadian oil supply and demand I mentioned a moment ago. For the purpose of this meeting, however, I think I should try to provide you with my personal speculative view of incremental production potential in Canada if the appropriate fiscal and economic climate were created. To do so I will assume that there is an improvement in the climate for tar sands development such that the only

101

constraints become those imposed by normal business prudence. Given the foregoing assumption, it is believed that four new tar sands plants - two based on mining and two on in-situ production techniques, could come on stream between 1987 and 1995. These plants would have a combined production capacity of 450,000 B/D by 1995.

I will further assume that there is fairly early action by the federal government to introduce legislation that will encourage the industry to expend money on frontier exploration and development, to the extent it is available. By analogy with the history of the producing regions, one can expect that at some point in the exploration history of each frontier region a major oil discovery will be made that will set off a wave of exploration eventually leading to full discovery and development of the region's oil potential. The Bent Horn discovery in the arctic islands could well represent the trigger oil find there. It seems a fair speculation that the Beaufort-Delta and east coast offshore regions' share of projected frontier drilling might produce a major catalytic discovery in one or both of those regions by 1985.

After the initial trigger discovery in each region, it is postulated that exploration activity will intensify and that the process of proving up oil will proceed at about the same rate, in proportion to reserve potential, as was the case in the producing regions. Analogy with field size distribution in Alberta suggests that for this exercise the estimates of reserve potential for the various frontier regions should be discounted by 30 per cent in the onshore and shallow offshore areas, and by 60 per cent in deep or ice-infested water, to allow for fields that will be uneconomic to develop.

For the arctic islands it is believed that field facilities and the necessary marine transportation system to move Bent Horn or other oil could be in place by 1983. To introduce some risk discounting let us assume that, of the Beaufort-Delta and East Coast offshore regions, only the latter will prove to have significant oil resources. In view of the technological advances required before production systems will be available for this region, and the time needed to deploy the required transportation systems, initial production for this region appears unlikely before about 1993.

Using the foregoing as a guide, one can prepare schedules of reserve additions for the frontier areas in question which can be converted to production on the basis of typical life indices. This process results in a projection of initial frontier production at 170,000 B/D by 1983, growing to 900,000 B/D by 1995.

Figure 15

CANADA

CRUDE OIL DEMAND VS PRODUCTION 1975-1995

SPECULATIVE HIGH SUPPLY SCENARIO

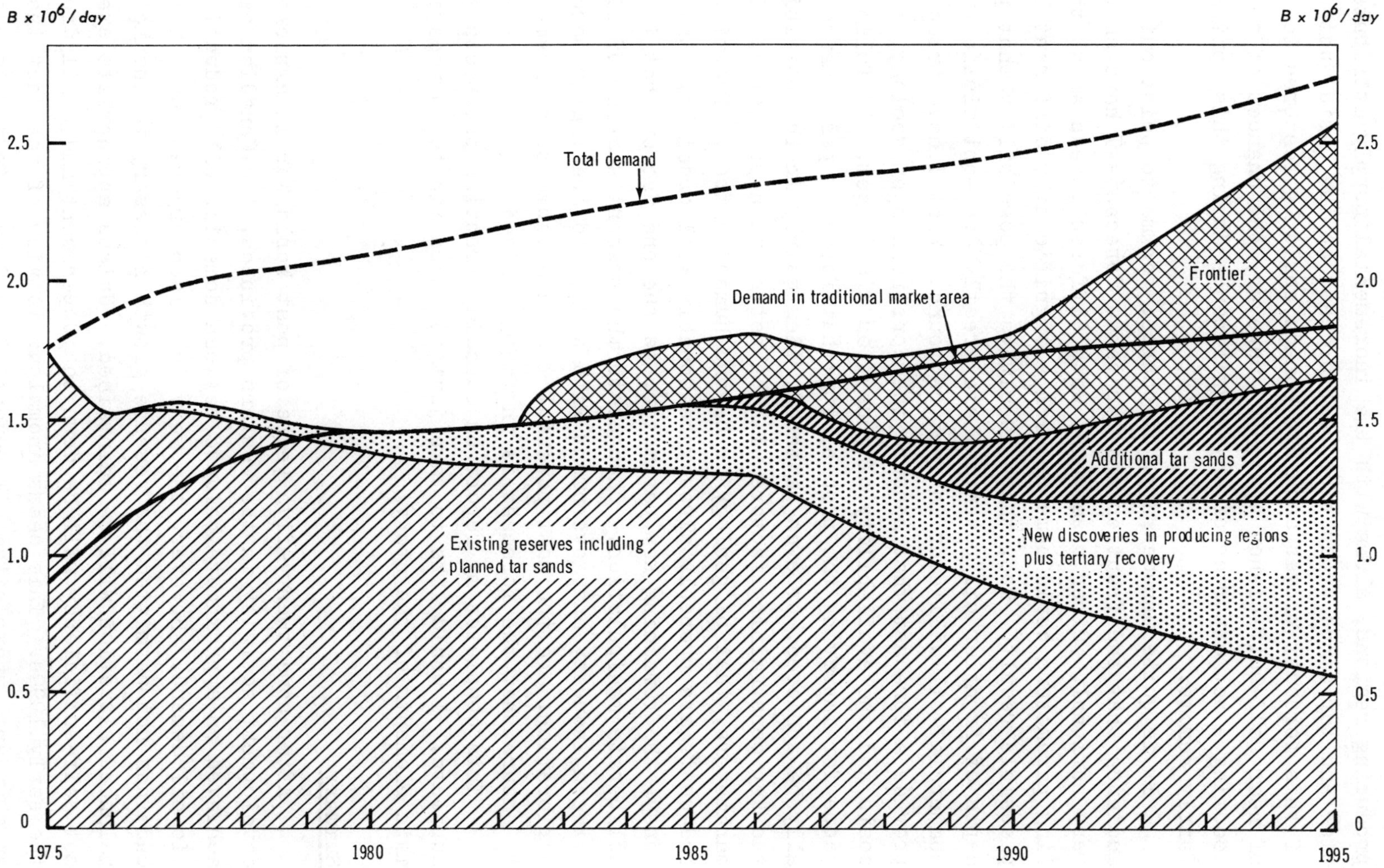

Figure 15 shows Canada's oil requirements versus supply from the producing regions, along with the incremental supply attributed to frontier and tar sands production, as forecast on the basis just discussed. If this forecast proves accurate Canada could be restored to a position of near self-sufficiency by 1995. However, the chances of the country becoming a net exporter of any significance during this century, appear fairly remote.

It is quite unnecessary, of course, for me to point out to this audience that the frontier and tar sands forecasts I have shown you are very notional. Other observers, with different ideas about the controlling parameters could come up with a significantly different but still plausible result. Too, there is always the possibility that present concepts of frontier oil potential are greatly over-optimistic and that no commercial oil will ever in fact be found. What I have tried to do is to provide a forecast that honours the constraints of geology, technology, costs and the availability of money, as I see them. It follows that the scope for improving the probability that forecast performance will actually be achieved, or even bettered, lies mainly in the possibility that the key constraining factors will be better than projected. Of these parameters, that of geology is beyond human control. Frontier technology and costs are probably already reasonably well ordained. The supply of money, the remaining key parameter, is the one factor that is subject to deliberate manipulation. Changes to the present sharing of oil revenue between industry and government with a view to freeing up more money for exploration, are the most powerful tool available to policymakers to try to ensure that the forecast is in fact realised.

This completes my report on Canadian frontier prospects and associated subjects. I would like to take just one more minute to recapitulate the main elements of the story.

Summary

After some twenty-five years of post World War II exploration and development activity in the western provinces, the Canadian oil industry achieved an approximate balance between domestic oil production and domestic oil demand. However, the very low discovery rate over the last decade has not permitted the reserve growth necessary to ensure that the balanced situation will be maintained. Unless appropriate steps are taken net self-sufficiency will be down to 50 per cent and still dropping, by the early 1990s. This seems anomalous in view of the fact that there is an abundant, unexploited oil potential in Canada in the form of geologically inferred reserves in the lightly explored frontier regions, and

very large tar sands deposits. There are, admittedly, immense technological, logistic and economic barriers to proving up these resources, a
circumstance that even now is just becoming well appreciated by the oil
industry, and perhaps not yet by the public and governments.

To date, policy reaction in Canada to rapidly increasing energy
prices has focused mainly on the opportunity presented to enhance government revenues and on the need to guard against oil industry capture of
perceived windfall profits. It is to be hoped that concern will shift
from these areas to concern for providing the oil industry with the incentive to work and leaving it with the money to do so. Under favourable
circumstances the nation might again approach net self-sufficiency some
twenty years hence.

Appendix

PROGRAM LISTING

```
1.              $WATORV
2.              DIMENSION COST(17),CPI(17),PGAS(17),
3.        C     AMPG(17),VM(17),YD72N(17),RU(17),
4.        C     ES(17),PCR(17),EFF(17),THETA(17),GAS(17),GASETU(17),
5.        C     PGAS11(17),PGAS77(17),YD(17),GASTAX(17)
6.              REAL NPOP(17),NPCR(17),MPG(17),MPGST(17),MPGEPA(17)
7.              DIMENSION GMMB(17),VMMM(17),APRMM(17),HPEA(17),GCPM(17)
8.              INTEGER YEAR(17)
9.              DO 5 I=1,17
10.             READ(5,*)  YEAR(I),CPI(I),PGAS(I),NPOP(I),RU(I),
11.       C     EFF(I),PGAS11(I),HPEA(I),YD(I),MPGST(I)
12.       5     YD72N(I)=(YD(I)*1.287/CPI(I))/NPOP(I)
13.             READ(5,*)(GASTAX(I),I=1,17)
14.       C     INITIALIZE VARIABLES
15.             AMPG(1)=13.71
16.             VM(1)=1017.9
17.             GASBTU(1)=9.2777
18.             ES(1)=63446.0
19.             THETA(1)=4627.72
20.             PCR(1)=109824.0
21.       C
22.             NNN=0
23.             NPCR(1)=8701
24.             GAS(1)=74.25
25.       C
26.             WRITE(6,1)
27.       1     FORMAT (12X 'THIS IS AN INTERACTIVE GASOLINE DEMAND MODEL')
28.             WRITE(6,2)
29.       2     FORMAT (12X,'***********************************************')
30.             PRINT 690
31.       690   FORMAT (12X 'CHOOSE OIL PRICE:  CONSTANT REAL DOLLARS (1),
32.       C     2% ANNUAL PRICE INCREASE (2), OTHER (3)')
33.             READ(8,*) NN
34.             PRINT 691
35.       691   FORMAT (12X 'CHOOSE MPG STANDARD:  EPCA STANDARD (1),
36.       C     NO STANDARD (2), OTHER (3)')
37.             READ (8,*) NSN
38.             IF (NSN.EQ.3) PRINT 699
39.             IF (NSN.NE.3) GO TO 82
40.       699   FORMAT (12X 'TYPE MPG STANDARD:  1977-1990')
41.             READ (8,*) (MPGST(I),I=4,17)
42.       82    IF (NN.EQ.2) GO TO 40
43.             IF (NN.EQ.3) GO TO 380
44.             DO 14 I=1,17
45.             PGAS(I)=PGAS11(I)
46.       14    CONTINUE
47.             GO TO 40
48.       380   PRINT 381
49.       381   FORMAT (12X 'TYPE GASOLINE PRICES, 1976-1990. 1975=$.561')
50.             READ(8,*)(PGAS(I),I=3,17)
51.       40    PRINT 385
52.       385   FORMAT (12X 'DO YOU WANT TO CONSIDER A GAS-TAX? (YES=1,NO=2)')
53.             READ(8,*)NTX
54.             IF (NTX.EQ.1) GO TO 386
55.             GO TO 45
56.       386   PRINT 387
57.       387   FORMAT (12X 'DO YOU WANT GAS-TAX THAT IS IN THE PROGRAM OR
58.       C     YOUR OWN?  (PROGRAM = 1,OWN =2)')
59.             READ(8,*) JTX
60.             IF (JTX.EQ.2) GO TO 388
```

```
 61.            GO TO 390
 62.   388      PRINT 389
 63.   389      FORMAT (12X 'TYPE YOUR OWN GAS-TAX')
 64.            READ(8,*)(GASTAX(I),I=1,17)
 65.   390      DO 391 I=1,17
 66.            PGAS(I) = PGAS(I) + GASTAX(I)
 67.   391      CONTINUE
 68.        C   *****************************BEGIN CALCULATIONS***********
 69.    45      DO 10 J=2,17
 70.            I=J-1
 71.            GCPM(J)=(PGAS(J)/CPI(J)/AMPG(I)
 72.            VM(J)=NPOP(J)*EXP(1.37205-.2251*ALOG(GCPM(J)+.63176
 73.        C   *ALOG(YD72N(J))+.005931*RU(J)-.9525*ALOG(HPEA(J))
 74.        C   +.306*ALOG(PCR(I)/NPOP(I)))
 75.            NPCR(J)=NPOP(J)*EXP(20.9943-4.1101*ALOG(ES(I)/NPOP(J))+
 76.        C   1.27339*ALOG(VM(J)/NPOP(J))+3.31406*ALOG(YD72N(J))-.071945
 77.        C   *RU(J)
 78.            PCR(J)=.934363*PCR(I)+NPCR(J)
 79.            ES(J)=NPCR(J)+.934363*.92097*ES(I)
 80.            MPG(J)=EXP(3.4096+.721484*ALOG(PGAS(I)/CPI(I)+.278516
 81.        C   *ALOG(EFF(J)))
 82.            IF (NSN.EQ.2) GO TO 190
 83.            IF (MPGST(J).GT.MPG(J)) MPG(J)=MPGST(J)
 84.   190      CONTINUE
 85.    30      THETA(J)=NPCR(J)/MPG(J)+.934363*.92097*THETA(I)
 86.            AMPG(J)=ES(J)/THETA(J)
 87.            GCPM(J)=(PGAS(J)/CPI(J))/AMPG(J)
 88.            VM(J)=NPOP(J)*EXP(1.37205-.2251*ALOG(GCPM(J))+.63176
 89.        C   *ALOG(YD72N(J))+.005931*RU(J)-.9525*ALOG(HPEA(J))
 90.        C   +.306*ALOG(PCR(I)/NPOP(I)))
 91.            GAS(J)=VM(J)/AMPG(J)
 92.            GASBTU(J)=5.248*GAS(J)/42
 93.            PGAS77(J)=PGAS(J)*1.838/CPI(J)
 94.            GMMB(J)=GAS(J)/(42*365)
 95.            APRMM(J)=NPCR(J)/1000
 96.            MPGEPA(J)=MPG(J)*1.141
 97.        C
 98.        C
 99.        C   UNIT 11 - TCB117
100.        C   UNIT 14 - BS2111
101.        C   UNIT 13 - BS217
102.        C   UNIT 17 - NCB117
103.        C
104.        C
105.        C
106.    10      CONTINUE
107.            IF (NN.EQ.3) GO TO 382
108.            IF (NN.EQ.2) GO TO 54
109.            IUNIT=14
110.            GO TO 56
111.    54      IUNIT=13
112.            GO TO 56
113.   382      CONTINUE
114.            IUNIT=17
115.    56      WRITE(11,156)
116.   156      FORMAT(//6X,'GMMB',4X,'APRMM',4X,'PGAS77',5X,'MPG'/)
117.            WRITE(11,5556)(GMMB(MM),APRMM(MM),PGAS77(MM),MPG(MM),
118.        C   MM=1,17)
119.            WRITE(IUNIT,157)
120.   157      FORMAT(//8X,'VM',10X,'NPCR',10X,'PCR',11X,'MPG',11X,'AMPG',
```

```
121.      C   12X,'GAS',10X,'MPG EPA',8X,'PGAS'/)
122. 5556      FORMAT(4X,F7.2,2X,F7.2,2X,F7.4,2X,F7.2)
123.          WRITE(IUNIT,58)(VM(M),NPCR(M),PCR(M),MPG(M),AMPG(M),GAS(M),
124.      C   MPGEPA(M),PGAS(M),M=1,17)
125.   58     FORMAT(F14.3,F14.4,F14.4,F14.7,F14.7,F14.4,F14.5,F14.5,2X)
126.      C***************WRITE OUT RESULTS***************
127.   60     PRINT 692
128.  692     FORMAT ('DO YOU WISH TO PRINT THE EXOGENOUS VARIABLES?
129.      C   (YES=1,NO=2)')
130.          READ(8,*) LLL
131.          IF (NN.EQ.2) GO TO 42
132.          IF (NN.EQ.3) GO TO 383
133.          WRITE(10,41)
134.   41     FORMAT(///12X,'CRUDE OIL REAL PRICE CONSTANT')
135.   49     IF (NSN.EQ.1) GO TO 1001
136.          GO TO 101
137. 1001     WRITE(10,1000)
138. 1000     FORMAT(///12X,'EPCA EFFICIENCY STANDARD ASSUMED')
139.          GO TO 5000
140.  101     IF (NSN.EQ.2) GO TO 2001
141.          GO TO 102
142. 2001     WRITE(10,2000)
143. 2000     FORMAT(///12X,'NO EFFICIENCY STANDARD ASSUMED')
144.          GO TO 5000
145.  102     IF (NSN.EQ.3) GO TO 3001
146. 3001     WRITE(10,3000)
147. 3000     FORMAT(///12X,'ALTERNATE EFFICIENCY STANDARD ASSUMED')
148.          GO TO 5000
149.   42     WRITE(10,43)
150.   43     FORMAT(///12X,'CRUDE OIL REAL PRICE INCREASES AT 2%/YEAR')
151.          GO TO 49
152.  383     WRITE(10,384)
153.  384     FORMAT(///12X,'THIS IS AN ALTERNATIVE CRUDE OIL PRICE RUN')
154.          GO TO 49
155. 5000     IF (NTX.EQ.1) GO TO 1002
156. 2002     WRITE(10,2003)
157. 2003     FORMAT(///12X,'NO GAS-TAX ASSUMED')
158.          GO TO 48
159. 1002     WRITE(10,1003)
160. 1003     FORMAT(///12X,'GAS-TAX ASSUMED')
161.   48     IF (LLL.EQ.2) GO TO 53
162.          WRITE(10,31)
163.   31     FORMAT(///30X,'EXOGENOUS VARIABLES')
164.          WRITE(10,32)
165.   32     FORMAT(30X,'*********************')
166.          WRITE(10,33)
167.   33     FORMAT(///8X,'YEAR',10X,'CPI',9X,'NPOP',11X,'YD',
168.      C   11X,'RU'/)
169.          WRITE(10,34) (YEAR(II),CPI(II),NPOP(II),YD(II),RU(II),
170.      C   II=1,17)
171.   34     FORMAT(8X,I4,3F14.3,F14.5)
172.          WRITE(10,35)
173.   35     FORMAT(//8X,'YEAR',9X,'HPEA',8X,'MPGST',11X,'EFF',
174.      C   10X,'PGAS'/)
175.          WRITE(10,36) (YEAR(II),HPEA(II),MPGST(II),EFF(II),PGAS(II),
176.      C   II=1,17)
177.   36     FORMAT(8X,I4,F14.5,F14.2,F14.5,F14.6)
178.   53     WRITE(12,99)
179.   99     FORMAT(//8X,'YEAR',10X,'PGAS',8X,'PGAS77',
180.      C   10X,'NPCR',11X,'PCR'/)
```

```
181.              WRITE(12,91)(YEAR(MM),PGAS(MM),PGAS77(MM),NPCR(MM),
182.        C     PCR(MM), MM=2,17)
183.              WRITE(12,98)
184.      98      FORMAT(//8X,'YEAR',10X,'ES',13X,'MPG',8X,'MPG EPA',9X,'THETA',
                  10X,
185.        C     'AMPG'/)
186.              WRITE(12,92)(YEAR(MM),ES(MM),MPG(MM),MPGEPA(MM),THETA(MM),
187.        C     AMPG(MM), MM=2,17)
188.              WRITE(12,97)
189.      97      FORMAT(//8X,'YEAR',8X,'GCPM',8X,'VM',11X,
190.        C     'GASAUTO',8X,'GASBTU'/)
191.              WRITE(12,93)(YEAR(MM),GCPM(MM),VM(MM),GAS(MM),
192.        C     GASBTU(MM),MM=2,17)
193.      91      FORMAT(8X,I4,2F14.5,2F14.4)
194.      92      FORMAT(8X,I4,F14.4,F14.7,F14.7,F14.5,F14.7)
195.      93      FORMAT(8X,I4,F14.8,F14.3,F14.4,F14.4)
196.              WRITE(12,11122)
197. 11122        FORMAT(/////////)
198. 9999         STOP
199.              END
200.              $DATA
201.              $ASSIGN 11 TO 8000/9999 OUTPUT
202.              $ASSIGN 14 TO 6000/6999 OUTPUT
203              $ASSIGN 13 TO 7000/7999 OUTPUT
204.              $ASSIGN 15 TO 8000/8999 OUTPUT
205.              $ASSIGN 8 TO TERMINAL INPUT
206.              $ASSIGN 10 TO 5000/5099 OUTPUT
207.              $ASSIGN 12 TO 5100/5999 OUTPUT
208.     1974   1.478    .5280    211.9   5.62   1.        .528   36.6   980.8    0.
209.     1975   1.612    .562     213.5   8.48   1.13      .562   36.1  1071.9    0.
210.     1976   1.727    .60      215.3   7.5    1.1639    .60    36.1  1188.8    0.
211.     1977   1.838    .6438    217.4   6.8    1.19882   .66    36.1  1306.3    0.
212.     1978   1.938    .707     219.5   6.5    1.23478   .74    36.0  1442.3   15.8
213.     1979   2.037    .771     221.5   5.9    1.27182   .78    35.9  1571.3   16.6
214.     1980   2.14     .852     223.6   5.6    1.31      .82    35.8  1711.8   17.6
215.     1981   2.236    .892     226.7   5.2    1.35      .85    35.7  1844.5   18.8
216.     1982   2.335    .943     227.8   4.9    1.39      .89    35.5  1981.4   20.1
217.     1983   2.434    .995     230.0   4.9    1.43      .93    35.4  2124.0   21.5
218.     1984   2.531   1.048     232.2   4.8    1.47      .97    35.3  2282.6   22.8
219.     1985   2.629   1.092     234.5   4.6    1.51     1.00    35.1  2438.5   24.1
220.     1986   2.728   1.147     236.7   4.6    1.55     1.04    35.0  2592.3   24.1
221.     1987   2.827   1.203     239.0   4.6    1.59     1.08    34.8  2742.6   24.1
222.     1988   2.925   1.261     241.3   4.8    1.63     1.12    34.7  2894.2   24.1
223.     1989   3.018   1.309     243.7   4.8    1.67     1.15    34.6  3052.1   24.1
224.     1990   3.107   1.368     246.0   4.8    1.71     1.19    34.5  3205.0   24.1
225.    0  0    0   0    0        .05  .1   .15     .2   .25  .3    .35   .4
226.    .45  .5   .5     .5
227.    $STOP
```

AN ECONOMIC ANALYSIS OF BRITISH NORTH SEA OIL SUPPLIES

by

C. Robinson and C. Rowland

This paper provides an economic analysis of the supply of British North Sea oil along the following lines:

1. A broad assessment of future North Sea production and its effect on the pattern of United Kingdom energy supplies.
2. The profitability of North Sea oil and how it may be affected by price variations.
3. Possible effects on profitability of government depletion control.
4. Balance of payments effects and their implications for depletion control.

The data on which the paper is based relates to June 1978: references to more detailed studies on earlier data bases are given at the end.

FUTURE SUPPLIES[1]

As a background to the rest of the paper it is useful to begin with an order-of-magnitude assessment of future oil supplies from the British sector of the North Sea which is then used for an economic analysis of some of the variables to which such supplies will be sensitive.

Any discussion of future supplies should be prefaced by acknowledging our present state of relative ignorance about the North Sea. There are already more than enough people estimating reserves and annual supplies without making the appropriate qualifications. A large area is as yet unexplored, especially north of the 62nd Parallel, in the Norwegian sector

---

1) For further discussion of future supplies, see Colin Robinson and Jon Morgan "North Sea Oil in the Future: Economic Analysis and Government Policy", Macmillan for Trade Policy Research Centre, London, July 1978, Chapter 3 and Colin Robinson and Jon Morgan, "The Economics of North Sea Oil Supplies", The Chemical Engineer, June 1977.

generally and to the west of Britain; as oil production is still in its infancy, there is insufficient operating experience even to make reliable reserve estimates for the fields under development. Exploration activity, recovery rates from fields discovered and depletion profiles will all be sensitive to prices and price expectations, to costs and taxes, about which one can only make guesses.

Consequently, anything one says about offshore supplies must inevitably be very tentative. The brief review of supply prospects which follows deals first with the outlook to 1980: this seems relatively clear at present, though some past operating accidents and delays are a reminder that development plans even for the next few years are by no means certain to be achieved. Then we turn to the longer term outlook which is very uncertain.

Oil production from the United Kingdom sector should build up rapidly in the next few years so that by around 1980, even allowing for delays in existing production programmes, there is a good chance that it will reach 100 million tonnes (about 2 million barrels per day) and be somewhat in excess of United Kingdom oil consumption. Natural gas output will also be increasing: the British Gas Corporation should have supplies of about 170 million cubic metres (6,000 million cubic feet) per day by 1980 and output may increase further in the 1980s.

Within a few years the North Sea should be supplying a large part of the United Kingdom's energy. In 1977, the North Sea provided almost all natural gas consumed in Britain (apart from small imports from Algeria); as Table 1 shows, natural gas output was equivalent to approximately 19 per cent of primary energy consumption. North Sea oil production began in the second half of 1975 and grew in the following two years also to the equivalent of some 19 per cent of primary energy. By 1980, the increase in oil output to around 100 million tonnes (170 million tonnes coal equivalent) and the rise in gas output should raise the share of North Sea production in total energy consumption to around 70 per cent; the other 30 per cent will come mainly from indigenous coal and nuclear electricity. Net imports of energy should have dwindled to zero by this time, though there will most probably still be substantial crude oil imports, balanced by exports of North Sea oil which is light and low in sulphur: the United Kingdom pattern of oil product demand will be met more economically by trading North Sea oil for heavier crudes.

To illustrate approximately what supplies may be in the longer term, we begin by making a very general assumption - that future combinations of costs, prices, taxes and other variables affecting profitability will be such that oil fields already discovered will be developed in accordance

Table 1

UNITED KINGDOM INLAND PRIMARY ENERGY CONSUMPTION

	Million tonnes coal or coal equivalent	
	1977	1980 Forecast
Coal	123	112 ± 10
Oil(*)	137	140 ± 10
Natural gas	63	80 ± 5
Nuclear	14	21 ± 4
Hydro	2	2
	339	335 ± 10
Output as percent of energy consumption:		
North Sea gas +	19	23 ± 1
North Sea oil	19	48 ± 5
	38	71 ± 6

*) Excluding non-energy products.

+) Including some imports of Algerian LNG and, in 1980, output from both the British and Norwegian sectors of Frigg.

Source: For all 1977 data: Monthly Digest of Statistics, May 1978, Tables 7.1 and 7.3.

with present ideas on depletion and that exploration will continue into the 1980s. Depletion profiles over time for fields already known and for fields assumed to be discovered in the course of future exploration efforts are then specified. The outcome is intended as a "surprise free" projection which can be used as a benchmark for testing the effects of such events as price variations or the imposition of depletion control.

The results are illustrated in Figure 1 which shows three production estimates. The lowest is for 22 "established commercial" fields either finalising development plans or already under development which appear to contain over 1.4×10^9 tonnes (11×10^9 barrels) of recoverable oil. Most of these seem likely to give acceptable rates of return, given present cost and price expectations and with existing tax rates. Actual supplies are likely to be substantially higher, since there are other known fields which, unless there is a sharp increase in the cost: price ratio or some other serious adverse changes in economic circumstances, should give rates of return of the same order as the 22 fields and will therefore most probably be developed also.

114

Figure 1

ESTIMATED UK NORTH SEA OIL OUTPUT 1976-1995 : SURPRISE - FREE PROJECTION

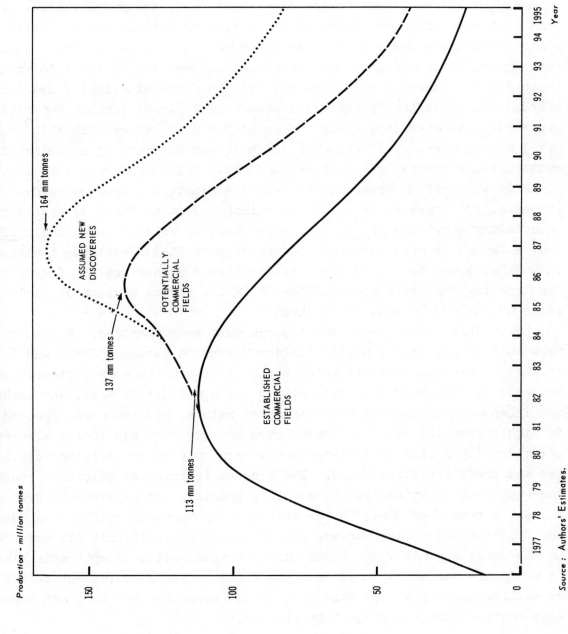

Source : Authors' Estimates.

To illustrate the effect of bringing in these fields, the "established plus potential" estimate includes another 19 "potentially commercial" fields with estimated recoverable oil reserves of about 0.6×10^9 tonnes (4×10^9 barrels).

Finally, some assumptions are made about future exploration activity on prospective structures in the area considered, the probabilities of finding fields of given sizes and the timing of future discoveries[1] in order to estimate newly discovered reserves. According to these estimates, further exploration may eventually reveal another 1.0 to 1.1×10^9 tonnes (7 to 8×10^9 barrels) of recoverable oil in finds of slightly smaller average size than the "potentially commercial" fields (around 200 million barrels). The aggregate figure of about 3.0×10^9 tonnes (23×10^9 barrels) of recoverable oil reserves which is our estimate of established commercial, potentially commercial and newly discovered reserves, is very close to the British Government's latest estimate for areas already licensed.[2] However, it should be pointed out that there is considerable uncertainty about the timing of discoveries, as well as about their size, and, although as much published information on field development plans as is available has been used there is inevitably a good deal of judgment in constructing depletion profiles for North Sea fields at such an early stage of the development of the area.

Two other points about the diagram must be emphasized. First, it refers only to oil in the United Kingdom North Sea between 55°50' and 62°N so that it excludes any oil which may be found in the more northerly areas, the whole of the Norwegian sector, the Irish and Celtic Seas, and onshore: nor is any associated and non-associated natural gas included. Second, it is highly speculative. One cannot even be very sure how fields already being developed will be depleted and we can only guess at future discoveries and their depletion rates. The diagram is intended only as a rough-and-ready way of attempting to quantify possible future events. Even so it can be concluded that, in the absence of government controls on output, the United Kingdom like Norway, should become a significant oil exporter in the early and mid 1980s (when United Kingdom oil consumption is likely to be in the range 100 to 120 million tonnes a year, equivalent to 2 to $2\frac{1}{2}$ million barrels a day) though it is not clear for how long net exports will continue nor how large they will be.

---

1) Future discoveries include developments resulting from finds made after end 1977.

2) Development of the Oil and Gas Resources of the United Kingdom, Department of Energy, 1978.

The above is a fairly mechanical kind of analysis in which the broad assumption is made that future cost, price and tax conditions will favour both the development of fields already discovered and continued exploration. However, one of the important questions concerns the extent to which future supplies are a function of economic variables. We now consider how supplies might be affected by price variations and by depletion control, using for the analysis some computer programmes developed for the economic assessment of North Sea oil fields.

OIL PRICES AND PROFITABILITY(1)

Fears are often expressed that a decline in world oil prices would seriously affect the exploitation of North Sea oil. If OPEC prices were substantially reduced - either deliberately or as a result of price-discounting - it appears that North Sea profitability might decline and that supplies might be cut. Similarly, further big price increases might be expected to stimulate North Sea activity and raise supplies. One cannot, however, draw simple conclusions about price effects, which will vary according to the stage of development of a field: the effect turns on the traditional micro-economic distinction between fixed (unavoidable) and variable (avoidable) costs. Price reductions are considered first in the analysis below.

Consider a North Sea oil field which has been discovered, on which appraisal work has been done, production platforms and pipelines installed - in fact, most of the capital which needs to be invested in the field has been spent, as it normally would be in the first few years after the find. From the producing company's point of view, such expenditure is then "sunk" or unavoidable. In considering the future output programme its prime concern is to ensure that revenues exceed costs of operating the field and yield a return as high as could be obtained elsewhere. Since capital expenditure is committed, the principal quantitative criterion to use in future planning of the field is whether <u>additional</u> (avoidable) expenditure from any point of time onwards would yield a positive Net Present Value. Our calculations indicate that even if world oil prices were to fall substantially, it would still be worth proceeding with

_____

1) See Robinson & Morgan <u>op. cit.</u>, Chapters 4 and 6. The cost estimates used in this paper are more up to date than those in Robinson & Morgan.

the development of most North Sea oil fields so far discovered because likely operating costs(1) are low relative to probable revenues.

Figure 2 illustrates the effect for four fields with varying characteristics. For each field we estimate the North Sea oil prices necessary to achieve an after tax internal rate of return (IRR) of 20 per cent. The cut off rate of 20 per cent in money terms (which, according to our calculations, approximates to a 12 to 14 per cent real rate of return) is admittedly somewhat arbitrary, but it seems unlikely that companies operating in the North Sea would be prepared to undertake projects expected to yield lower returns: they might require a higher return but our general conclusions turn out to be very robust with respect to the cut off rate which is used.(2)

We first calculate the price, which, if maintained for the whole of a field's life, would just yield 20 per cent IRR after tax - in other words, an estimate of costs including "normal" producing profit. In the case of Forties, for example, this price is estimated to be about $4\frac{1}{2}$ per barrel - the point of intersection between the Forties curve and the vertical axis. For Dunlin the required price to achieve 20 per cent IRR appears to be more than twice as high - approximately $10 per barrel - and the Brae price appears to be as much as $16 per barrel (off the diagram).(3) The mid 1978 parity price of North Sea oil (based on the $12.70 per barrel f.o.b. price of Light Arabian Crude and allowing for transport and quality premia) is around $13.70 per barrel. As one moves to the right along the horizontal axis in the diagram the prices indicated are those which would achieve 20 per cent IRR from that point in the project onwards - ignoring all previous costs and revenues. Naturally, since capital expenditures are concentrated during a few years early in the project, the effect up to about years 6 to 7 (year 0 being project start) is to remove large quantities of expenditure but not to change project revenues significantly: consequently the calculated required price drops sharply. After years 6 to 7,

---

1) There is considerable uncertainty about the future size of operating costs and what portion will be truly variable. The calculations which follow are only intended to indicate approximately what the effects of price changes might be.

2) For example, see Colin Robinson "The Economics of North Sea Oil and Gas", Erdol Erdgas Zeitschrift, International Edition, May 1976 and Robinson and Morgan, op. cit., Chapter 6.

3) These very large differences in required prices are, of course, a consequence of the considerable variations in the characteristics of North Sea oilfields. The cost estimates for Brae are particularly speculative because no firm development plan for this field has yet been announced.

the required price stabilizes or increases a little (because of increasing unit operating costs). For example, Forties which, according to our calculations requires a North Sea oil price of around $4½ per barrel to earn 20 per cent DCF on the whole project, would after five or six years of development need a price of only around $1½ to 2 per barrel - that is, approximately the pre-October 1973 price - to earn 20 per cent on avoidable expenditures from then on. Even rather high-cost fields such as Brae and Montrose seem to be similarly placed, though the required prices are higher - about $3½ to 4½ per barrel after some six years into the project.

The purpose of this analysis is to quantify something which is intuitively clear, not to give precise figures of the prices at which various rates of return are achieved. Given the concentration of capital spending in the early years of a field's development, even a large decline in OPEC prices, once development is underway, might have a relatively small effect on the output of that field. There might be some tendency for recovery rates to fall because of cost-cutting, but the essential point is that once a large proportion of capital has been committed, returns on incremental expenditures seem likely to appear high to producers, as compared with alternative investment opportunities, even at very low prices. One might therefore draw the general conclusion that any likely declines in world oil prices will not greatly affect North Sea oil output in the next few years since most fields already under development should go ahead almost regardless. The producers are unlikely to walk away at any prices one can foresee. It might reasonably be expected, therefore, that output from the 22 established commercial North Sea fields - the lowest line in Figure 1 - should be insensitive to price fluctuations. That is, output of 100 million tonnes per year (2 million barrels per day) or more in the early 1980s should be little affected by a price decline.

It is quite a different story with fields on which development has not yet begun and those still waiting to be discovered - the higher curves in Figure 1. In such cases, the avoidable costs of potential producers include not only operating costs but capital costs of development and, in the case of fields not yet discovered, exploration costs as well. To try to assess the effect on exploration and development activity we have calculated for six of the potentially commercial fields, how returns on total projects might be affected by price increases or decreases.

A number of sensitivity analyses were performed using various future price trends and taking both Internal Rate of Return (IRR) and Net Present Value (NPV) criteria (after tax). Table 2 gives some of the conclusions.

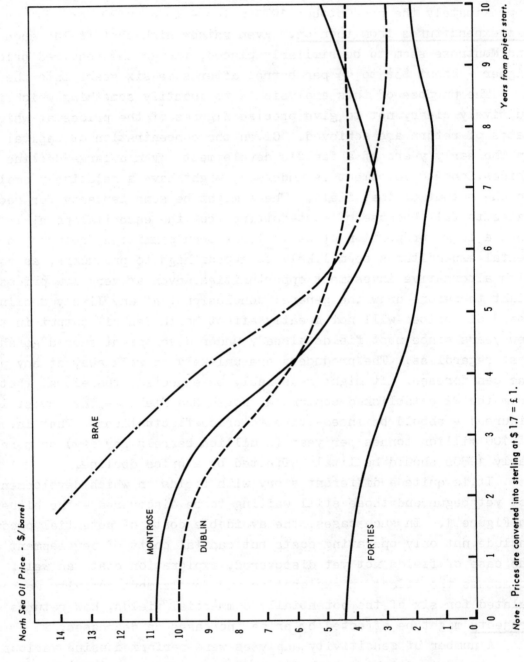

Figure 2

PRICE AT WHICH 20 % AFTER TAX IRR IS ACHIEVED
FROM DIFFERENT POINTS ONWARD IN PROJECT LIFE

North Sea Oil Price - $/barrel

Years from project start.

Note : Prices are converted into sterling at $ 1.7 = £ 1.

Source : Authors' estimates.

The three scenarios used for the North Sea oil price(1) are as follows:

Limited Price Decline (LPD): The current $13.7 per barrel price falls steadily to $13 in 1980 and then increases by about 3 per cent per annum to 1985 and by 7 per cent per annum thereafter.

Crude Price Indexation (CPI): A 5 per cent per annum rise through to 1985 is followed by 7 per cent per annum increases in the price.

Use of Monopoly Power (UMP): Immediate large price rises push the price to approximately $19 by 1980, and another price increase of 25 per cent in 1985 occurs. In the other years a 7 per cent per annum increase is assumed.

Unsurprisingly, the UMP scenario generates relatively high returns for all the offshore projects considered, and with price developments along these lines the six potentially commercial fields would all be viable. In these conditions of sharply rising prices the higher schedules in Figure 1 might be underestimates of the oil production from the North Sea. However, examining the results for the CPI scenario and bearing in mind that returns are expressed in current terms, shows a considerable difference. How some of the fields (Alwyn, Hutton and Lyell) look much poorer prospects and, consequently, the eagerness of the companies to develop these discoveries compared to the situation of higher price would dwindle. The impact of a limited price decline (LPD) reduces the same three fields' returns to levels at which most companies would not consider them worthwhile investment projects.

Because of this effect on expected total project rates of return of a price decline, there would very likely be a noticeable reduction in exploration and development activity in the North Sea under the LPD scenario compared with a situation of constant prices in real terms (approximated by our CPI scenario). The magnitude of the supply response is not easy to predict, especially since the price fall would reduce oil company returns on production worldwide and would alter price expectations. The calculations of rates of return will not bear precise interpretation because of uncertainties about future costs, taxes and revenues, nor can one know what rates of return various companies will expect. Moreover, company response to price changes will depend partly on what sizes and types of fields companies expect to find in the future. Nevertheless, one can be fairly sure that there would be a significant reduction in the output levels shown by the potentially commercial and new discoveries supply

---

1) See Robinson and Morgan, op. cit., pp. 121-123 for further explanation.

Table 2

ESTIMATED PROFITABILITY OF NORTH SEA OIL

| | Estimated Recoverable Costs | Unit Costs* ($/barrel) | | Returns on given Price Scenarios | | | | | |
| | | | | UMP | | CPI | | LPD | |
	Million Tonnes	Capital	Operating	IRR %	NPV+ £m	IRR %	NPV+ £m	IRR %	NPV+ £m
Larger Sized Potentially Commercial Fields									
Alwyn	40	2.8	8.3	29.4	130	20.2	2	15.1	-96
Crawford	45	2.5	3.2	40.2	255	30.4	113	24.9	50
Toni/Thelma	50	2.2	3.0	35.6	245	27.5	103	22.8	36
Smaller Sized Potentially Commercial Fields									
Andrew	20	2.5	4.5	36.7	103	27.2	35	21.7	8
Hutton	35	3.9	3.4	29.7	129	21.6	17	17.3	-28
Lyell	35	3.5	3.3	25.6	62	19.2	-7	15.9	-34

Note: All costs and rates of return are quoted exactly as calculated - they are subject to significant error margins for reasons given in the text.

*) The unit cost figures are total (undiscounted) costs ÷ total output. They are given merely to
indicate the scale of expenditure involved. All calculations of course use discounting. Costs
are calculated in £s and then converted to the more familiar $/barrel units at $1.7 = £1. The
initial calculations are made in sterling to avoid distorting the value of capital allowance
for tax purposes.

+) At a 20 per cent discount rate.

Source: Authors' estimates.

lines in Figure 1 if prices were to drop in the near future. Other things being equal, the later the price decline the less would be its effect on supplies, since more capital would have been committed. Possibly, a continued price slump (unlikely as this may be) would result in a supply curve only fractionally above the output of the established commercial fields, unless the government decided to exploit fields thought to be submarginal by the oil companies.

DEPLETION CONTROL, PROFITABILITY AND SUPPLIES(1)

In the surprise-free assessment of future British North Sea Oil supplies in section 1 above, it was assumed that there would be no government control of company depletion policy. However, a system of detailed regulation of company production plans was instituted by the Petroleum and Submarine Pipelines Act of 1975 and there are other possible means of controlling output (for example through the 1976 Energy Act or the licence award system): one must assess to what extent this may influence profitability and supplies, either directly through government-imposed changes in depletion profiles or indirectly through its effect on company attitudes towards North Sea investment.

The producing companies now have to seek Department of Energy approval for output programmes specified within rather narrow limits and recently the Department of Energy has apparently decided it will approve development plans in stages - for example, it will initially only approve the build-up period before the "plateau" is reached.

The government has power subsequently to vary company programmes already approved within a range which it determines though any increase which it wants may not exceed the output which would be obtained from drilling one additional well. Reductions from specified programmes are supposedly constrained by guidelines set out in the House of Commons in December 1974 by Mr. Eric Varley (then Secretary of State for Energy) though it is not clear whether the "Varley Guidelines" are consistent with later government actions such as the Energy Act of 1976. Most of the discussion about depletion control in the United Kingdom has used rather naive "national interest" arguments in support of reducing output below company plans. Although the theoretical basis for such arguments is

---

1) Further details and references are in Robinson and Morgan, op. cit., Chapter 7. Critical comments on depletion controls are in Colin Robinson, "A Review of North Sea Oil Policy", Zeitschrift für Energie Wirtschaft, September 1978.

extremely weak, this paper is concerned only with summarizing what would be the effects of downward depletion control on company profitability and company investment policy.

The essence of such conclusions as can be quantified is that the impact on company after tax DCF returns of production cuts within the Varley Guidelines would probably be fairly small.(1) Government-imposed output reductions would be unlikely to reduce project internal rates of return (in money terms) more than some 3 or 4 percentage points as compared with a situation of no controls. This might adversely affect the exploitation of some marginal fields but on the face of it would not appear likely significantly to reduce the profitability of oil from the North Sea. If the government were to impose development delays as well as production cuts, however, profitability would be more severely affected.

Perhaps more important, some of the effects of depletion control are likely to be less tangible. First, there is the additional uncertainty which seems to be the inevitable accompaniment of political involvement in output decisions. The calculations above of the impact on profitability of output regulation only demonstrate ex-post effects on the assumption that existing guidelines are maintained. However, the companies cannot be sure precisely how the Varley Guidelines will be applied, nor are they certain how long they will remain in force. In other words, by assuming that the guidelines remain and that control is imposed in particular ways, our calculations may assume away one of the principal effects of depletion control - the generation of more uncertainty than in an unregulated situation. Then there is the effect of regulation on the efficiency of North Sea operations. Presumably if companies have to alter production programmes there will be real costs involved in doing so, in terms of extra production costs and management time occupied in discussions with the regulatory authority and coping with revised plans. For such reasons it is possible that the presence of a regulatory system may have some adverse effect both on the willingness of companies to invest in the North Sea and on the efficiency of their activities. One must, however, distinguish between the effects of having a regulatory system and the effects of government-imposed changes in output as compared with company plans. The system is already in being in the sense that companies are having their production programmes examined by the Department of Energy. Nevertheless, despite the popular belief that there are "national interest" grounds for reducing output in the 1980s below company programmes one cannot assume that such reductions will actually be imposed. Whatever one thinks of national

_____

1) See Robinson and Morgan, op.-cit., pp. 143-144.

interest arguments, for reasons given in section 4 below it may not be very realistic to think that any British Government in the foreseeable future will be willing to impose output reductions. The United Kingdom might suffer the disadvantages in terms of increased uncertainty for the producers of having a detailed regulatory apparatus even though that apparatus leads to little or no change in company output programmes.

IMPLICATIONS FOR THE BALANCE OF PAYMENTS AND THE BRITISH ECONOMY(1)

As Britain substitutes indigenous oil for crude oil imports which would otherwise have been made and becomes for a time a net oil exporter, a potential balance of payments gain will emerge. For most of the fields now being developed North Sea oil production costs, though high in comparison with Middle East costs, are still lower than the world oil price: consequently, the use of British resources in offshore oil production and associated activities (as compared with employing these resources elsewhere) should lead to an increase in real GNP and thus to a potential balance of payments improvement.

To measure this potential gain one can compare a future "North Sea state" (NS) with a future "no North Sea state" (nNS), assuming that in each case the level of factor employment is the same. For each future year, various North Sea oil output levels and various world oil prices are assumed and the implications for the current and long-term capital account balance of payments are calculated with the same series of computer programmes used in the analyses mentioned earlier in this paper: the inputs are estimates of recoverable reserves, production rates, capital costs for established commercial, potentially commercial and newly discovered fields defined as in Figure 1. Present tax and royalty rates are assumed to continue unchanged. Obviously there is great uncertainty about future costs, outputs, prices and tax rates but - as in all the research presented in this paper - the intention is to provide estimates which define the general order of magnitude of the impact of North Sea oil.

The payments effect is very sensitive to one's assumptions about a number of key variables, especially North Sea output and world oil prices. To indicate this sensitivity each of the three price scenarios used on page 120 is combined with each of the three production cases in Figure 1. Separate calculations are then made of the estimated saving on visible trade in oil, the import content of equipment and services for North Sea

1) Detailed calculations and explanation are in Robinson and Morgan, op. cit., Chapter 8, although on an earlier data base.

activities, United Kingdom exports of such equipment and services, interest on overseas loans and profits accruing to overseas companies, and inflows and outflows of long-term capital. The exchange rate is assumed throughout to be the same in NS as in nNS (although in practice the exchange rate might well appreciate in NS) so that in effect the balance of payments benefits are assumed to accrue to foreign exchange reserves.

Figure 3, which assumes for illustrative purposes the LDP price scenario and the established commercial, potentially commercial and newly discovered fields (the highest curve in Figure 1), demonstrates how dominant is the visible trade saving ("oil trade effect"). "Non oil trade effects" (mainly imports of equipment and services in the 1970s and interests and profits outflow abroad in the 1980s) offset the visible trade effect to a certain extent throughout, and there is a moderate net outflow of capital in the 1980s, but all the calculations show substantial overall balance of payments benefits.

The aggregate effect shown in the diagram is approximately £8,000 to £10,000 million (in current prices, not real terms) in the late 1980s but it is possible to obtain much smaller or much larger gains by varying the assumptions about prices and output. After examining a large number of such sensitivity analyses our judgment is that future potential balance of payments gains (current and long-term capital account in NS versus nNS) are likely to be of the following orders of magnitude.

		£000 million (current prices)
1980	4 to 6	(about $2\frac{1}{2}$% to 4% of GDP)
1985	6 to 12	(about 2% to $4\frac{1}{2}$% of GDP)

By any standard such potential gains appear very substantial. Furthermore, they are benefits not available to other Western European countries (apart from Norway) which will continue as importers of OPEC crude. Nevertheless, one must be extremely cautious in interpreting the results since the advent of North Sea oil in no way guarantees any actual balance of payments gain or indeed any lasting benefit to the economy. Our results turn on a comparison between two future states - one with (NS) and without North Sea oil (nNS). But it is extremely difficult to know what differences, other than the presence or absence of the North Sea, there might be between the two states.

The estimated potential gain to the balance of payments is essentially an increase in real GDP stemming from the improved use of factors of production. Not all the GDP increase need flow into the balance of payments: the United Kingdom might take the gain in the form of higher investment or

Figure 3

BALANCE OF PAYMENTS EFFECTS : L P D PRICE SCENARIO INCLUDING ESTABLISHED COMMERCIAL, POTENTIALLY COMMERCIAL AND NEWLY DISCOVERED FIELDS

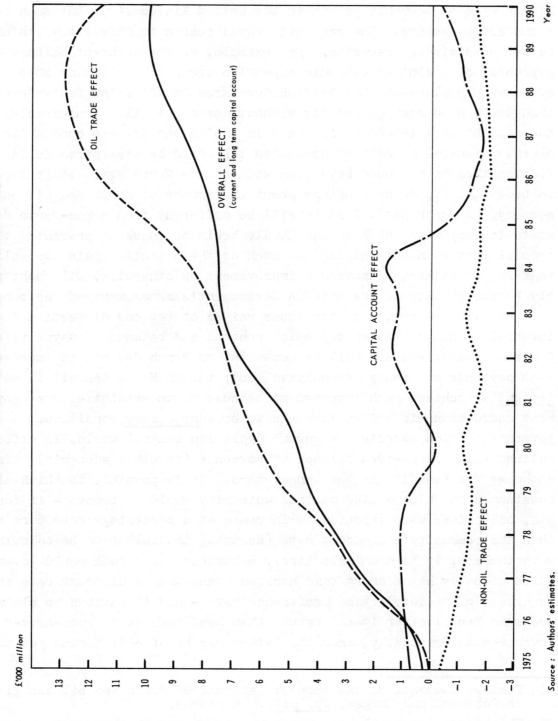

Source : Authors' estimates.

higher public or private consumption than would have occurred without North Sea oil.(1) Those who are particularly optimistic see much of the gain flowing into increased investment, thus leading to a considerably higher rate of economic growth in the United Kingdom.(2) But such conclusions are premature. One can, with equal reason at this stage, think of rather pessimistic scenarios. For example, as the apparent balance of payments constraint on economic expansion seems to be lifting at a time of heavy unemployment, the British Government might allow faster expansion than would have seemed possible without North Sea oil: conceivably, though there might be a temporary increase in public and private consumption, much of the potential benefit of North Sea oil could be dissipated in faster inflation than there would have been without the North Sea. It is important to beware of simple conclusions about the impact of North Sea oil on the economy. A North Sea oil state will be different from a non-North Sea oil state in many ways which we can hardly begin to guess at present: the crucial factor in determining how much of the potential gain is realised in terms of balance of payments improvement or otherwise, will most probably be the quality of the British Government's management of the economy.

    Finally, we return to the issue raised at the end of Section 3 - the interrelationship between depletion control and balance of payments effects. For some years, Britain will be dependent on North Sea oil to improve overseas payments and repay accumulated debt; though North Sea oil is not sufficient to achieve such improvement because of uncertainties over government macro-economic policy it seems to be a necessary condition. A government which exercised downward depletion control would, in effect, be relinquishing short-term balance of payments (or other economic) gains in exchange for benefits in the longer term. It is possible to think of circumstances in which a wise central authority would so behave - if for example, oil prices were expected to increase at a percentage rate more rapid than the community's discount rate (assuming the latter to be calculable). Nevertheless, it is much less likely behaviour for a real world government which probably has a short time horizon - and thus a discount rate greatly in excess of society's time preference rate - and is anxious to claim economic benefits for itself rather than hand them on to successor-governments, especially since the latter may be of a different political

1) Further comments on the economic effects of North Sea oil are given in Robinson and Morgan, op. cit., Chapter 9.

2) This would presumably be a faster rate of growth than would otherwise have occurred but not necessarily faster than in the 1960s and early 1970s since North Sea oil costs will be higher than the price of imported crude pre-October 1973.

party. Both major political parties in Britain have argued that depletion
control is a matter for the 1980s, not the late 1970s. When one looks at
the prospects for the British economy, however, one wonders whether even
in the 1980s governments will be holding down North Sea oil production.
Whatever one's opinions on the desirability of depletion control (upward
or downward) it is quite possible that for many years to come, far from
trying to reduce North Sea output, British governments will be working
hard behind the scenes in an endeavour to induce the companies to maximise
production.

THE FUTURE SUPPLY OF INDIGENOUS OIL AND GAS IN WESTERN EUROPE: WITH SPECIAL REFERENCE TO THE NORTH SEA

by

P.R. Odell

INTRODUCTION

My paper is directed mainly and specifically towards what I think to be a real possibility for energy independence in Western Europe for the period 1980-2000 based on that quite remarkable resource phenomenon - the North Sea oil and gas province: and to the institutional difficulties which are involved in getting the enormous hydrocarbon resources of that major new province effectively developed. If the opportunities and the challenges associated with the exploitation of the North Sea's resources are accepted then other Western European offshore oil and gas developments will follow as night follows day and so extend energy independence well into the twenty-first century - without worrying too much about nuclear power and other high resource cost alternatives.

What has happened to date in the North Sea depends essentially on the international oil companies for they provide the dominant element in the exploration for and the production of the province's oil and gas resources. Indeed, they have been allocated - directly and indirectly - well over 70 per cent of all the North Sea acreage allocated (including an even higher percentage of the best blocks); they have discovered all the major fields found so far (except one, viz. Brae) together with most of the minor fields; and they are responsible for almost all the fields in production or under development towards production.

This is so in spite of the interest shown in the North Sea both by new oil companies formed with the North Sea in mind and by non-oil companies in Britain and Norway deciding to diversify into a new field of endeavour. Even the rapid growth of state oil companies - particularly the British National Oil Corporation and Norway's Statoil - seems unlikely to diminish very much the role of the international oil companies in determining just how far and how fast the North Sea oil and gas province will be developed. Thus, the motivations of the international oil companies in respect of the North Sea opportunities - in relation, that is, to alternative opportunities which they have elsewhere - together with these companies' responses to the petroleum legislation of the countries surrounding the North Sea have largely determined the progress towards, and still condition the prospects for, the development of this major new oil and gas province.

THE RESOURCE BASE

Progress and prospects - though dependent in the final analysis, as we shall show later, on the "behaviour" of the companies (and the governments) - depend in the first instance on the resource base. Knowledge of this is, in the initial period of exploration limited and speculative and the hypotheses on the likely occurrence of oil and gas are based largely on geological analogies. The size and complexity of the North Sea basin constituted a major difficulty in the early evaluations. Size, though a simple concept, was consistently underrated such that the North Sea came to be seen as a kind of European backyard in which prospects for exploitation were limited and over which the main concern was that false hopes should not be raised about its potential. However, what has already happened in terms of successful exploration and what remains to be done in this respect (and this adds up to an exploration effort that will be several times the magnitude of that already made) should be seen in the context of a potentially petroliferous North Sea province which is almost exactly the same size as the petroliferous region of the Persian Gulf - a comparison which is illustrated in Figure 1. As in the Persian Gulf, where there are many different oil and gas "plays" (i.e. many potentially productive horizons in the underlying geological series), so also with the North Sea where the first decade of exploration has also clearly demonstrated that there are many sorts of potential reservoir rocks which are worthy of investigation, so opening up not only the likelihood of exploration in parts of the province which were earlier considered not worthwhile, but also the desirability of deeper exploration in areas previously investigated only by shallower wells.

Thus, both geographical scale and geological complexities underlie the amazing success to date in the exploration efforts in the North Sea and provide the basis for continuing efforts stretching out over the next twenty-five years at least. The success to date is pinpointed in

Table 1

NORTH SEA OIL AND GAS PROVINCE: OFFSHORE DISCOVERIES BY OCTOBER 1976

	Oil and Gas	Gas Only	Not Known	Total
1. Number of Discoveries - Total	65	50	57	172
a) In Southern Basin	5	39	25	69
b) In Northern Basins	60	11	32	130
2. Designation of the Discoveries - Total	65	50	57	172
a) Fields with Declared Reserves	38	38	-	76
b) Fields without Reserves Declaration	27	12	-	39
c) Discovery Wells (no other information)	-	-	57	57
3. Size Distribution of Declared Fields - Total	38	38	-	76
a) More than 2×10^9 bbl. oil or equivalent	4	1	-	5
b) $1 - 2 \times 10^9$ bbl. oil or equivalent	9	2	-	11
c) $0.5 - 1.0 \times 10^9$ bbl. oil or equivalent	10	3	-	13
d) Smaller fields	15	32		47
4. Production and Production Plans				
a) Number of Fields with Declared Reserves	38	38	-	76
b) In Production	8	9	-	17
c) Production Plans	16	7	-	23
d) No Plans for Production	14	22	-	36

Table 1 showing the number of finds that had been made in various classes up to the end of October 1976. The essential proof of the prolific nature of the basin lies in the fact of 172 discoveries to date of which about two-thirds (136) have already been designated as gas or oil and gas fields. Of this number, about 66 per cent have had reserves' figures declared for them. Moreover, of the 76 fields with declared reserves, 29 have more than 500 million barrels of oil (or oil equivalent), so making them "giant" fields in usual North American oil industry parlance.

NATURAL GAS

What the discoveries mean in terms of effective overall recoverable reserves is especially difficult to evaluate in the case of natural gas. For example, many of the gas-only fields lie in the Dutch sector of the

Table 2

NATURAL GAS RESOURCES OF THE SOUTHERN NORTH SEA BASIN
(excluding associated gas)

	Dutch Sector	British Sector	German Sector
Total number of Gas Discoveries	33	31	3
Number of Discoveries Declared Gas Fields	11	15	-
Governments' Declarations of Gas Reserves (10^9 m^3)	322(1)	878(2)	
Number of Fields in Production	2	7	-
Other Fields with announced Production Plans	8	-	-
Current (1976) Annual Production (10^9 m^3)	c.3	38	-
Estimate of 1980 Production from Fields currently on production or in development (10^9 m^3)	c.10	c.42	-
Likely Current Reserves for all fields in each sector at overall 90% probability (10^9 m^3)	1000+	1250+	100+
1980 Production Potential with Full Exploitation of the already discovered reserves (10^9 m^3)	40+	50+	5

Notes: 1) Arithmetic total of "proven" reserves of 11 declared
           fields.
        2) Arithmetic total of proven, probable and possible reserves
           of declared fields.

 + Based on all discoveries made and not just on declared fields.
++ Based on 20-25 year depletion periods for the fields.

North Sea (Figure 2). Here, unfortunately, there is obligation on
neither company nor government to give any fields' and reserves' information at all to the general public, so eliminating the possibility of
depletion policy making (and, indeed, energy policy making) which can
be seen to be justified by the rate of reserves' discovery. Table 2
details the gas fields, discoveries and reserves position for the southern
basin of the North Sea. In the Dutch sector there are at least 33 offshore discoveries(1) about which virtually nothing has been made known
to the public in terms of reserves figures for individual fields. The
Dutch Government has merely given a total figure for the sector's proven
reserves without indicating what is meant by "proven" or even the number
of fields whose reserves are included in the total! A specific request
to the Ministry of Economic Affairs to say how many fields were included

(1) There are also a few others about which not even an announcement of
    discovery has been made.

Figure 1

A COMPARISON OF THE SIZE OF THE PETROLIFEROUS
AREAS OF THE PERSIAN GULF AND THE NORTH SEA

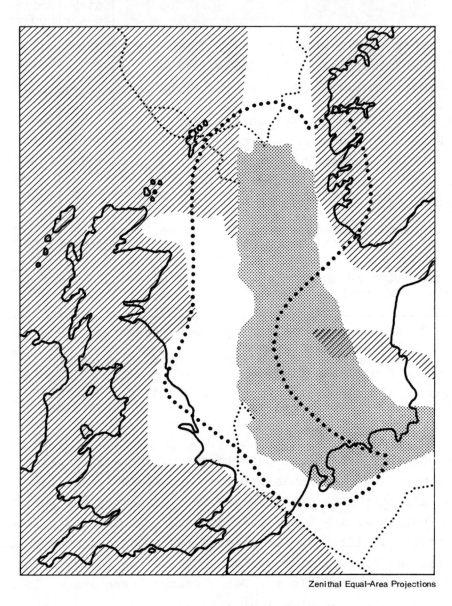

Zenithal Equal-Area Projections

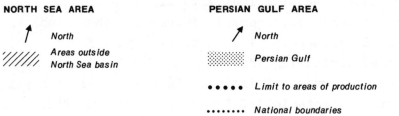

Figure 2

GAS FIELDS IN THE SOUTHERN BASIN OF THE NORTH SEA OIL PROVINCE

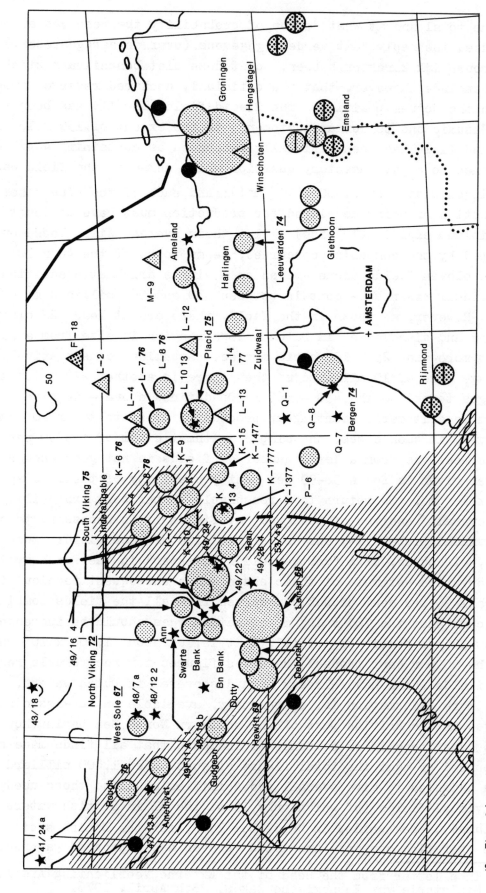

(See Figure 8 for the key to the map)

135

in the total and at what degree of probability the reserves are declared produced the reply "dat verdere gegevens (over de aardgasreserves) een vertrouwelijk karakter hebben, zodat deze niet beschikbarr zijn".(1)   One can conclude, however, that the officially declared reserves grossly understate the actual position. The Placid field (L10/11) has been forward sold (to Gasunie and to German customers) to the extent of 150×10^9 m^3; the Ameland field has been officially announced as containing 44×10^9 m^3 (in response to a parliamentary question which asked if the field was a second Groningen with over $2,000 \times 10^9$ m^3!); and each of the nine other fields in production or being developed for production must have at least 25×10^9 m^3 (a minimum exploitable size) whilst their average size, based on hints dropped by the companies concerned, seems to be of the 40×10^9 m^3. Thus these eleven fields alone – even on a simple arithmetic addition of their individual reserves – contain between 430 and 565 milliard m^3 of natural gas. However, as shown in the Table, there are at least 22 other gas fields and discoveries in Dutch waters ranging in size from a few milliard up to more than 25×10^9 cubic metres. Assuming, conservatively, an average size of 10 milliard m^3 then there is another 220×10^9 m^3 of natural gas to add to the $430 - 565 \times 10^9$ defined above – to give a total, by a simple arithmetic counting-up procedure, of up to 785×10^9 cubic metres. However, in such cases, the simple arithmetic sum of the 90 per cent probable reserves from a large number of fields is not good enough, as the answer given is for a 98-99 per cent probability of their being at least 785×10^9 m^3. To return to a reasonable 90 per cent probability figure for the overall reserves of the group of fields it is necessary to collate the individual probability curves for the size of each field: such curves are also, of course, of a "vertrouwelijk karakter" in the Dutch system but a conservative estimate of their shape and a subsequent recalculation of the overall 90 per cent probable reserves from all the fields combined would indicate that the 785×10^9 m^3 of natural gas should be increased by at least 25 per cent: to a total, that is, of about $1,000 \times 10^9$ cubic metres as the 90 per cent probable reserves figures for recoverable reserves from the fields which have already been discovered in Dutch waters.

More information is, fortunately, available on non-associated gas reserves in the British part of the southern North Sea basin though even in this case there is no evidence to indicate that all finds made have been announced or that the total figure of just under 1,000 milliard m^3 has been adjusted to give overall probabilities equal to those used for the individual fields (of which only 15 of the 31 in British waters have been declared). Meanwhile, in September 1975, for the Danish sector the D.U.C. (the operating company) announced "proven and unproven" gas reserves for

1) In a letter from the Head of the Externe Voorlichting Afdeling of the Ministerie van Economische Zaken, 28th April, 1976.

three of its discovered fields of between 24.3 and 41.9 milliard m^3. The much higher reserves of 69.1 milliard m^3 estimated for the three fields by the consultants appointed by the Danish Government again point to the continuing tradition of understatement of gas reserves by the companies which have made gas discoveries - a tradition in Western Europe dating back to the earlier traumatic understatement by Shell and Esso of the Groningen field reserves.

In Table 3, therefore, we find a contrast between the gas reserves' position as it was officially presented in 1975 and what seems likely to be the situation by the early 1980s if we move to a more reasonable approach to the calculation of reserves and if discoveries of new associated and non-associated gas reserves continue at the level achieved over the last few years. Indeed, the estimate of 10,500 x 10^9 cubic metres of recoverable reserves of gas by that date can now be considered as the minimum likely rather than the maximum possible. Moreover, beyond that date one can now also be confident that new discoveries - and new reserves in previously known discoveries - will continue to add to the gross available amounts of recoverable natural gas so that, even without the discovery of another major off-shore gas province somewhere around the much intended coastline of Western Europe, an ultimate resource base of at least 20,000 x 10^9 cubic metres does not seem to be unduly optimistic.

This size of resource base provides the scope for a steadily increasing rate of natural gas production in Western Europe for the whole of the period up to at least the end of the century from its present annual total of about 220 x 10^9 m^3 (some 250 m.t.c.e.). Thus viewed from the point of view of the gas resource base it would seem quite reasonable to assume that natural gas, the preferred fuel by most users in household, commercial and industrial energy markets - can continue to increase its share of the Western European energy market with the capability of reaching ultimately, a level of roughly the same order of magnitude as the 35 per cent share of gas in the United States energy market in the late 1960s/early 1970s.[1] This is an eminently reasonable view to take on the incorporation of natural gas into the Western European energy economy given a situation in which European natural gas in general, and North Sea gas in particular, is in much closer proximity to markets such that the (relatively-high) transport costs involved in taking gas to its consumers do not become an important element in its supply price. This advantage arising from the geography of the supply of and demand for gas in Western Europe is demonstrated in Figure 3 in which the European situation is compared with the

1) Including imports of natural gas principally from the Soviet Union and North Africa. Such imports, however, would not need to form more than one-fifth of the total gas used.

Table 3

WESTERN EUROPE: AN ESTIMATE OF ITS NATURAL GAS RESOURCES AND
PRODUCTION POTENTIAL IN THE EARLY 1980s

	Recoverable Reserves		Early 1980's annual production potential	Million of tons of coal equivalent(*)
	As declared in 1975	As developed by 1980		
	(x 10^9 m^3)	(x 10^9 m^3)	(x 10^9 m^3)	
On shore Netherlands	2 300	2 750	135	150
South North Sea – British Sector	1 000	1 250	60	75
South North Sea – Other Sectors	350	1 250	40	50
On-Shore West Germany	300	650	35	40
Austria, France, Italy, etc.	400	600	35	45
Middle North Sea Basin – United Kingdom/Norway	750	1 600	80	90
North North Sea Basin – United Kingdom/Norway	600	2 300	80	90
Rest of European Continental Shelf	50	350	15	20
	5 750	10 450	480	560

Source: For 1975 various national and oil companies estimates and Oil and Gas Journal,
29th December, 1975. Estimates for 1980 and the author's own.

(*) Conversion to coal equivalent based on known calorific values of the various gas supply sources.

much less favourable circumstances in the United States and the Soviet Union but in each of which natural gas has, nevertheless, been able to compete against alternative fuels in the main demand centres.

Thus, given an adequate gas resource base and the proximity of a readily available market for as much gas as can and will be produced, one can expect (company and government behaviour permitting) a continued rapid contirbution of natural gas to the Western European energy economy, particularly, of course, in the countries surrounding the North Sea from which the bulk of the resources will be available for the foreseeable future. The prospects for the supply and transportation patterns for natural gas by the early 1980s are shown in Figure 4.

OIL

Progress to date in the development of indigenous Western European hydrocarbons thus relates more to natural gas than they do to oil. However, the even more recent developments in the exploration for oil have produced even more exciting results in terms of the resource base potential than those of the somewhat earlier search for gas. As a result, the prospects for the medium to longer term future of oil production from the North Sea now exceed all earlier forecasts and expectations.

Table 4 presents the minimum likely position on reserves at October 1976 from already known oil fields. This shows that ultimately recoverable reserves from discoveries that have already been made are likely to be at least 42,000 million barrels - after making only quite modest allowances for the appreciation of fields with already declared reserves and for fields which have been announced as having been proven but about which no information on reserves has yet been given. It does not, however, make any allowance whatsoever for the additional 57 discovery wells which have been drilled (see Table 1). At least some of these will eventually prove to have recoverable reserves of oil and so push the sum total of discovered reserves to date above the 42×10^9 barrels' figure.

Table 5 then puts this real-world reserves' development position and prospects for development in the perspective of some of the estimates which have been made over the last few years about North Sea oil. Shell and B.P. both gave earlier estimates for 1976 and 1980 reserves which lie at a level of 50 per cent or less of what has occurred, or what can now be seen as most likely to happen in terms of reserves' development. These companies' figures even for the province's "ultimate" reserves range from 35 to 44×10^9 bbls - figures, that is, that indicate ultimate levels of

Figure 3

MAPS SHOWING RELATIONSHIP BETWEEN AREAS OF NATURAL GAS PRODUCTION AND AREAS OF ENERGY CONSUMPTION IN WESTERN EUROPE, USA AND THE USSR

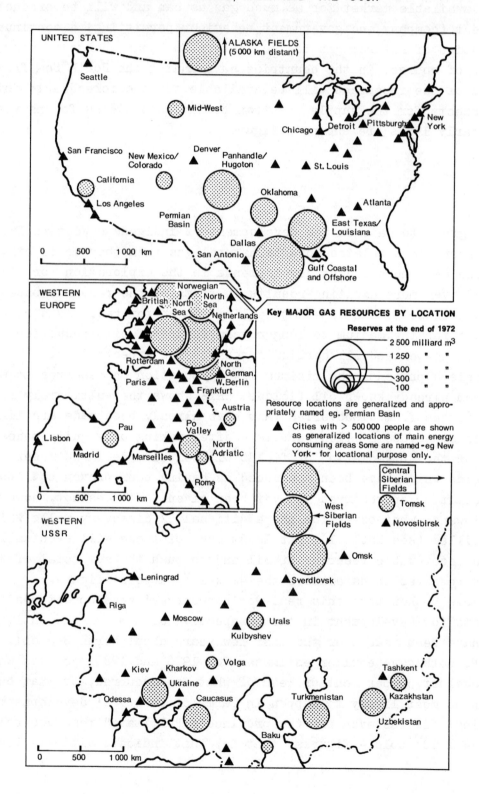

Table 4

NORTH SEA OIL RESERVES - 31ST OCTOBER, 1976

1. As Declared		million barrels
a) By Simple Addition of Declared Figures for 38 fields	about	22,000
b) After Adjusting to a Summed 90 per cent Probability (from the 38 fields assuming each is declared at a 90 per cent probability)	at least	29,500
2. On Extrapolation		
a) With Upward Revision following Production from the 38 fields (average 15 per cent)	at least	34,000
b) Addition of Reserves from the so far undeclared 27 discovered fields(*) with an assumption that these fields are on average two-thirds smaller than the declared fields		8,000
MINIMUM total(*) of North Sea Oil Reserves discovered to date	more than	42,000

*) In addition there have been - as shown in Table 1 - another 57 oil
(or gas) discovery wells many of which will eventually be declared as
oil fields. The 42×10^9 barrels of oil shown in this Table as already
having been discovered is thus a minimum likely figure - and not even
the most reasonable estimate. See below in Table 5 for details on
future estimates of North Sea oil reserves.

reserves which are less than those that are currently known to exist or
the existence of which can be confidently extrapolated from present
evidence. This, moreover, is in spite of the fact that most of the exploration work in the North Sea still remains to be done - including exploration work on some of the largest structures in the Norwegian sector
in blocks which have not yet been allocated to companies for such work,
in accordance with Norwegian conservation policies. Of the oil companies'
estimates only CONOCO's 1975 estimate of up to 67,000 million barrels of
ultimately recoverable oil remains credible and one is forced to the
conclusion that the oil industry's conservatism over the question of
reserves' evaluation (and/or its wish to protect its vested interests)
has once more led it to inappropriate pronouncements on the resource potential of a region whose future economic and even political survival
depends on a realistic evaluation of its indigenously available oil. In
this respect it should be noted that Table 5 also shows that our

141

Table 5

NORTH SEA OIL RESERVES: ESTIMATES AND FORECASTS COMPARED

	by 1976	by 1980	Ultimate
1. As Declared			
a) By simple Addition of Declared Figures for 38 fields	22	$(bbl \times 10^9)$	
b) After Adjusting to a Summed 90 per cent probability (from the 38 fields assuming each is declared at 90 per cent probability)	29+		
2. Declared plus Additional Discovered Reserves not yet Declared			
a) From upward revision of reserves in declared fields		34+	
b) From fields discovered but not yet declared (27 fields)		8+	
3. New Discoveries from 1976 to 1980		+10	
Totals	29+	±55	
4. As predicted by EGI Simulation Model (50 per cent Probable)(1)	17.75		
5. As Hypothesised by EGI Model for 1976 Reserves after appreciation plus discoveries 1976-1980 (50 per cent probable)(1)		48.6	
6. As Forecast by EGI Model on full development of the Province(1)			
a) 90 per cent probability			78
b) 50 per cent probability			109
7. As Forecast by Oil Companies			
a) By Shell in 1972(5)(2)	10-12	17.5	(35)
b) By B.P. in 1973(3)	-	-	38
c) By B.P. in 1974(4)	16-18	24-30	44
d) By Conoco in 1975(5)	-	-	45-67
e) By Shell in 1976(6)	23		35

Notes: 1) P.R. Odell and K.E. Rosing, The North Sea Oil Province: An Attempt to Simulate its Development and Exploitation, 1969-2029, Kogan Page Ltd., London 1975.

2) At E.I.U. International Oil Symposium, London, October 1972 (A. Hols, Head Production Division of Royal/Shell Exploitation and Production Co-ordination), Ultimate figure is 1975 forecast at Tønsberg Conference.

3) At Financial Times North Sea Conference, London, December 1973 (Dr. J. Birks, Director B.P. Trading).

4) At Annual Conference of the Society for Underwater Technology, Eastbourne April 1974 (Mr. H. Warman, Exploration Manager, British Petroleum Co. Ltd.).

5) At Conference on the Political Implication of North Sea Oil and Gas, Tønsberg, February 1975, (Mr. T.D. Eames, Oil Exploration Division, Conoco North Sea Ltd.)

6) Shell Briefing Service, Offshore Oil and Gas North West Europe, July 1976.

Figure 4
1980 PRODUCTION AND DISPOSAL OF NATURAL GAS IN WESTERN EUROPE

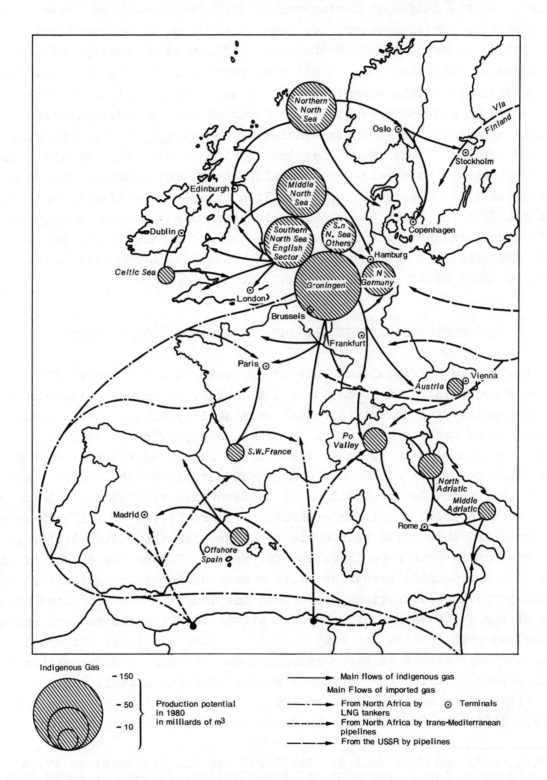

Indigenous Gas

— 150
— 50
— 10

Production potential
in 1980
in milliards of m³

→ Main flows of indigenous gas

Main Flows of imported gas

—·—·→ From North Africa by ⊙ Terminals
LNG tankers

------→ From North Africa by trans-Mediterranean
pipelines

— —→ From the USSR by pipelines

simulation model study(1) of the North Sea oil reserves - a model which was
generally criticised in the industry for its highly "optimistic" conclusions - is also under-predicting the rate of development of reserves of
the province. This, however, does not surprise us as we described it as
a conservative model in which the probabilities of discovery, of size of
field and of recoverability of oil etc. were also orientated to minimum
rather than reasonable opportunities. Thus, although it is possible
that the 1976 and 1980 results from the model may be underestimates
because of mistakes we made in simulating the <u>timing</u> of the exploration
and development effort (such that the model's results will within a few
years "catch up" with real-world developments), nevertheless the 90 per
cent probability figure of 78,000 million barrels of ultimately recoverable oil from the North Sea province is beginning to assume a high degree
of "reasonableness" when set against the more than 42,000 million barrels
which are likely to be recoverable from known fields and with another
57 discoveries still awaiting evaluation.

THE CONTRASTS IN THE OIL AND GAS RESERVES EVALUATIONS

Why should "officialdom" have so seriously underestimated the
resource potential? It seems to be due largely to an unwillingness to
try to quantify the future potential availability of the so-far undiscovered recoverable resources of the North Sea province: that is, because
of an unwillingness to do in Western Europe something which is common
place in other parts of the world and most notably North America where
the U.S.G.S.-directed evaluations of undiscovered recoverable oil and gas
are used as a prime input for estimating future levels of oil and gas production. And this quite reasonable procedure for the United States in
this respect is even more important for Western Europe where most of the
potential hydrocarbon resources still remain undiscovered given that most
of the petroliferous regions are around the long and much indented coastline of the continent in off-shore locations in which oil and gas exploration and exploitation has only recently become possible. Figure 5
shows the vast extent of these petroliferous offshore regions within the
total extent of which the North Sea constituted but a small part.
Within this context, the concept of total potential resource base
is very different from the way in which is has been conceptualised at the

---

1) P.R. Odell and K.E. Rosing: The North Sea Oil Province; an Attempt
   to Simulate its Development and Exploitation, 1969-2029, Kogan Page
   Ltd., London, 1975.

Figure 5
WESTERN EUROPEAN REGIONS OF OFF-SHORE OIL AND GAS POTENTIAL

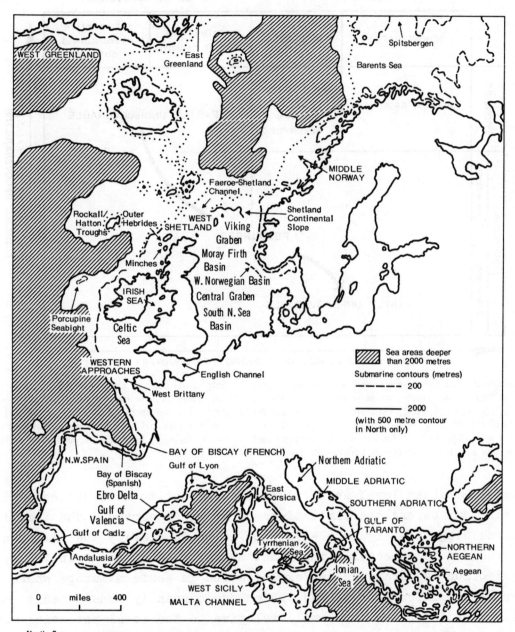

North Sea — *Off-shore areas where extensive oil and/ or gas deposits have been located and which are under active development*

IRISH SEA — *Off-shore areas where exploratory work has been done and where either initial drilling has indicated the presence of oil and gas or there are high expectations of hydrocarbons occurrence*

Andalusia — *Other interesting off-shore areas with relatively shallow water where there are geological and/ or geophysical expectations of oil and gas and where exploration will begin before 1985*

Figure 6
RESOURCE DIAGRAM

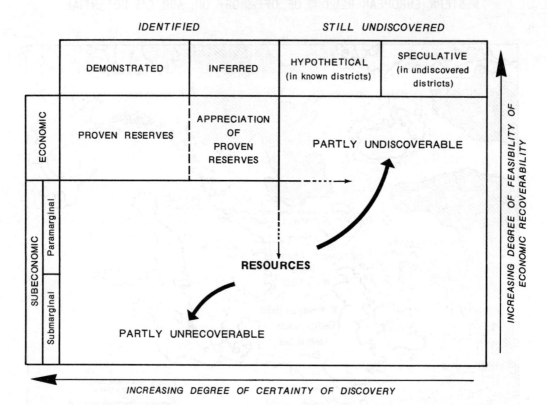

This illustrates the categories into which resources (of, for example, oil and gas) may be divided and the way in which geological and economic factors influence the relationships between the categories. In light of levels of knowledge, levels of technology and price levels, the dividing lines between the categories vary over space and time. The proven and inferred reserves of oil and gas in Western Europe at the present time are relatively small but all this means is that conditions to date have worked against their development. With changed conditions - of knowledge, technology, economics and politics - the resource base can be more effectively explored and developed.

official European level where planners seem not to have succeeded in reaching out beyond the limitations of proven reserves. The familiar resource diagram – Figure 6 – perhaps helps to clarify the differences involved and is especially important at the present stage in the cycle of petroleum explorations and exploitation in Western Europe where proven reserves and inferred reserves, which are currently considered to be economic to produce, are very small relative to the total potential which can be developed as knowledge increases, technology improves, politics change and economics give more emphasis to indigenous resource development.

To return to the North Sea, however, we can simplify the evaluation which is necessary to try to understand how its oil and gas resources are developing. This evaluation is presented in Figure 7 with its x and y axes representing "discovery" and "recovery" the complementary aspects of resource development.

The size of the resource base depends, in the first instance, on the number of fields and on their size. Equally obviously, in any province, there are a finite number of fields and how many of these are discovered is a function of the size of the investment in the exploration effort. The more of the fields that are discovered, the greater the quantity of reserves in the province - a development which is illustrated in the y axis in Figure 7. Increases in reserves figures thus depend first on the continuation of an exploration effort in the province over a long period of time during which the results of the continuing effort are expected to yield knowledge of successively smaller fields. However, in light of specific economic and political conditions, it is possible that the exploration effort will be terminated before all the prospects have been tested so that resources, which are discoverable in a technological sense, will remain undiscovered and so keep the reserves figures of the province at a lower level than would otherwise have been the case. The degree to which this phenomenon is likely to occur in respect of North Sea exploration remains uncertain. Many companies, however, have indicated that the much increased physical costs of developing North Sea fields coupled with the higher share of the revenues which now have to be paid to governments following the renegotiation of the terms of the concession arrangements (and which the companies see logically enough as constituting an additional set of costs affecting the viability of their operations) constitute good reasons as to why exploration should cease when all the larger structures have been tested. This might then exclude the search for all fields which are expected ultimately to yield less than 200 million barrels of oil (or the equivalent thereof of associated or non-associated natural gas). However, as such reservoirs are unlikely to contain much more than 15 per cent of the total reserves of the province (based on the usual statistical distribution of reserves by field size in a petroleum province) then the effect of this component in accounting for the difference between the simulated availability of reserves and the quantities expected by the oil companies will be relatively minor, unless, of course, other factors, such as government imposed limits on the amount of exploration (as in Norway) intervene to inhibit discoveries.

Apart from this influence, however, the much more important component in determining contrasts in estimates of reserves emerges from the other axis (the x axis) in Figure 7. This is a component which represents possible variations in the degree to which the resources of oil and gas which have been found in a set of fields are actually exploited. Of the oil and gas in place in any reservoir a certain percentage will be recoverable with a given technology over an economically relevant time

Figure 7

DISCOVERY AND RECOVERY: COMPLEMENTARY
ASPECTS OF RESOURCE BASE EVALUATION

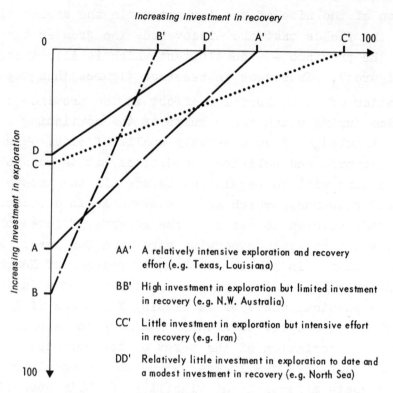

AA' A relatively intensive exploration and recovery effort (e.g. Texas, Louisiana)

BB' High investment in exploration but limited investment in recovery (e.g. N.W. Australia)

CC' Little investment in exploration but intensive effort in recovery (e.g. Iran)

DD' Relatively little investment in exploration to date and a modest investment in recovery (e.g. North Sea)

The proven oil reserves of any given petroliferous province depend, first, on how much investment is put into the exploration effort; that is, in testing all the possible occurrences of hydrocarbons in different sorts of structures in different horizons in the geological succession. Full exploration is a costly and a time-consuming process and most hydrocarbon provinces of the world have, as yet, only been explored to a limited degree. In the case of the North Sea, the exploration effort to date, as shown on the diagram, covers no more than one-third, at most, of the total exploration which is necessary fully to explore the province: and, this, if it happens at all, will be spread out over at least the next two decades.

Second, proven, recoverable reserves depend on investment in the facilities needed to recover the discovered oil. Fields may be "creamed" of their lowest-cost resources (with costs represented almost entirely by investment costs) or developed more intensively so as to push recovery towards the "limit" set by technology. The technology itself, of course, changes over time as does the commercial viability of recovering more or less oil from a field or group of fields. In this diagram North Sea investment in recovery is shown to be relatively higher than the current level of investment in exploration (line DD'). It is, however, still modest compared with what could be done for reasons which are discussed in this monograph.

period. This percentage figure will be a function of the level of investment made in the oil recovery system such that the more money that is spent, the more oil that will be recovered. Moreover, technology also improves over time and so increases the possible recoverability of oil from discovered fields, so enabling more oil to be recovered with additional expenditure on the development systems. Finally, of course, technological improvement also costs money and investment in this way may also be seen as a means of increasing the recoverability of oil from a field. Overall, therefore, one can hypothesise that the more investment that is made in developing a field more intensively and extensively the more oil that will be produced, so pushing out to the right in Figure 7 the recoverability component. By this means, too, total reserves are increased.

In my view the major difference between our simulation model's predictions of the reserves of the North Sea and the estimates of "officialdom" emerges from the way in which this component is being allowed to work itself out. What it boils down to is a difference between the possible and the likely recoverability of oil from a discovered field and this is a question - with both technical and economic aspects - which has not yet been anything like adequately explored in spite of its important policy implications for both companies and governments.

As shown previously (Table 1) a large number of oil and gas fields have already been found in the North Sea province: and their geographical distribution - by size and by type - as it was at the end of October 1976 is illustrated in Figure 8. There are almost 30 "giant" fields (using American parlance in which a "giant" is a field with over 500 million barrels of oil or oil equivalent) and Figures 9 and 10 locate and name most of these and also show their approximate shapes and geographical extent (important variables in the reserves recovery component).

In the case of each of these fields some part of their technically recoverable reserves will be recovered on the basis of an installed production system the decision on which will depend essentially on the operating company's evaluation as to how it can earn, at best, maximum and, at worst, sufficient profits. In other words the declared recoverable reserves of a field are not a fixed quantity: on the contrary, they are a highly variable element and are essentially a function of the investment decision which the operating company takes. The initial investment decision is, moreover, one which, in the unique circumstances of the North Sea environment, determines more or less once and for all what percentage of the technically recoverable reserves shall be recovered over the full production life cycle of the field. This is because, given the size, the shape and the deep-water location of the fields, the initial decisions on the number of platforms to be put on the field and the number of wells to be associated with them is the critical variable for defining the quantity of the reserves which will be recovered: that is, for defining the size of the field in terms of recoverable reserves.

How this works out in practice is illustrated in Figure 11, showing how, for a hypothetical field, one, two or three platforms can be located to deplete the reservoir; or, rather, the oil reserves of part or parts of the reservoir. In economic terms, moreover, each additional platform is less productive than the previous one (successive platforms produce

Figure 8

OIL AND GAS FIELDS IN THE NORTH SEA TO OCTOBER 31, 1976

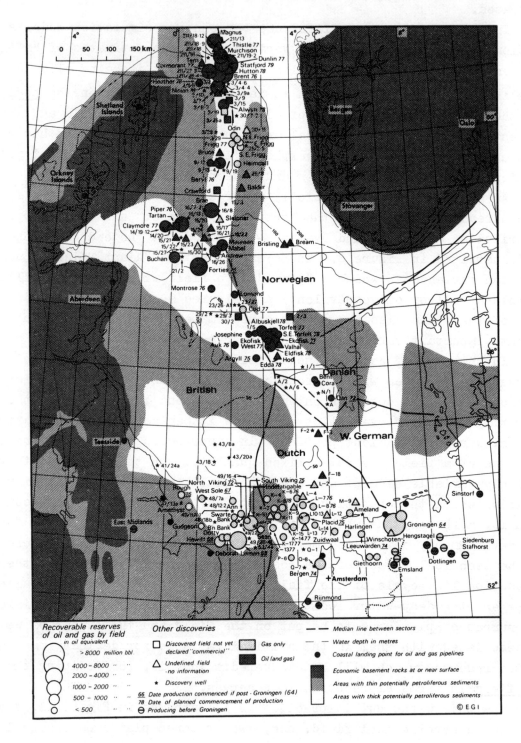

Figure 9

LARGE OIL FIELDS IN THE VIKING GRABEN

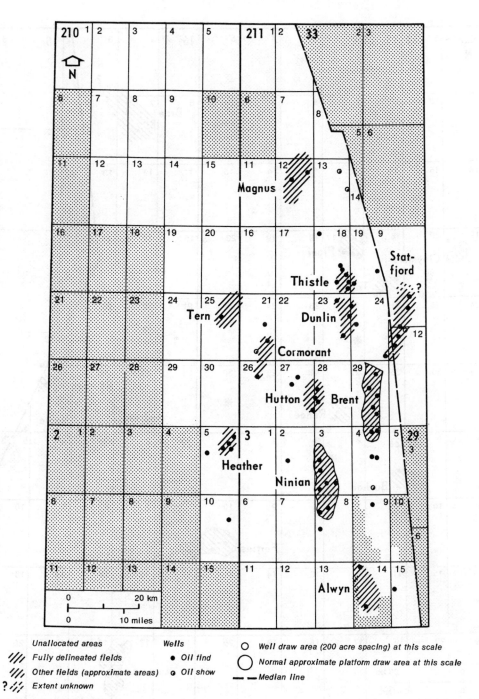

Figure 10

LARGE OIL FIELDS IN THE REST OF THE NORTHERN NORTH SEA BASIN

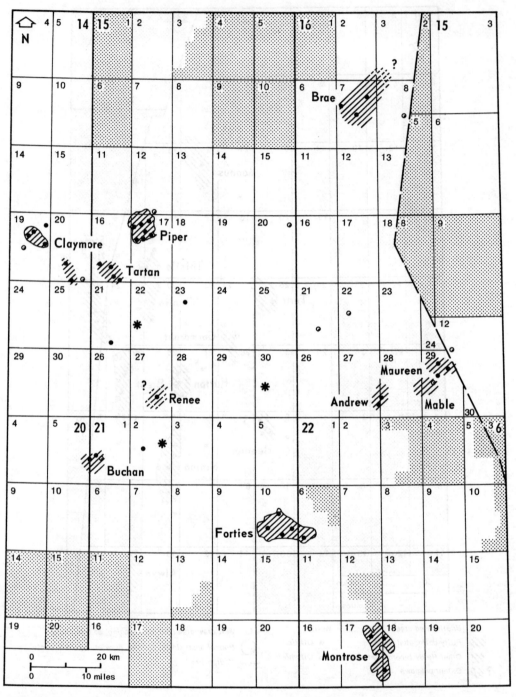

Figure 11
RE-ORGANISATION OF PLATFORM LOCATION WITH INCREASING SYSTEM SIZE

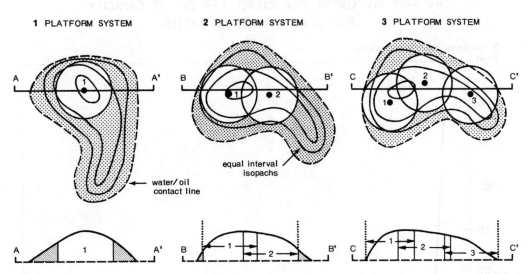

A hypothetical oil field is shown with a one, a two and a three platform system, respectively on the three maps of the field. Below each map is an appropriate cross-section diagram. If we compare the one with two platform system and the cross section A-A' with the cross section B-B', it can be seen that the introduction of the second platform has caused the relocation of the first resulting in the deepest oil bearing sands being shared between the two platforms for production purposes. The first platform now produces less oil than the one platform in the one platform system. Comparing the two with the three platform systems and sections B-B' and C-C', the same phenomena can be observed in respect of the location and productivity of platforms one and two.

decreasing quantities of oil but do not cost any less to install or to run) and so, as shown in Figure 12, there is a rising average unit investment cost curve as a field is more extensively and/or intensively developed.

Thus, the geography of any North Sea field; its technical characteristics; and the economics of the different production systems which can be developed to deplete it have important consequences for the field's unit production costs. Moreover, as the unit revenue curve can be taken to be horizontal (as it is not affected by the production decision), then there are also consequences for the unit profitability of production. In such circumstances we would argue that there is a high propensity on the part of the operating companies to play safe and thus to take exploitation decisions which mean that the fields are simply "creamed" of their lowest-cost-to-produce reserves. Thus some or even most of the technically recoverable reserves of a field become unrecoverable and so do not constitute recoverable reserves at all in the context of company evaluations on how to maximise their returns on investment in a field - or evaluations on how to achieve lowest acceptable rates of return on investment when compared with the opportunities which exist for nearly all the companies concerned

Figure 12
AVERAGE INVESTMENT PER BARREL PER DAY OF CAPACITY
WITH A FOUR PLATFORM SYSTEM

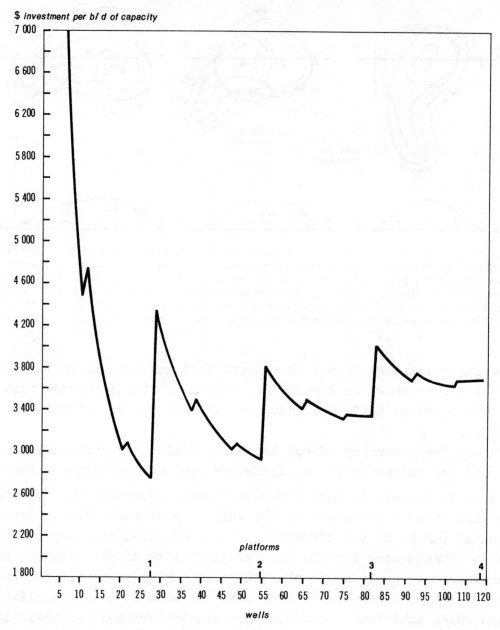

For the installation of this system on the field four separate platforms each with an ultimate capacity for handling 27 wells are required. Within each platform there is again the falling unit cost curve as average productivity increases with the increasing number of wells - except that again costs of expanded platform facilities at 12 and 20 wells per platform creates the upward kinks in the curves. The installation of each additional platform, however, reduces overall productivity and there is a jump back to a higher unit investment cost - for example, from the minimum $3,350 when full productivity is achieved from the third platform to $4,000 as the first well on the fourth platform comes into production. Note the steadily increasing unit cost from the most productive situations on Platform 1 to Platform 2 ... to Platform 4: that is, from $2,750; to $2,950; to $3,350; and to $3,700 with the four platforms.

Table 6

FORTIES, MONTROSE AND PIPER FIELDS: CONTRASTS IN BENEFITS FOR THE UNITED KINGDOM
ARISING FROM COMPANY AND COUNTRY OPTIMAL DEVELOPMENTS OF THE FIELDS

Fields / Platform Systems — Characteristics	Forties		Montrose		Piper		Total of the Three Fields		
	4 Platform	6 Platform	1 Platform	3 Platform	1 Platform	3 Platform	Companies' Optimum	Country's Optimum	% Difference
a) Quantity (million barrels) of Recoverable Oil in Economically Relevant Time Period (Companies' Estimate of oil Recoverable)	1935 (1800+)	2135	194 (c.160)	416	649 (642)	965	2758	3516	27.5%
b) Production of Oil in 1980(1) (million bbl)	221	255	21	42	71	100	313	397	26.9%
c) Peak Year Production of Oil	1980/81	1981	1979/82	1981/82	1979/82	1981/82	-	-	-
d) Flow of Government Revenues in 1980(1) (million $)	2138	2310	140	282	738	983	3016	3575	18.4%
e) Foreign Exchange Value of 1980(1) oil (million $)	3210	3697	320	628	1071	1494	4601	5819	28.6%
f) Present Value of all Government Revenues (milliard $)	7.04	8.07	0.46	0.71	2.45	3.54	9.95	12.33	23.9%
g) Present Value of Total Volume of Oil Produced (milliard $)	13.1	15.1	1.34	2.67	4.82	6.82	19.26	24.59	27.7%
h) Capital Investment in 1974 1975 1976 1977 (million $)	367 397 361 277	367 443 500 599	42 98 84 28	85 213 215 162	93 217 186 62	113 279 274 191	502 712 632 367	565 935 988 950	12.5% 31.3% 56.5% 158.7%
i) Number of Platforms - Built by 1976 / Under Construction in 1976	4 0	4 2	1 0	2 1	1 0	2 1	6 0	8 4	33.3%

1) 1980 has been selected for illustrating the contrasts in government benefits to show the relatively near future importance of the differences arising from the systems and not because the differences reach their peak in 1980. Indeed, relatively the gap continues to widen throughout the 1980s but, in terms of the absolute difference between the size of the benefits the peak is reached in 1984/5.

for the investment of the same money in other activities or in oil producing activities elsewhere in the world.

It is this set of facts, we conclude, that forms the principal reason for the contrast between the simulated level of North Sea reserves (and in which simulation model a primary assumption was that all reserves which were discovered and technically producible would be produced) and the much lower levels which are being defined by the companies. The problem has been fully defined and analysed in a recently completed study.(1) Here I offer simply the conclusion of the study in respect of three North Sea fields (Forties, Piper and Montrose) which were analysed in detail in respect of possible alternative production systems, their productivities and associated profit levels and their revenues to government and contributions to foreign exchange earnings. The results of the study are summarised in Table 6 demonstrating quite unequivocally the great difference between what has been decided by the companies in terms of the reserves they have decided to produce from these three fields and what might have been achieved if alternative production systems had been installed on the fields with the objective of achieving production levels which would have maximised the economic returns to the United Kingdom. The difference in terms of the reserves figures is of the order of 27.5 per cent: the difference from the point of view of the benefits to the United Kingdom economy is just as important, ranging from 23.9 per cent in terms of the present value of the future flows of tax revenues to over 50 per cent in respect of the present value of investments in the production systems; a factor which is of great significance for job creation in the economy in general and in the oil producing areas in particular.

THE IMPLICATIONS OF THE DIVERGENT EVALUATIONS OF THE NORTH SEA'S RESOURCES

The establishing of a set of hypotheses for the divergent evaluations of the oil (and gas) resource base is thus a complex task: but there does seem to be a rational explanation for the divergence – in terms, basically, of who is making the evaluation and for what purposes. What emerges from the analysis, however, appears to be equally important from the point of view of the idea that the North Sea could provide almost enough oil and gas to make Western Europe almost independent of the

1) P.R. Odell, K.E. Rosing and H. Vogelaar: Optimal Development of the North Sea's Oil Fields: a Study in Divergent Company and Government Interests and their Reconciliation, Kogan Page Ltd., London, 1976.

Figure 13

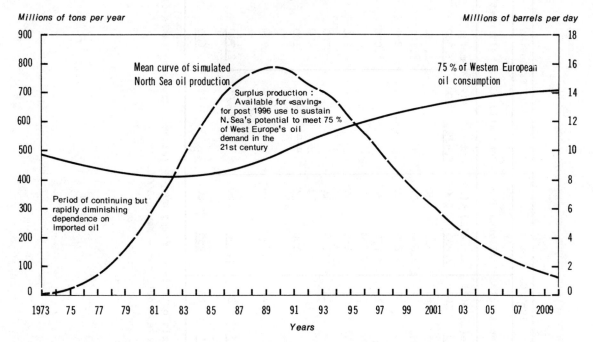

Millions of tons per year .... *Millions of barrels per day*

Mean curve of simulated
North Sea oil production

Surplus production :
Available for «saving»
for post 1996 use to sustain
N. Sea's potential to meet 75 %
of West Europe's oil
demand in the
21st century

75 % of Western European
oil consumption

Period of continuing but
rapidly diminishing
dependence on
imported oil

Years

Mean curve of North Sea oil production compared with expected Western European oil consumption in light of the impact of the oil crisis, the much higher oil prices and policies which aim to substitute other energy sources for oil. The comparison of North Sea production potential with 75 per cent of oil demand instead of with total expected demand is in recognition of the fact some parts of Western Europe can be more effectively served from other oil producing areas and that some heavier crudes will have to be blended with the lighter North Sea crudes in order to achieve the right "product mix".

rest of the world for its required supplies of energy in the period 1980-2000.

This is shown in respect of oil in Figure 13 indicating the possibility that the North Sea could provide at least three-quarters of Western Europe's oil needs over this period. And from Table 7 it is seen that the prospect of a rapidly rising rate of production of North Sea oil and gas (with contributions from other off-shore areas as well) alters quite

fundamentally the energy outlook for the continent. 80 per cent self-sufficiency in energy supplies – without any unreasonable expansion of expensive nuclear power, giving a residual requirement for imported oil amounting to no more than 150 million tons a year, emerges as the prospect for the economy given a rate of growth of energy demand of about 4 per cent per annum over the whole period (except for a somewhat lower rate over the period 1975-80).

However, as stated in the Introduction, North Sea progress and prospects depend on the international oil companies which are, indeed, the only existing institutions capable of developing the potential of the

Table 7

CONVENTIONAL AND ALTERNATIVE ESTIMATE ON WESTERN EUROPE'S ENERGY SUPPLY 1975-2000

	1975 Expected Approximate Use		1980 Estimates Conventional		1980 Estimates Alternative		1985 Estimates Conventional		1985 Estimates Alternative		2000 Estimates Conventional		2000 Estimates Alternative	
	mmtce(1)	%(2)	mmtce(1)	%(2)	mmtce(1)	%(2)	mmtce(1)	%(2)	mmtce(1)	%(2)	mmtce(1)	%(2)	mmtce(1)	%(2)
Total energy	1,580	100	2,225	100	1,850	100	2,870	100	2,200	100	5,425	100	2,300	100
Oil Total	920	59	1,500	66	785	42	1,845	64	800	36			1,350	41
1. Indigenous	30	2	50	2	380	20	195	7	600	27			1,100	33
2. Imported	890	57	1,450	64	405	22	1,650	57	200	9	not		250	8
Gas Total	215	13	265	12	575	31	385	14	750	34			1,050	32
1. Indigenous	200	12	215	10	500	27	300	11	600	27			800	24
2. Imported	15	1	50	2	75	4	85	3	150	7			250	8
Coal Total	400	26	280	12	310	17	310	11	350	16	available		500	15
1. Indigenous	360	23	205	9	230	12	220	8	250	11			400	12
2. Imported	40	3	75	3	80	5	90	3	100	5			100	3
Primary Electricity	45	3	210	10	180	10	330	12	300	14			400	12
Total Indigenous	635	40	680	31	1,290	69	1,045	37	1,750	80			2,700	81
Total Imported	945	60	1,575	69	560	31	1,825	63	450	20			600	19

Sources: OECD, EEC and various national estimates prior to the oil crisis of future energy supply patterns from the basis of the conventional estimates. The alternative estimates are the author's own - March 1974.

1) Million metric tons coal equivalent.
2) Columns do not necessarily add to 100 because of "rounding".

158

North Sea on the scale required for this level of contribution to the
West European energy economy. Unfortunately, however, these companies
are already tied to - and/or willing participants in - a system, viz.
that of the Organisation of Oil Producing and Exporting Countries (OPEC),
whose power centre is external to Western Europe and this inhibits their
being willing or able to make anything like the efforts needed in the
North Sea to achieve the level of developments which the resource base
itself certainly makes possible.

THE NORTH SEA, THE COMPANIES AND THE INTERNATIONAL OIL SYSTEM

Thus, even those oil companies which have been eminently successful
in finding North Sea oil fields have little motivation to try to ensure
that the oil from the fields they have found will provide them with all
or most of the supplies they require for their British and European
markets. This is because their evaluation of the opportunities they have
for recovering oil from North Sea fields is made along with their evaluation of the cost of, and the profit to be made from, alternative sources
of supply. This alternative oil is, of course, that which is available
from the countries that have traditionally supplied oil to Western Europe,
viz. the countries which now constitute the member countries of OPEC.
And, contrary to general belief, the international oil companies - the
same companies, that is, that are almost entirely responsible for developments in the North Sea - have, in the aftermath of the 1973/4 oil crisis,
achieved a good working relationship with the OPEC countries and are able
to continue to make a substantial profit in importing oil from them.
Indeed, in the middle of 1976, as shown in Table 8, the weighted average
up-stream company profit margin on crude oil delivered to the refineries
and crude oil import terminals in Rotterdam was no less than $1.15 per
barrel.(1) This profit per barrel made on importing OPEC oil to Europe
has, moreover, several special qualities which make it even more attractive. In the first place it is profit which can be virtually free of tax
for two reasons; first it is earned off-shore and can, if necessary, be
accounted for in tax-havens; second, under existing legislation in the

_____

1) Though the weakness in some oil product markets in Western Europe
   causes some products to be sold below "cost" (i.e. below the import
   cost shown in the Table plus refinery and distribution costs) so
   reducing the final overall profit level to below this figure, it is
   evident that such "downstream" losses would occur no matter where the
   oil came from. Thus, the equivalent to this $1.15 per barrel profit
   on OPEC oil must be made from the production of North Sea oil to make
   it at least as attractive to the companies as OPEC oil.

Table 8

THE COST AND PRICE OF OPEC OIL - AND OIL COMPANIES' MARGINS

(May 1976)

From:	Price f/ton(1) delivered Rotterdam	Price $/ton(2)	Freight(3) Cost $/ton	Netback(4) Price/ton ($)	Netback(5) Price/bbl ($)	Cost of Crude(6) oil to Company $/bbl	Company Margin $bbl
Libya	257.31	98.98	1.84	97.14	12.71	12.32	0.39
Nigeria	265.01	101.35	2.89	98.46	13.36	12.46	0.90
Iran	251.47	97.25	6.46	90.79	12.46	11.29	1.17
S. Arabia	254.17	97.76	6.24	91.52	12.42	11.03	1.39
Kuwait	253.70	97.33	6.32	91.01	12.57	11.08	1.49
United Arab Emirates	263.67	100.39	6.23	95.16	12.52	11.60	0.92

Weighted Average of Company's Profit Margin $1.15

Notes: 1) From Dutch foreign trade statistics, May 1976.

2) Converted from Gulden at rate of F.2.60/$.

3) Based on assumed use of mixture of time charters (at WS51) and spot charters (at WS25).

4) Delivered price per ton Netherlands less the freight cost.

5) Price per barrel based on the per ton price divided by a factor for °API crude from the different countries.

6) Based on Crude oil prices given in the Petroleum Economist for May 1976. Where both buy-back and equity oil costs are shown for a country, a 50:50 split has been assumed for companies' offtake.

160

"mother" countries of the international major oil companies (viz. the United States, Britain and the Netherlands), profits earned on operations in the traditional producing countries are free of additional tax obligations at home because the taxes already paid in the producing countries (amounting up to $10 per barrel) make a large tax credit position available to the companies concerned. This element itself makes the apparent per barrel profit really worth up to twice as much as the $1.15 per barrel for were this profit to be made on the production of North Sea oil then, even after payment of royalties and petroleum revenue or other special oil and gas production taxes, it would still be subject to normal corporation tax at a rate in the United Kingdom of 52 per cent and in Norway at the rate of 49.5 per cent.

But there is a third - and perhaps an even more important component - to add to the equation which the companies have to solve in trying to determine the relative profitability to them of either importing OPEC or producing North Sea oil. This arises from the fact that the companies do not need to make any new investment whatsoever to secure and to handle oil imports from OPEC sources. As far as the production itself is concerned - together with the production development facilities required to maintain and/or increase the output potential - the OPEC countries themselves now take care of the investment required given their nationalisation of the companies' assets. In as far as technical and managerial help is required from the companies to keep production going and/or to develop new potential, then the companies collect a fee - either in cash or in the form of oil at a specially discounted price. Such fees are also highly profitable in themselves so adding another favourable element to the companies' interests in maintaining and even increasing the flow of OPEC oil.

A similar situation favouring traditional flows of oil exists with respect to the capital investment by the companies in downstream facilities in Western Europe. The demand for oil in Western Europe is currently running at almost 30 per cent below the quantity that was by now expected to be in demand. However, investment plans and projects initiated in the early 1970s are still coming on line and adding to the capacity of the West European oil industry to handle supplies. Not only is such capacity not really needed, in light of the present and near future overall supply/demand situations, but most of the projects are, moreover, of a kind and in locations related to the expectation of increased flows of OPEC oil from the Persian Gulf and Africa.(1) Much of this investment cannot, of course, be served by oil from the North Sea which, indeed, requires its own new infrastructure of terminals, re-built or new refineries and new distribution facilities to enable it to be incorporated into the Western European energy market.

In light of this situation, any greater than really necessary change in the geographical pattern of oil supply to Western Europe - such as would occur if there were too large and too rapid a rate of substitution of oil from traditional sources by oil from the North Sea - would have a further serious adverse effect on the economic viability of the oil companies' operations in Western Europe. And this would be in a situation in which the much reduced demand for oil has in itself already caused them serious problems of retrenchment in their operations, with consequential higher unit costs for processing and distribution and a squeeze on profit margins. Still higher prices to meet the higher costs would then cause a further set-back in the development of demand. In other words, both the demand situation overall in Western Europe and the location and type of the industry's existing infrastructure in the region run counter to any enthusiasm on the part of the international companies for too much North Sea oil at any price, let alone for too much produced at low rates of return on investment which do not match their expectations from alternative investment opportunities elsewhere in the world.

Thus, the decline which we have hypothesised in the average internal rate of return on the increasing levels of investment which are required to increase the output of oil from a particular field will, sooner or later - and probably sooner rather than later - bring them to a cut-off point for their investment in the field. Thus, if oil company policy aims at maximising the returns on the limited amount of money that it has available for investment, then this is most likely to be achieved by limiting the degree to which any particular field is developed.

Given the way in which the oil supply/demand situation has developed in Western Europe, there are currently no restraints on a company acting in this way in respect of its North Sea oil reserves. This is unlike the

1) One such example is the £60 million crude oil import terminal on Anglesey: this was planned in the late 1960s and built between 1972 and 1975 so that Shell's Stanlow refinery could expand and secure its increased crude oil needs by means of 500,000 ton tankers bringing oil from the Persian Gulf to the new deep-water terminal. Now the expansion is not needed (because demand has not developed). Moreover Stanlow will soon be running to a large degree on North Sea crude oil which will be brought from the Shetland Islands in small tankers that can use the pre-existing import facilities at the refinery. The terminal will become an expensive white elephant.

United States or Canada where, in both cases, national policies lay down that the pattern of oil and energy supply is a matter for national determination so that energy imports in general and/or oil imports from particular sources can be kept under control. By contrast, oil companies in Western Europe continue to enjoy the right to determine the energy supply pattern to the region. They can, in other words, "off-load" the oil which they are committed, or which they want, to take from OPEC countries in Western European markets, irrespective of how much oil they could, for example, recover from the North Sea developments. Thus, the exploitation of North Sea oil is placed in jeopardy because so many of the developmental decisions belong to exactly the same international oil companies which, as shown above, have a commercial interest in the maintenance of oil imports to Western Europe. Or perhaps it would be more correct to say that the development of the North Sea's recoverable reserves has been placed in jeopardy because of the fact that the companies are still allowed to take supply production and refining decisions in Western Europe that are virtually free of relevant and meaningful government interference.

And matters must remain like this until action is taken by the governments and the other institutions concerned in Western Europe to secure the divorce of the indigenous oil industry from this degree of external control over its affairs. Such a divorce can only become absolute as the international oil companies are required to take their decisions on the North Sea in the light of the countries' best interests rather than in the context of the best interests of the international oil system.

THE EXPLOITING COMPANIES AND THE GOVERNMENTS OF WESTERN EUROPE

Unfortunately, misunderstanding - or even perhaps deliberately misleading statements - on the nature and probable size of the North Sea oil and gas resource base, and its possibilities for influencing the medium-term outlook for the Western European energy economy, have detracted attention away from the fundamental issues involved in achieving the full exploitation of North Sea fields. Thus, whilst much effort has been directed by Britain, Norway and the Netherlands to ways of toughening up the conditions under which the companies have been allowed to operate (so that the countries concerned could secure a larger share of the economic rent arising from the recovery of each barrel of oil that the operating company determined to produce), none of these countries has taken any action which obliges the international companies to produce all the recoverable oil and gas from each field as quickly as possible (or, indeed, at all) in order to have additional energy available as alternatives to imported oil. And neither have these three producing countries, nor any Western European oil importing country, yet offered guarantees on the profitability of the oil that could be produced by the producing companies, in order to get them to develop their fields more intensively than they will do in light of their own calculations of the optimal situation in the open-market conditions of the Western European energy economy.

Indeed, the very contrary is happening - from two points of view. First, the main potential customers for North Sea oil, viz. France and West Germany - appear to view with suspicion the idea that they should underwrite or guarantee the production of high-cost North Sea oil. Instead they prefer to keep their energy supply options orientated in two other directions. Firstly to the protection and the financing of extra-ordinarily expensive and large programmes of nuclear power development and secondly to special deals and/or greatly enhanced trading relation-ships with the Arab oil producing countries of the Middle East seen, in a psychological sense, "as a natural extension of Western Europe" and with which therefore such trading relationships to secure essential oil sup-plies seem to be preferred over the idea of special arrangements with the potential oil-producing countries of Western Europe.

Second, the oil and gas producing countries of Western Europe are themselves not doing very much - if, indeed, anything at all - to persuade their neighbouring countries to change their policies in this respect. Indeed, the Netherlands, Norway and Britain all appear to be convinced that their resources of oil and gas are so scarce that they need to be protected against the rapacious demands of their neighbours. Such resour-ces, it is argued, are better left unproduced now for the sake of future generations. The validity of such arguments of inevitable scarcity are very much open to question and appear to emerge mainly out of too un-questioning an acceptance of oil company sponsored views on the size of the reserves (ignoring the fact that the companies do have a good commer-cial motivation to minimise rather than maximise their estimates as to how much oil is recoverable) and out of a serious misinterpretation of the nature of the resource base.

Here we are not concerned with the validity or otherwise of the scar-city hypothesis as such, but only in terms of its validity with respect to the impact that it has on the way governments look at the development decisions on North Sea oil and gas reserves. Thus we may, for purposes of our argument, accept that a country may wish to prevent the discovery

or the development of a field or fields because it wishes to save the
potential reserves for the future. However, what one cannot accept is
that a limited rate of development of a discovered field is inevitably
better for a country than an expanded rate, such as could be achieved if
there were more and earlier investment in its exploitation.

This is unacceptable because the proposition leads inevitably to the
conclusion that it is better for a company to cream a second field of its
easiest and lowest-cost-to-produce reserves, instead of developing the
first field to a degree which produces more-than-the-cream of the reserves
through a higher investment and an expanded production plan.

Moreover, in the too general and too ready acceptance of the idea
that each field should be developed slowly, the strong possibility that
this is by no means an automatic conservation of resources approach is
usually ignored. Indeed, government controls on the rate at which a
field's resources are recovered are prescribed as a conservation measure.(1) This is hypothesised on the grounds that a 1-platform installation on a field will take out less oil in the short-term than a 2-platform
installation (and so with 2 instead of 3; and 3 instead of 4 etc.) and
hence, ipso facto, save oil for the future. It is, however, possible to
argue quite a contrary hypothesis. The minimum-number platform system
will have its platform(s) located on the field such that the maximum
possible oil can be produced through the system in the shortest possible
time. This must be done in order to enhance the present value of the
cash flow in a situation in which the producing company is probably

discounting future revenues at discount rates as high as 25 or even
30 per cent. This sort of system reduced the probability of recovering -
or eliminates the possibility of recovering entirely - the previously
estimated technically recoverable oil reserves in the field. This results,
for example, from the depressuring of the field to a greater than desirable
level or from the loss of oil reserves as a result of the irregular advance of the oil/water contact zone cutting off reserves from future recovery efforts. However, once the initial platform(s) - location decision
has been taken, then every other location for a platform on the field becomes sub-optimal with reference to the original technically recoverable
reserves. Thus, even if there are later additional installations on the
field, then, in the absence of technological change or major change in the
economic environment, less oil will be produced than if a system with a
greater number of platforms had been installed in the first instance.

_____

1) The British Government, for example, now has reserve powers to limit
   the rate at which a field is depleted in the interests of the conservation of the field's resources.

In other words, a development decision which initially involves the installation of fewer platforms and wells on a field than are needed to produce all the technically-recoverable oil from a field in an economically relevant time-period is likely to be counter-productive in terms of national policy concern for the conservation of resources. In a multi-field province such as the North Sea there may be something to be said on grounds of conservation (if not of economics) for not producing some of the fields which have been discovered or, even better, for not discovering so many of the fields in the first instance! But neither on grounds of conservation nor on grounds of national economic considerations is there any reason to allow, let alone encourage, a field which is in the process of development not to be developed to its maximum possible extent - to a degree of development, that is, which lies beyond that which is optimal for the oil company.

This is so because a decision which means the development of a field only up to the level of investment and hence of production which is optimal for the company concerned, instead of to the maximum possible level of recovery (at or approaching the level of the technically producible reserves), will have an adverse effect on the degree to which the exploitation of the oil benefits the economy of the country concerned. In the short-term this will arise because jobs and profits which can be generated from oil development related activities will be less than they could be. In the medium- to the longer-term, the benefits to the country will be reduced in terms of the size of the government revenues arising from oil production, of the contribution of the production of oil to the GNP and of the balance-of-payments' effects of the ability to substitute oil imports by North Sea production and/or the creation of an oil export potential.

We thus hypothesise an _inevitable_ and _normal_ divergence of company and government interests in the development decision on a field. On the one hand, one must recognise the validity, from the oil company's point of view, of the strictly commercial methods used by the company to calculate the optimum recoverability of oil from a field in light of competing alternative opportunities _at the international level_ for the available investment funds at the moment in time when the development decision on the field concerned has to be made. On the other hand, the government has an interest in maximising the returns to the country from a field by way of ensuring that its development contributes to the growth of the GNP, to the creation of employment and of local regional multiplier effect, to the balance-of-payments' situation and to government revenues.

To date, however, the divergence of interests has been presented and analysed solely in terms of how the parties concerned differ over the sharing of the benefits to be gained from the development of a field. In such arguments the size of the field is a constant in the equation. As a result there has been much mutual recrimination between the parties and only an uneasy and unstable modus vivendi has so far been established as a result of the negotiations, as well as of "shows of force" by both sides (with threats of nationalisation on the one hand matched by threats of the withdrawal of the international oil industry from the North Sea on the other). We would argue differently: to the effect, that is, that the justifiable commercial decision of the oil company on the development of a field will, normally, lead to a degree of development of the field that fails to maximise what the government could get for the benefit of the nation were a different and more intensive development plan for the field implemented.

If, as should be self-evident in the light of the previously described impact of the initial development decisions on the future recoverability of oil from a field, the government's interest can be shown to lie in maximising the recoverability of oil from a field over a finite period of time (of, say, 20 years) then, in order to persuade the company concerned to do what the government believes to be necessary to maximise recovery, flexibility in the arrangements for taxing the production of oil from the field and/or the possibility of the government contributing to the financing of the field's development have to be considered.

Such a government contribution - direct, by way of an investment in the field's development costs when a more intensive development than that which is optimal for the company is required, or indirect, by way of a modified tax regime - is necessary in order to create conditions in which the company concerned is at least no worse off when it is obliged to install a more complex production system. This more complex system will aim to maximise the degree to which the technically recoverable oil is recovered in an economically meaningful time period and will replace the company's preferred less complex system which, however, on the basis of the company's own evaluation of costs, prices and taxes, appears to provide the company with its optimum investment strategy as far as the development of the field is concerned. Thus, the company will need to be compensated for being forced to pursue a commercially sub-optimal development plan.

A full analysis by government of each of the several possibilities for developing a field is a requirement in order to establish the degree of divergence of interest that exists between the company seeking to optimise its return on investment and the "best" solution from the point of view of the government and country. We have tried to show elsewhere that it is possible to establish the degree of divergence of interests involved in a field development. The degree of divergence established can then provide the basis for an agreement on the degree to which government needs to give financial help to the company in order to eliminate the divergence of interests. Such financial intervention by government is required in order to make it worthwhile for the company to forgo its preferred recovery system and to establish another system which enhances the degree of recovery of the resources within an economically-significant time period. By means of such help the company can be assured of a return on its investment with the more intensive development which is no less than the return it would have earned with its preferred system of development of the field.

From the government's point of view the share of the capital costs which it meets, or the amount of tax which it remits, in order to protect the company's investment at a given minimum level, does not give the government any equity interest in the field. What its help does is to enable the field to be produced more extensively and intensively than would otherwise be the case. As the field is so exploited more oil is produced more quickly so that more jobs and a larger multiplier effect are generated in developing the field (so increasing GNP). The present value of the oil from the point of view of its contribution to the balance of payments is enhanced and there are ultimately more taxes to be collected from the revenues generated by the sales of more oil over a shorter time period. Government itself can determine if these expected additional benefits justify the financial help it must give in the first place to secure the more intense development of a field. If the benefits are considered to justify the assistance, then the assistance can be given; if not, then the help need not be given and the field's development will remain at the level as evaluated by the company.

The research which we have done to date indicates that agreements between government and companies to produce the kind of results described above are possible. It is not "wizardry": it is essentially a function of the contrasting risks and present values which are attached to the possibilities of the future production of oil from a given field by the two parties concerned. In essence, the companies are moved from their propensity to "cream" the fields in a risk minimisation/profit maximisation approach to the exploitation of the resources. Instead, by means of government help they are encouraged and enabled to undertake the technically more difficult and economically more risky task of getting a

greater percentage of the technically recoverable reserves of oil out of
the reservoir more quickly so that its value can make a positive contribution to the needs of the nation and so that a proper conservationist
attitude towards technically available oil can be established.

Moreover, it is only with the divorce of the potential for the
production of North Sea oil and gas reserves from the control mechanisms
of the international oil companies that we can hope to reduce the uncertainties in the energy sector of the European economy to manageable
proportions by maximising the production of indigenous resources. The
essential element in achieving this is, as argued in this paper, an
effective co-operative effort between governments of European countries
and the international oil companies. This, ironically, however, can
only be possible when the latter have been brought under control - in
respect, that is, of eliminating their continuing opportunity to exercise,
jointly with OPEC, the right to determine the supply and price of energy
in Western Europe. When this has been done - within a necessarily inter-
ventionist framework of the kind that Europe has built up over the years
in respect of other, and less-important, sectors of the economy - will it
be possible effectively to realise the opportunities offered by the North
Sea oil and gas reserves. This is a development which is as important to
Britain, Norway and the Netherlands - the important producing countries
of today - as it is to the other nations in Western Europe, whose energy
needs can be more securely and more cheaply met out of the North Sea's
resources than they can from their continued reliance on insecure oil
from the OPEC countries.

This analysis may be construed as a criticism of the international
oil companies. But this is not so, for these companies justifiably, from
the point of view of their own interests, seek to achieve the most profitable set of operations possible. And they have to do this in relation
to governments and to inter-governmental organisations. On the one hand,
they have to deal with OPEC which, in its wisdom and as a result of many
years of experience with the international oil companies, has created very
effective constraints on the companies' activities. It has, however, also
ensured the continuity of highly profitable operations for the companies.
In Western Europe, on the other hand, it is the lack of the right sort of
governmental constraints on the companies which enable the latter to pursue policies which lead to the inevitable divergence of what they do from
what they should be doing for Europe's well-being and security. And this
is done without the companies, in their turn, really knowing if and how
they are going to be able to make adequate profits out of activities in

Western Europe. In no respect is this gap between governments and companies of more importance in 1976 than in respect of decisions which affect the speed and the degree of development of the North Sea's considerable resources of oil and gas. Too many of the provinces' known or expected recoverable reserves remain unnecessarily unexploited basically because the international oil companies and European governments have not yet found a modus vivendi which makes the development of them a top priority of common interest to both parties. As a result the energy supply potential from the indigenous oil and gas resources of Western Europe remains much more limited than it need be with the consequential dangers of a continued too high a dependence on imported OPEC oil and/or the need for an economically unsustainable and environmentally unacceptable degree of development of nuclear power.

BIBLIOGRAPHICAL NOTE

This paper is based on the author's many years of interest in and research on the international oil industry in general and the North Sea in particular. The details of the background to his views and hypotheses can be found in the following publications:

1.   Oil and World Power, Penguin Books, London. First published in 1969 and now in its fourth Edition, 1975.

2.   Natural Gas in Western Europe, e.f. Bohn, Haarlem, 1969.

3.   "Against an Oil Cartel", New Society, 11th February, 1971.

4.   "Europe and the International Oil Industry in the 1970s", Petroleum Times, January 1972.

5.   "Europe's Oil by 1980", National Westminster Bank Review, August 1972.

6.   "Indigenous Oil and Gas Developments and Western Europe's Energy Policy Options", Journal of Energy Policy, Vol. 1, No. 1, June 1973.

7.   "European Alternatives to Oil Imports from OPEC Countries: an Energy Economy based on Indigenous Oil and Gas", in Energy Policy Planning in the European Community, (Ed. J.A.M. Alting von Geusau), Sijthoff, Leiden, 1975.

8.   The North Sea Oil Province: an Attempt to Simulate its Exploration and Development, 1969-2029, (with K.E. Rosing), Kogan Page, London, 1975.

9.   The West European Energy Economy; Challenges and Opportunities (1975) Annual Stamp Memorial Lecture, on the application of economies and statistics to current problems, to the University of London, Athlone Press, London, 1975.

10. "The World of Oil Power in 1975", The World Today, Vol. 31, No. 7, July 1975.

11. "On the Optimisation of the North Sea Pipeline System" (with K.E. Rosing and H.E. Beke-Vogelaar), Journal of Energy Policy, Vol. 4, No. 1, March 1976.

12. The Western European Energy Economy: The Case for Self-Sufficiency 1980-2000, H.E. Stenfert Kroese N.V., Leiden, 1976.

13. "Europe and the Cost of Energy: nuclear power or oil and gas", Journal of Energy Policy, Vol. 4, No. 2, June 1976.

14. "The EEC Energy Market: Structure and Intergration" in Economy and Society in the EEC: Spatial Perspectives, R. Lee and P.E. Ogden, Eds. Saxon House Books, Farnborough, 1976.

15. Optimal Development of the North Sea's Oil Fields: A Study of Divergent Government and Company Interests and their Reconciliation (with K.E. Rosing), Kogan Page, London, (to be published 28th November, 1976).

SPANISH OIL PROSPECTS

by

R. Querol

INSTITUTIONAL FRAMEWORK

Spain's industrialisation during the 60s and early 70s carried with
it a very significant increase of energy consumption, which averaged about
9 per cent/year between 1964 and 1973. Since most of the additional energy
was used in the form of petroleum, the average increase in oil consumption
was as high as 15 per cent/year during the same period.

The 1973 crisis slowed Spain's economy considerably. Together with a
much smaller increase in GNP/year (average 2.7 per cent in 1974/1977) came
a slower growth of energy consumption (3.8 per cent) and oil consumption
(1.9 per cent) during the same period. Figures 1 and 2 show the oil consumption and energy consumption in Spain since 1959. In 1977, Spain consumed about 98 million tons of coal equivalent, including about 43 million
tons of oil. Since only about 1 million tons were produced in the country,
most of the oil was imported and contributed largely to the huge deficit
in the balance of payments (about 3 billion dollars/year in 1974/1977).
82 per cent of the crude imported in 1977 came from the Middle East,
13 per cent from Africa and 5 per cent from other countries.

There are ten refineries in Spain with a present total capacity of
63 million tons of oil/year. They process the crudes for internal consumption and, in small measure, for export.

Nationalisation of marketing took place in 1927, with CAMPSA as the
agency administering the monopoly.

HISTORY OF EXPLORATION

Scientific methods in petroleum exploration had little or no application before World War II; generally, shallow wells were drilled near oil-
sand outcrops or seeps. Petroleum exploration rights were granted under

173

Figure 1
OIL CONSUMPTION

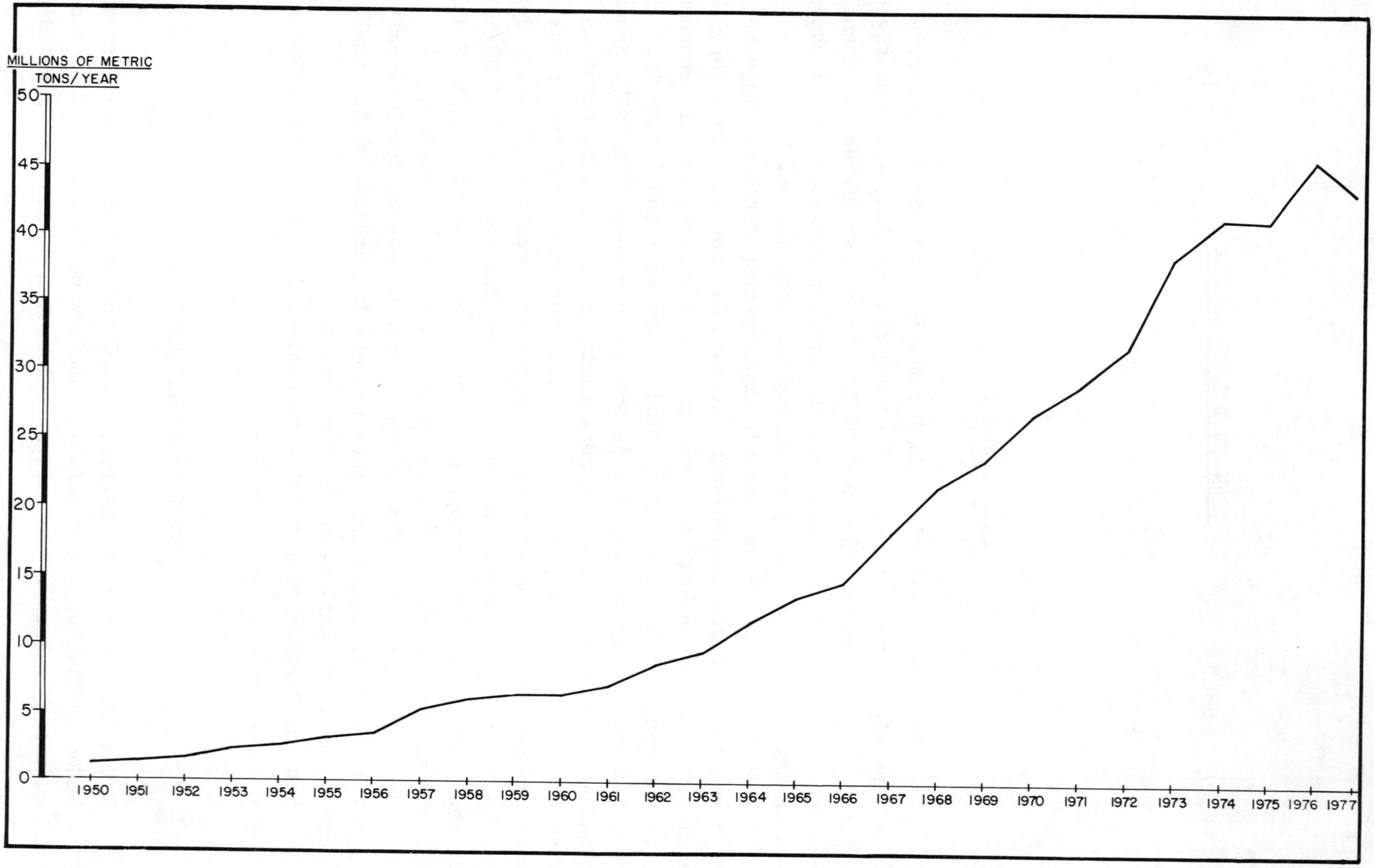

MILLIONS OF METRIC TONS/YEAR

50 45 40 35 30 25 20 15 10 5 0

1950 1951 1952 1953 1954 1955 1956 1957 1958 1959 1960 1961 1962 1963 1964 1965 1966 1967 1968 1969 1970 1971 1972 1973 1974 1975 1976 1977

174

Figure 2

ENERGY CONSUMPTION IN SPAIN

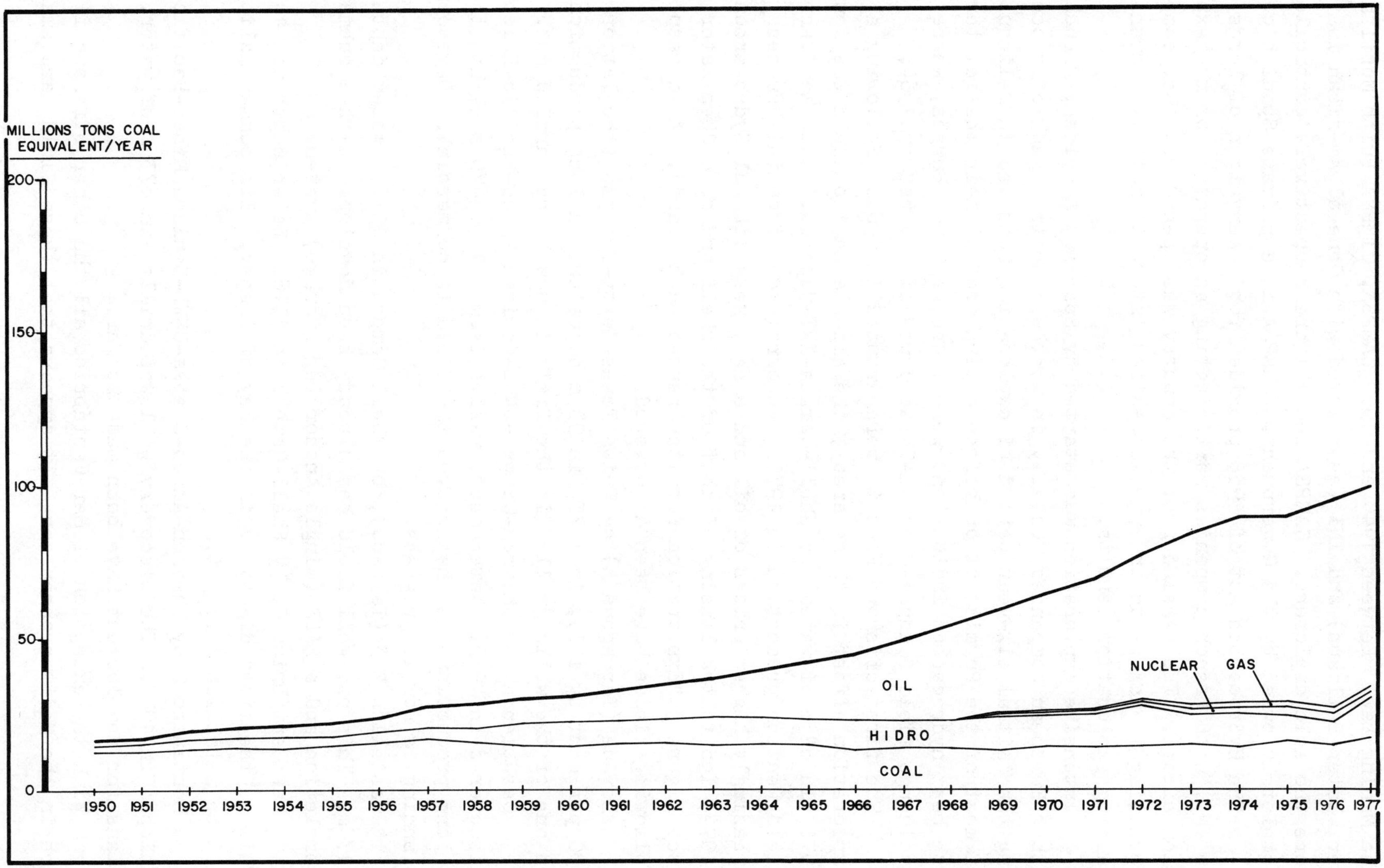

the Mining Law. Between 1940 and 1959 CAMPSA, CIEPSA (with Mobil as partner, later Deilmann) and INI (associated with General American in 1952) were the sole explorers. CAMPSA and INI are respectively controlled and completely owned by the Government. CIEPSA is a private Spanish company.

The Hydrocarbon Act of 1958 provided for reservation of large areas to INI, with French companies participating as operators of the exploration effort. The remainder of the country was opened to other companies, resulting in major- to intermediate-sized international oil companies obtaining exploration permits.

Intensive exploration was started by Spanish, American, French and other European companies, chiefly in northern Spain. Lack of success during the initial six-year period of exploration resulted in relinquishment of most of the Permits at or before expiration of their initial terms. The best hydrocarbon indications were found north of Burgos, where the Ayoluengo field (Campsa-Chevron-Texaco) was discovered in 1964.

The first offshore Permits were granted in 1965. Following a moderate exploration effort in these areas, the Amposta Marino field was found in 1970 (in 62 m of water) by Shell-Campsa-INI-Coparex. This Mediterranean field went on production in 1973. Encouragement from this success, increasingly higher prices of oil and a new, very liberal Hydrocarbon Act (1974) led to the leasing of most of the shelf and some deep water areas (see Figure 3 "Exploration Permits Granted Each Year"). As a result the following fields have been discovered.

Dorada (Tarragona E) by Union Texas-Getty-Eniepsa (the latter owned 100 per cent by INI), in 1975, in 95 m of water. A long production test on one of the wells drilled in the field is now being carried out.

Casablanca, by Chevron-Ciepsa-CNWL-Denison Mines-Pacific-Eniepsa, discovered in 1975. Water depth varies from 121 to 470 m within the field. A long production test has proven the field is commercial. Development planning is now in progress.

Castellón B 5 (Tarraco), by Shell-Campsa in 1976. Water depth is 117 m. This one-well field has already been developed with a submarine completion and a SALS (Single Anchor Leg Storage) System.

Mar Cantábrico C, by Shell-Campsa in 1976. Water depth was 147 m. This is the first discovery in the Bay of Biscay. Its commerciality has not be established as yet.

Montanazo C by Chevron-Amoco-Ciepsa-CNWL-Denison Mines-Pacific-Eniepsa in 1977. The discovery well was drilled in 673 m of water. No plans for development have been made so far.

With the exception of Mar Cantábrico all the other offshore fields are in the Mediterranean sea. It appears that these fields are small and

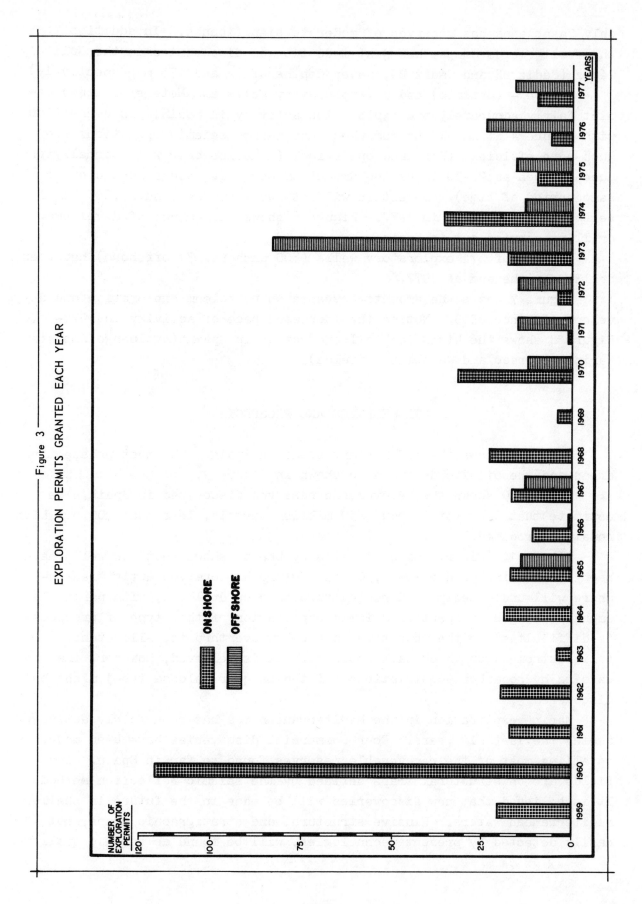

Figure 3

EXPLORATION PERMITS GRANTED EACH YEAR.

ONSHORE

OFFSHORE

NUMBER OF
EXPLORATION
PERMITS

YEARS

177

only Casablanca has reserves of moderate significance. In addition to these oil fields Campsa has just made two gas discoveries in the Gulf of Cádiz (Cádiz B2 and Cádiz B3, water depths of 92 and 133 m respectively).

Figure 4 (Seismic) and 5 (Exploratory Wells and Meters Drilled) depict fairly accurately the exploration activity in Spain. In connection with Figure 4 it should be remarked that marine seismic operations are much more efficient than land operations (a marine crew will normally produce as much as 25-30 times the amount of coverage than a land crew in the same period of time). Seventeen wildcats were spudded and 27,685 km of seismic were produced in 1977. Figure 6 shows the amount of development drilling carried out in Spain.

A total of 371 exploratory wells (300 onshore, 71 offshore) had been drilled by the end of 1977.

Figure 7 shows the amounts invested on petroleum exploration and development since 1959. Notice the increased pace of activity in 1975-77. Figure 8 shows the historical oil production in Spain (Ayoluengo Amposta Marino, Tarraco and Casablanca fields).

OIL RESERVES AND PROSPECTS

Figure 9 shows the sedimentary areas in Spain. The most prospective areas and the oil fields are also shown in Figure 9.

Figure 10 shows the recoverable reserves discovered in Spain. We estimate that they total about 250 million barrels, less than 200 of which remain unproduced.

Figure 11 (Distribution of Wells by Depth) shows that few wells have been drilled in Spain below 3,500 m. Future exploration activities onshore will undoubtedly be directed towards deep drilling, with particular emphasis in gas prospects. Seismic exploration of this type of prospects is difficult since the objectives are below overthrusts, allochtonous (olistostromes) or thick salt deposits. It is believed, however, that rewarding hydrocarbon accumulations (of the Lacq or Malossa type) might be discovered.

Marine exploration in the Mediterranean Sea has been fairly active during the last six years. Four commercial discoveries have been made, totalling most of the recoverable reserves found so far in Spain. The fields are on structural traps defined by the seismic reflection method. It is believed that new discoveries will be made in the future in shallow Mediterranean waters. Elusive structural and stratigraphic traps, not easily detected by present technologies, will be found as seismic quality

Figure 4

SEISMIC

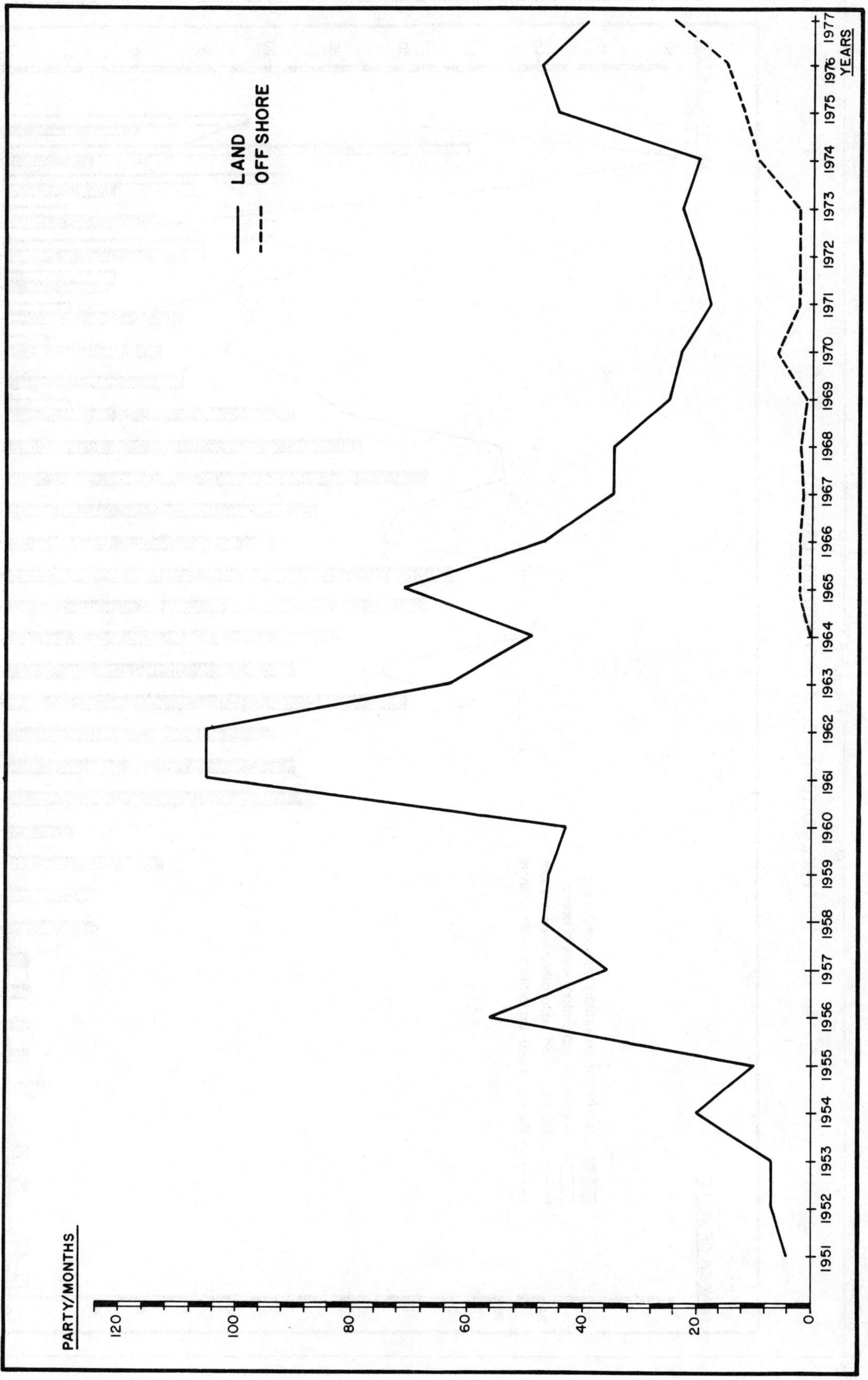

PARTY/MONTHS

LAND

OFF SHORE

YEARS

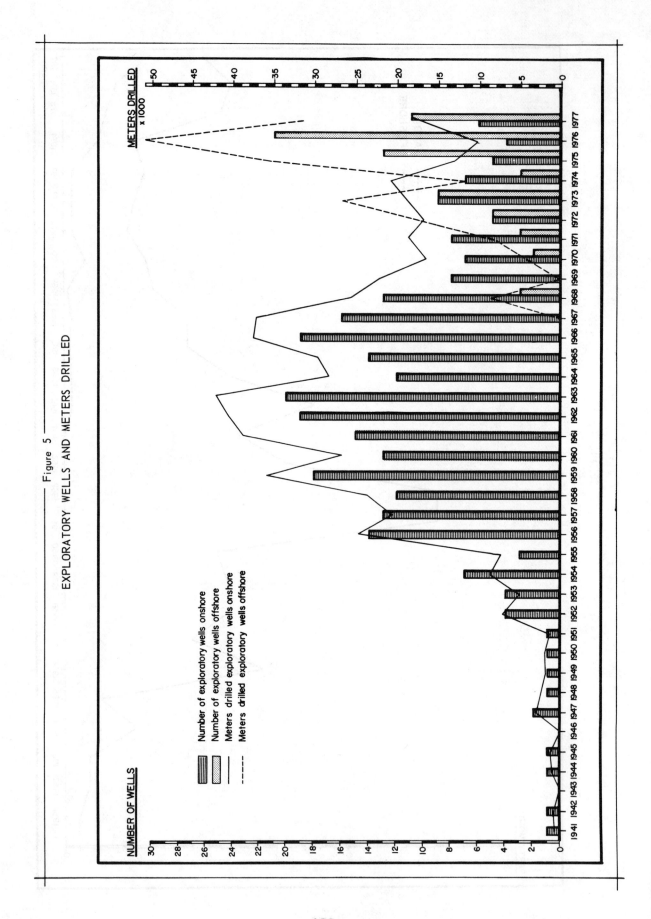

Figure 5

EXPLORATORY WELLS AND METERS DRILLED

Figure 6

DEVELOPMENT WELLS AND METERS DRILLED

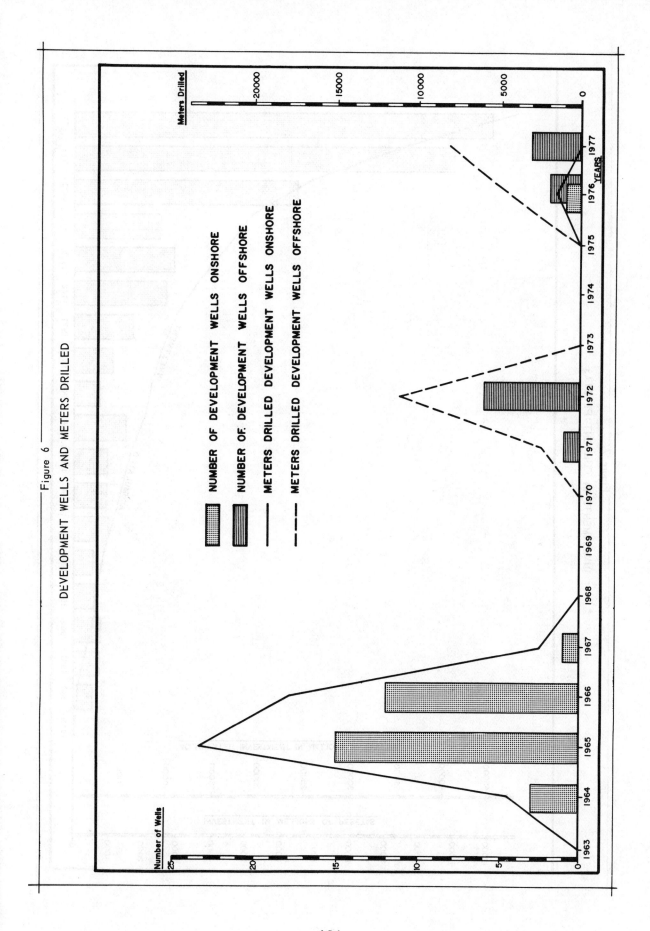

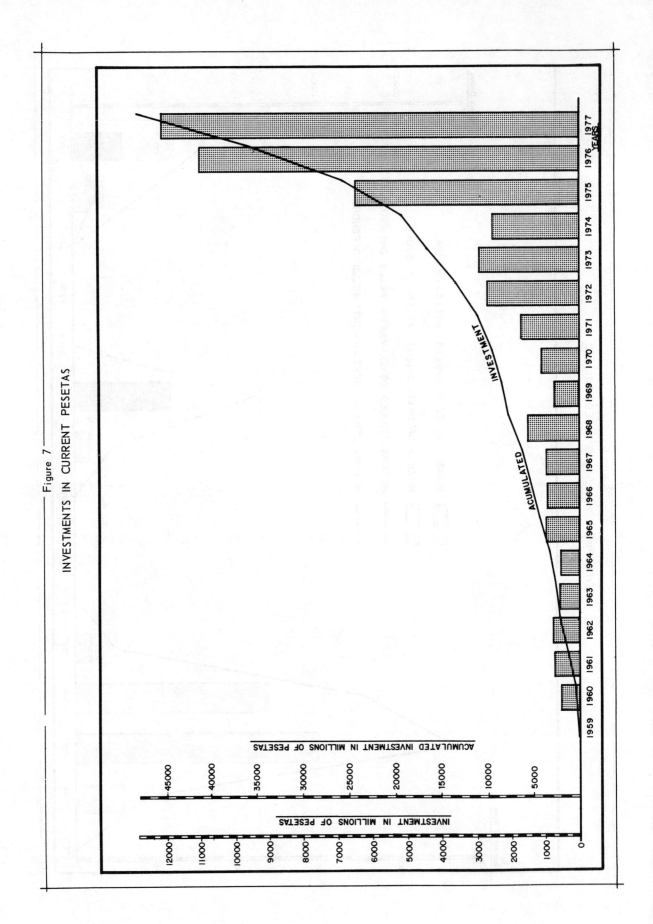

Figure 7

INVESTMENTS IN CURRENT PESETAS

Figure 8

OIL PRODUCTION

Figure 9

PROSPECTIVE AREAS IN SPAIN

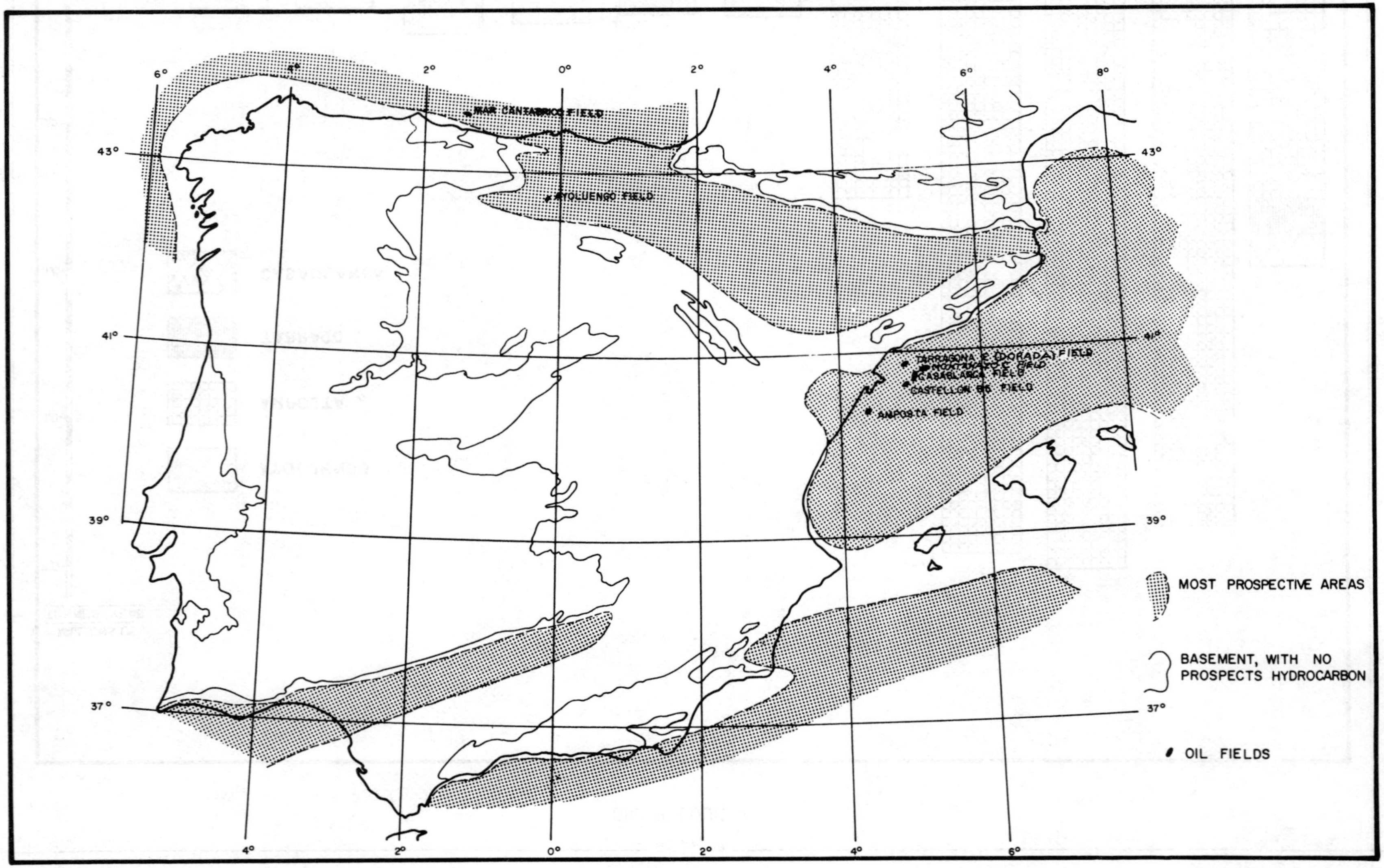
MAR CANTABRICO FIELD
AYOLUENGO FIELD
TARRAGONA E (DORADA) FIELD
CASABLANCA FIELD
ANPOSTA FIELD
6°
4°
2°
0°
2°
4°
6°
8°
43°
41°
39°
37°
MOST PROSPECTIVE AREAS
BASEMENT, WITH NO PROSPECTS HYDROCARBON
OIL FIELDS

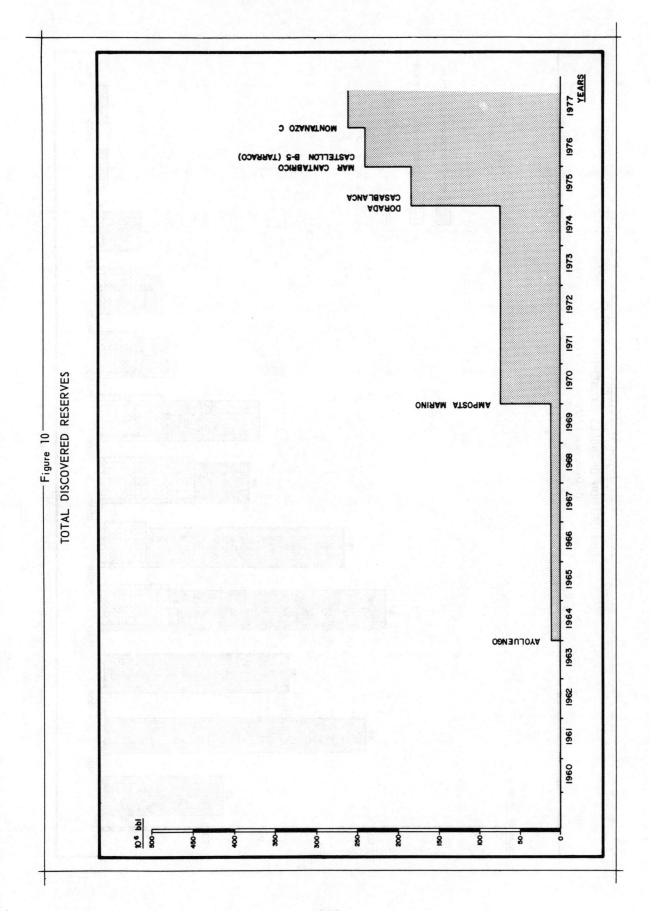

Figure 10

TOTAL DISCOVERED RESERVES

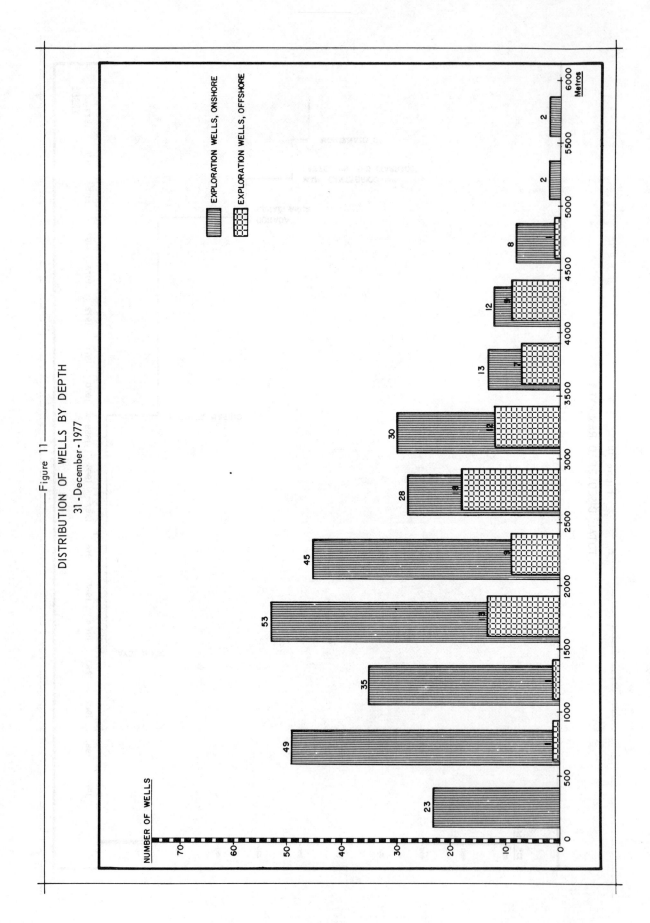

— Figure 11 —

DISTRIBUTION OF WELLS BY DEPTH

31 - December - 1977

EXPLORATION WELLS, ONSHORE

EXPLORATION WELLS, OFFSHORE

NUMBER OF WELLS

Metros

improves. In the near future the Mediterranean continental shelves should
provide most of the new oil in Spain, although discoveries will probably
be of modest size and productions should remain at level of a few million
tons/year during the 80s.

The deep waters in the Spanish Mediterranean are now beginning to be
explored. Eniepsa-Phillips-Getty have recently drilled a well there in
1,144 m of water. Good encouragement was provided by the Montanazo C discovery. The seismic method indicates that thick sedimentary basins exist.
The largest of them extends northeast and south of the Balearic Islands in
waters of up to 2,600 m. From a geological point of view the area has good
hydrocarbon potential. Technology will determine if and/or when the possible discoveries would become commercial.

Two recent gas discoveries by Campsa in the Gulf of Cádiz appear encouraging. Further appraisal work is required to determine the significance of the reserves.

An oil discovery was made in the Bay of Biscay in 1976 (Mar
Cantábrico C2, which tested 7,000 bopd at 1,450 m). An appraisal well
(Mar Cantábrico C 3), tested oil but it apparently proved that reserves
were not commercial. It is believed that prospects exist in the Bay of
Biscay, but development will be costly due to difficult metereological
conditions, complicated geology and associated mechanical problems.

PRODUCTION AND DEVELOPMENT COSTS

The only offshore field that has been developed is Amposta. A drilling and a production platform were installed in 61 m of water. The oil
is produced from a SBMS into a tanker. Lifting costs are in the order of
$1-2/bbl increasing as the field is depleted. Development costs (in
1971-72) were in the order of $750 per initial bopd.

Castellón B 5 produces from a submarine completion through a SALS
system into a tanker containing separating facilities. It appears that
Casablanca will be developed from a conventional fixed platform while
Dorada will be produced from a mobile platform through a multiple riser
system. Probable lifting and development costs are $1-2/bbl. and $1,500/
bopd in case mobile platforms and SBMS or SALS are used. If fixed platforms and pipelines are installed, lifting costs might be reduced to
$1/bbl but development investment would go up to $5,000/bopd.

FORECAST OF FUTURE DISCOVERIES AND PRODUCTION

From what has been indicated above it appears that oil prospects on-shore are small; gas reserves of significant size at great depths could however be discovered, particularly in Northern Spain.

It appears reasonable that ultimate recoverable reserves in Spanish waters down to the 200 m isobath would be in the order of 1 billion bar-rels; admittedly this is only an educated guess. As far as the deep waters go little can be said except that thick sedimentary basins are present; they presumably contain good source and reservoir rocks.

Estimating future production is always difficult. Figures of 2 mil-lion tons for 1980, perhaps increasing slightly in 1985 appear reasonable. Notwithstanding the aggressive development of energy sources other than oil (particularly the ambitious nuclear programme) it is obvious that Spain will be strongly dependent on imported oil during future decades.

STABLE GROWTH AND ENERGY STRATEGY
TWO APPROACHES

by

T. Ikuta

STABILITY IN THE MIDST OF INSTABILITY

In the context of the energy problem, 1977 may be regarded as the
year in which stability existed in the midst of instability. This sounds
paradoxical enough, but the fact is that such a condition did prevail both
in the world economy and the Japanese economy.

When I say "instability" I mean, first of all, the unstable condition
in the world economy generally. Japan and other major industrialised
nations tried hard to stabilize their economies, but the prospects for the
stable, non-inflationary growth which they tried to achieve remained uncertain. In fact, so many uncertainties existed in the world economy as
well as in the Japanese economy that it was extremely difficult to make
any reasonably accurate prediction about the future. So this underlying
uncertainty in the world economy was the first factor that contributed to
the instability in the energy situation.

A second element of instability was the continuing uncertainty in the
global oil supply situation. Petroleum, of course, is the most important
source of energy, but whether we would be able to secure adequate oil supplies in the future remained a big question.

The biggest reason for that was the growing instability in the political and military situation in the Middle East. Something quite unexpected could happen anytime in this highly volatile region, as the surprise
visit to Israel by Egyptian President Anwar Sadat amply demonstrated.
Since that historical event the political situation in the Middle East has
been evolving rapidly. But exactly where the current peace effort is heading is not clear. There is no telling, for example, whether a final peace
settlement will be reached within the broad framework of Geneva peace talks.
The Sadat journey to Jerusalem has raised hopes for peace but it has also
caused deep splits within the Arab countries.

It is not only the confrontation between Israel and Arab nations which lies behind the unstable situation in the Middle East. Relations between Iran and Saudi Arabia, for instance, contain seeds of trouble although their ties have improved lately. More serious, however, are the military relations between Iran and Iraq which is supported by the Soviet Union. Considering that the Soviet-American rivalry in the Middle East is reflected indirectly in these relations between Iran and Iraq is a question that has a critical bearing on the Mideast situation.

This question is not unrelated to the reversal of Iran's oil pricing policy that took place prior to the OPEC meeting held in Caracas last December. The Shah of Iran, Pahlevi, changed the position which his country had maintained all along - that the price of oil should be raised by a wide margin. Thus he came out against any oil price increase and called for a price freeze. As a result, Iran joined the ranks of the "moderates" at the Caracas meeting and even played the leading role in their drive for a temporary price freeze. That meant a complete switch in the Iranian oil pricing policy. At previous OPEC meetings, Iran had always proposed a large price increase while Saudi Arabia had called for either a price freeze or a moderate raise. And pricing decisions by OPEC had represented compromises made between these two leading oil-producing countries.

Why did Iran reverse its position? The presumed reason is a military one - that is, the security of Iran itself. Iran, of course, is in a very difficult position military, so that it probably has no alternative but to rely on the United States. The price Iran paid for increased United States support - in the form of new arms sales to Iran - was its decision in favour of a price freeze. That is how the policy reversal is generally explained. In fact, political and military moves are now closely related to the price and supply of oil.

The situation described above will probably continue in years ahead. The Middle East is one region where the East-West relationship is subtly reflected. It is in this region that the diplomatic and military strategies of the United States and the Soviet Union come into play. So the Mideast situation is likely to remain in a flux as long as the rivalry between these two superpowers continues in this region.

Turning back to the energy problem, such political and military instability is expected to continue, at least in the time framework in which energy supply and demand projections are made. So the politico-military uncertainty in the Middle East was the second factor that contributed to the instability in the energy situation.

A third element of instability was the existence of various problems in the development of alternative energy sources. Most of these problems date back to the oil crisis of late 1973, which prompted major industrialised nations to launch new energy development programmes designed to reduce their independence on petroleum. However, these programmes faced considerable difficulties in the subsequent years - for economic and social reasons, with the result that their implementation was seriously hampered. The current trend is still for a continuing slowdown in the development of alternative energy.

This is particularly true of nuclear energy, which had involved difficult problems even before the oil crisis visited upon us. Difficult problems were also involved in the development of other alternative sources of energy. Liquefied coal, tar sand, oil shale and heavy oil - all these could replace oil completely as liquid energy. But it is quite unlikely that any of these potential substitutes for oil will be commercially developed before global oil shortages develop. As for solar energy and other types of energy of the future, the situation is simply hopeless - in the sense that they will become available only in the distant future.

What about energy conservation? In the United States, for instance, efforts in this direction are making no substantive progress. The energy bill introduced by President Jimmy Carter to the Congress, a measure which incorporated a new oil conservation policy, was drastically altered in the Senate. It was not clear whether the United States would have any effective energy programme in the foreseeable future.

Development of alternative energy and expansion of utilisation are aimed, of course, at resolving the problem of possible oil shortage on the supply side, whereas conservation is a means of coping with the same problem on the demand side. So, looking back on the past year, there was little prospect that energy supplies would be expanded sufficiently to meet this growing demand. This, then, was the third factor that contributed to the instability in the energy situation.

In summary, we had three elements of instability - namely, (1) the stagnation in the world economy and increased global economic uncertainties, (2) the unstable or dangerous political and military situation in the Middle East, and (3) uncertainties over the development and utilisation of alternative energy.

But stability did exist in these unstable situations. That stability was the lack of serious problems in the global energy supply and demand situation. At least that is how the situation stood in 1977. Take oil for example. The world market remained generally weak, creating the impression that there is no such thing as energy crisis, or that no serious energy shortage is likely to occur. This is the meaning of "stability in the midst of instability". But this seemingly contradictory situation is not desirable, not because we do not want stability but because we paid an extremely high price for that stability.

First, we paid a high price - in the form of the global economic stagnation - for stability in the energy supply, particularly in the oil supply situation. Energy experts often commit the mistake of believing that stability in the energy supply is an end in itself. In other words, they often forget that such stability is meaningful only when it contributes to economic growth and human welfare. It is not desirable, therefore, to stabilize the energy supply and demand balance at the expense of economic stagnation. We should try first to eliminate the underlying instability in the energy situation, even if we have to sacrifice the temporary stability. Short-term problems would arise in this case, but governments should deal with such problems by adopting effective policy instruments, in order to prevent temporary disruptions in the energy supply.

While the world energy supply situation remained relatively stable throughout 1977, oil consumption in the United States continued to expand rapidly. The United States was the only major industrialised nation in the non-Communist world that experienced a solid economic recovery. Most of the increased domestic oil demand was met by increasing oil imports. However, this aggravated the United States balance of international payments, thus causing a huge trade deficit. The deficit problem became a major irritant in Japan-United States trade relations and applied downward pressure on Japanese exports to the United States. It also affected the world economy significantly.

Rising oil consumption fueled by strong domestic economic growth was a phenomenon confined largely to the United States in 1977. But oil use in other major oil-consuming countries is also expected to increase if the world economy shows faster recovery in the future. If this happens, much of the increased consumption will have to be met by expanding oil imports. This will probably bring further strains into the relationship between OPEC and oil-consuming countries. I believe there is a strong possibility that Japan will tread the same path that the United States followed in 1977 if the Japanese economy is launched into the orbit of stable growth.

DILEMMAS IN THE ENERGY PROBLEM

I have explained how and why the world energy situation remained stable in 1977 despite its underlying instability. This conditions has created dilemmas in the energy problem in most countries of the world.

In my view there are two major dilemmas. One dilemma is this: there are no immediate problems in the energy supply and demand situation, particularly in the area of petroleum, although we have paid a very high price in the form of worldwide economic stagnation. But, as many experts predict, the present supply and demand position of oil will be reversed later than 10 years from now, for example, if the world economy achieves a degree of stable, non-inflationary growth. In other words, demand will exceed supply. I think this is a reasonable prediction. Consequently, we have to begin preparations now to cope effectively with an anticipated energy crisis, or an oil crisis, to be more precise. We must act now because we need a certain amount of lead time in which to carry out the necessary measures to promote energy development and conservation.

Of course, experts are fully aware of all this, but general public are not. It is extremely difficult to convince the general public that, although oil is currently available in abundance we are going to suffer serious shortages in the not-too-distant future, or that an oil crisis is at least a theoretical possibility under a set of conditions. Given this lack of public awareness, it is difficult to ask them to make sacrifices now. So this is the first dilemma in the energy problem.

The other dilemma is related to the problem of natural resources, or their long-term availability. The earth abounds in various energy resources including petroleum. But this is one thing. Whether or not we will be able to use them at reasonably low cost is quite another matter. So the physical abundance of energy resources does not necessarily mean that we do not have to worry about their actual availability. This is the second dilemma in the energy problem.

According to previous estimates, the ultimate recoverable reserve in the world is about 2 trillion bbl. Moody's estimate puts the sum of proven and prospective reserves at 1.1 trillion bbl. Of this total 300 billion bbl. has already been consumed. So we have 800 billion bbl. left. In addition, there is a discoverable reserve of 930 billion bbl. So these figures given by Moody also add up to about 2 trillion bbl. As for discoverable reserves, estimates vary from 600 billion bbl. to 1.4 trillion bbl. The figure of 930 billion bbl. is the median value of these different estimates.

According to Moody, 28 different studies have been made with respect to oil reserves. Moody's estimate comes in between the lowest and highest of these calculations, so that it may be considered reasonable. It is one thing that we have about 2 trillion bbl. in ultimate recoverable reserves. How much of these reserves will be actually recovered, or when or how will this become possible is quite another question.

Table 1

WORLD OIL RESERVES (J.D. MOODY)(*)

Proven reserves + prospective reserves

     = 1.1 trillion bbl.
       (0.3 trillion bbl. already used;
        0.8 trillion bbl. still underground)

Discoverable reserves

     = 0.6 ~ 1.4 trillion bbl.
       (0.93 trillion bbl.)(*)

Remarks: Median value of 28 different estimates.

DISTRIBUTION OF UNRENEWABLE WORLD RESOURCES

North America	45%
Russia, China	30%
Other regions	25%

Remarks: Oil resources in the Middle East account for
        only 1 per cent of the total.

*) J.D. Moody: "Possible Exploration Activities and
    Ultimate Petroleum Reserves", 1977.

We must also consider other energy resources. Coal and other substitutes for oil, such as tar sand, oil shale, heavy oil, natural gas and uranium ore also abound in various parts of the world. Forty-five per cent of such non-renewable resources are deposited in North America, 30 per cent in Russia and China and the remaining 25 per cent in other regions. It would seem significant that most of these resources exist in three of the world's great powers.

But the existence of such huge quantities of energy resources is, in itself, not as important as it may seem. What is more important is whether or not these resources can be used at reasonably low cost. It is their actual availability that really counts. If the mere presence of energy resources under the ground means that they will become entirely available – that is, if a huge reserve can bring about a large increase in supply, then why did the price of oil jump four times in 1973? This is a question Moody asks. The question is quite relevant because oil resources in the Middle East account for only 1 per cent of the global total of non-renewable energy resources. These do not include, of course, renewable energy such as solar heat and hydropower. Yet the countries holding only 1 per cent of the unrenewable energy reserve succeeded in raising their oil prices 300 per cent. This is something quite unthinkable if we follow the above-mentioned line of reasoning. Yet, in reality, those countries continued to raise their prices and there are no signs that they will stop doing so on the longer-term basis. All this means that the size of a reserve does not really matter if it cannot be exploited on a profitable basis. We must take into account various constraints that exist on the supply side. So it is basically wrong to think that we do not have to worry about future energy shortages because we have enough resources. Such a simplistic view often confuses the debate on energy problems.

I have explained two dilemmas in the energy problem. To sum up, one dilemma is that, while the present energy situation is relatively stable the future has some serious problems in store for us. The other dilemma is that, while we have plenty of energy resources we cannot necessarily use them as much as we want to. Energy problems have been discussed recently against the background of these dilemmas. Particularly in Japan a great deal of discussion has taken place during the past year. It is, of course, useful to discuss a problem from many angles. But it seems to me that some of the arguments made by experts in this country missed the point completely - so much so that for the general public the energy problem became something very difficult to understand.

ECONOMIC SCALE AND ENERGY DEMAND - PRICE ELASTICITY

One critical view on the general energy outlook is that energy demand is not likely to increase as much as it is projected to. Energy demand is discussed in two ways. First, it is argued that the growth of energy consumption can be curbed considerably through conservation. This argument has been advanced, for example, by Japan's political parties, which have their respective energy plans. One thing these plans have in common is that very high targets are set for energy conservation.

As shown in Table 2, the conservation ratio for 1985 is set at 5.5 per cent in the reference case and at 10.8 per cent in the accelerated policy case of an interim forecast by the supply and demand division of the Advisory Committee for Energy. The ratio was 9.4 per cent in the recommendations submitted by the same committee in August 1975. Later, that figure was revised to 5.5 per cent and 10.8 per cent respectively, to bring it up to date. If the energy conservation ratio is set at such high levels much of the energy problem will be resolved.

The energy conservation ratio is related, of course, to total energy demand. Since the demand volume is very large, even a fractional change in the ratio translates into a considerably large reduction in demand. In

Table 2

ENERGY CONSERVATION TARGETS (1985)

August 1977 estimate	June 1977 estimate	
	reference case	accelerated case
Industry 6.9%	6%	11%
Transport 18.4%	2%	10%
Household 13.5%	6%	14%
Other 2.4%	2%	3%
Total 9.4%	5.5%	10.8%

Notes: 1) 1985 conservation ratios vs. 1973 figures.
        2) Prepared (partly estimated) from data com-
            piled by the Advisory Committee for Energy.
        3) Conservation ratios set at 15-16 per cent
            in the energy plans of political parties.

the case of the energy plan prepared by the Japan Socialist Party (JSP),
for instance, the conservation ratio is set so high that demand for electricity from nuclear power plants will be reduced to zero. Other parties
are not much different from the JSP in that they all project high conservation ratios. But none of them explains in concrete terms why it is possible to attain such high conservation targets. In my view, it will be
difficult to achieve the 10.8 per cent target in the accelerated policy
case. It will be no easy task to hit the 5.5 per cent target in the reference case, either. But my feeling is that at least that much conservation must be achieved.

The other way to look at energy demand is to discuss it in more
theoretical terms. The argument here is that energy demand probably has
greater price elasticity than is generally thought. This is an important
question which deserves full consideration.

The view that energy demand probably has considerable price elasticity
is supported by the expectation that energy supply may also exhibit some
price stability. In reality, however, the price elasticity of energy supply is almost nonexistent as far as production in the OPEC area is concerned. In other words, production will not increase if the price is
raised. This is clear enough from what I have already explained regarding
the problem of availability and politico-military problems involved in the
oil supply situation. In the case of oil produced in the United States,
however, some price elasticity can be expected. The reason is because production is likely to increase if the domestic price is raised to a more
appropriate level.

As for alternative energy, such as tar sand, oil shale and liquified coal, considerable price increases will probably accelerate production. The basic policy of OPEC is to raise the price of oil to the point where that price equals the marginal cost of alternative energy development. OPEC pursues this policy, because by doing so they can expect the supply of alternative energy to increase and oil demand to decrease. And a slow-down in oil demand will enable them to conserve their oil resources for a longer period. So I do not think that the supply of alternative energy has no elasticity at all. But price movement will have to be rather rigid. As for OPEC's oil policy, I would say that it is completely rigid.

There is little empirical data in Japan and other countries concerning the price elasticity of energy demand. So, few analyses have been made on this subject. It can be said with a reasonable degree of certainty that actual elasticity is considerably small, at least within a certain initial margin of price increase.

Table 3

PRICE ELASTICITY OF ENERGY DEMAND

(prepared from data compiled by the Advisory
Committee for Energy, Esso Standard, etc.)

Short term	−0.1
Long term	−0.2 ～ −0.3

As shown in Table 3, the price elasticity of energy demand is about −0.1 on the short-term basis. But the value increases on the longer-term basis, probably to the range of −0.2 and −0.3. This is because long-term price changes cause adjustments in industrial and economic structures. Elasticity analyses made in the past are based on data collected when energy prices were on the decline. So we do not have sufficient data as yet regarding elasticity in periods of rising prices. For this reason it is difficult to make an accurate approach to this question.

In my view, however, the price elasticity of energy demand is considerably small. The reason is that energy consumption has increased despite the quadrupling of oil prices and the resultant general rise in energy prices.

Energy consumption figures are compared with Table 4. On a sector-by-sector basis, consumption in mining and manufacturing dropped 5.3 per cent in 1975 from 1973. On the other hand, energy use in the transport and household-commerce sectors increased about 13 per cent. This shows, of course, that energy demand expanded despite the sharp rise in oil and energy prices. Thus the price elasticity of energy demand showed positive values. In other words, demand increased despite price hikes. This runs counter to the textbook principle of supply and demand. But in the mining and manufacturing sector demand did decline. Why did all this happen? To obtain a valid answer we should consider what happened in the other sectors. In the agriculture, forestry and fishing industry, demand increased 8 per cent. It expanded 7 per cent in other manufacturing fields, and 4~2 per cent in electric power, gas and food processing. By contrast, demand dropped considerably in chemicals, cement, etc., metal products and

Table 4

ENERGY CONSUMPTION BY SECTOR (1973 vs. 1975)

(data compiled by the Institute of Energy Economics)

Mining and manufacturing	− 5.3%	Gas	+ 3.4%
Energy	+ 4.4%	Food manufacturing	+ 2.1%
Transport	+12.5%	Chemicals	−16.9%
Household and commerce	+13.1%	Cement, etc.	−15.8%
(industrial sector breakdown, major fields only)		Nonferrous metals	−13.2%
		Paper-pulp	−10.9%
Agricultural & fishing ind.	+ 8.4%	Metal products, machinery	− 9.2%
Other manufacturing	+ 7.1%	Steel	− 8.1%
Electric power	+ 4.5%	Railways	− 2.0%

machinery, and steel. One conclusion that can be drawn from this is that the drop in energy consumption in mining and manufacturing was due to a kind of income effect rather than the price effect. In other words, structural changes in the industrial sector must have been responsible for the reduced energy use. To put it another way, the rise of energy prices must have eroded the competitiveness of the industries involved. But I do not think that this was a decisive factor in the decline of energy consumption. Let me take the Japanese steel industry as an example. Steel production has dropped since 1973, but this is not the result of the higher energy price but because demand has dropped due to the changing circumstances at home and abroad. The same thing goes for other industries.

The majority view of energy experts here and abroad is, therefore, that energy demand has very small price elasticity, if any, but that it does have considerable income elasticity. Energy demand projections can prove wrong if too much emphasis is put on changes in demand or supply due to price increases. I think income or structural changes should be given primary consideration in order to make accurate demand forecasts.

RETHINKING ENERGY-CONSERVATION

Turning to energy conservation again, some experts believe that conservation in the narrow sense of the word - meaning technical <u>conservation</u> or reduction of unit energy consumption - can help a great deal in cutting back on overall use. The effect of such conservation varies, however, depending on how it is approached. To take a particular section of a production process, this effect will be considerably large. But the effect on the entire production process or the average effect for all industries is moderate.

Table 5 shows how unit energy consumption in major industries changed in the period from 1973 to 1975. Considerable efforts were made in all these industries to cut back on energy consumption in order to cope with the higher oil price. In the steel industry, however, unit consumption increased in nominal terms. This is due to a drop in the rate of capacity utilisation and a deterioration in the quality of coking coal. So, adjusted for these factors, the real unit consumption index stands at 99, as against the base figure of 100 for 1973. This unit consumption dropped in the steel industry in real terms, but only to a marginal degree. This means that conservation had only very small effect in reducing unit consumption. In the case of integrated steel companies who operate blast furnaces, however, this effect is a little larger than is indicated in the Table. But in the steel industry as a whole, including open-hearth and electric-furnace steelmakers, the effect of conservation is marginal. The situation is not much different in aluminium smelting, but in the cement industry unit consumption dropped more than 6 per cent. In the paper-pulp industry the unit ratio increased. Thus the results of energy conservation are scattered. But I am not saying that we do not have to make conservation efforts. On the contrary, we must step up efforts in this direction. In fact, we need an effective conservation policy because that is what the Japanese economy badly needs. But conservation is not a panacea. It doesn't provide any magic formula that would resolve all energy problems once and for all. In the industrial sector, for instance, it is not likely that the energy conservation ratio will increase markedly.

At this point, let me consider the larger question of putting the Japanese economy back on the course of stable growth. This question is central to the energy problem. More specifically, the question at stake

Table 5

ENERGY UNIT CONSUMPTION IN MAJOR INDUSTRIES(*) (1973 vs. 1975)

(data compiled by the energy conservation
division of the Advisory Committee for Energy)

		(1973 = 100)
Iron and steel	nominal	108
	adjusted	99
Aluminium smelting		100.6
Cement		93.7
Paper-pulp		104.8

*) Unit energy consumption means the amount of energy
used to produce one unit (quantitative unit) of
product.

is what sort of economic structure we will have after the economy is put
back on the orbit of stable growth. One significant development that
should be noted in this connection is that production in the secondary industry has declined since the oil crisis, while that in the tertiary industry has expanded. Thus there has been a shift in the relative importance of the secondary industry, with the tertiary industry taking up much
of the slack in the manufacturing sector. And this trend is expected to
continue. But I don't think this is the way to put the Japanese economy
on the course of stable growth.

Consequently, an upturn in the secondary industry is needed for the
continuing growth of the Japanese economy, although the kind of rapid
growth we have experienced in the past is out of the question. In other
words, if the Japanese economy is to achieve stable growth, energy demand
will have to increase again as a result of structural changes in Japanese
industry. Of course, the growth in energy demand will be curbed because
further progress will be made in energy-saving improvements or in technology development investments that will also help to reduce energy consumption. But the resultant drop in energy demand will have to be more than
offset by a general increase in such demand.

On this assumption, the rate of conservation in energy demand is
estimated at 6 per cent in the industrial sector in the reference case and
at 11 per cent in the accelerated policy case. However, the 11 per cent
conservation in the latter case seems too high. My feeling is that even
if considerable efforts are made the most we can achieve will probably be
about 6 per cent in the reference case. So I am pessimistic regarding the
target set in the accelerated policy case.

Table 6 shows how energy consumption changed relative to industrial production in major industrialised countries in the 10 years from 1965 to 1975. In other words, it indicates the elasticity of energy use to industrial output. More specifically, it shows much energy consumption changes when the industrial production index changes by one point. In Japan's case, the elasticity value for the 10-year period under review was 0.883, which

Table 6

ELASTICITY OF ENERGY CONSUMPTION TO INDUSTRIAL PRODUCTION IN MAJOR COUNTRIES (1965 vs. 1975)

(data compiled by the Institute of Energy Economics)

United States	1.425	France	0.499
Italy	1.220	United Kingdom	0.437
West Germany	0.926	Japan	0.883

is not very high. This compares with 1.425 in the United States, 1.220 in Italy and 0.962 in West Germany. The relatively low rate of increase in Japan's energy consumption seems to contradict the fact that the rapid expansion of the Japanese economy was supported mainly by the growth of heavy and chemical industries, which consume huge quantities of energy. In my opinion, energy consumption in Japan increased less rapidly than industrial production because of extensive rationalisation investments, including those designed to reduce energy use in these industries. Such investments were made in order to maintain the competitive position of Japanese industrial products on world markets.

In this connection I would like to point out that Japanese industries generally are more efficient than similar industries in other countries. A case in point is the steel industry. An efficient industry is like a man who is physically fit. This means that such an industry finds it extremely difficult to raise the efficiency further, just as it is very difficult for a trim person to reduce his weight further. So I think we should not underestimate the level of future energy consumption in the industrial sector.

As for the transport and household-commerce sectors, energy demand has been rising despite price increases. This also indicates that it will be extremely hard to reduce the demand in these sectors. Take the transport sector for example. At least in principle, energy conservation is to be promoted through a shift from individual transportation to mass transportation or from motor vehicles to railroads. In reality, however, not much progress is being made in this direction. This means that the structure of Japan's transport sector is so rigid that it permits little change. Mass transportation by means of railways or trucks and buses does not necessarily offer much economic advantage. People drive their own cars instead of taking buses, or goods are carried by trucks rather than by railways, not simply because doing so is fashionable these days but because some hard economic considerations are involved. It is such economic factors that make for a rigid structure in the transport sector. So it will be no easy task to alter this structure.

Particularly in Japan's case, any serious attempt to increase the weight of railroad transport will come up against the difficult problem of reconstructing the financially ailing Japanese National Railways (JNR). It is easy to speak of the need to alter the structure of the transport industry in favour of railroad service. But I do not think this will be accomplished, unless we come to grips with the problem of the JNR.

Regarding the household-commerce sector, energy demand also has been growing. The basic reason for this is, of course, because consumer life in Japan has improved in both quantitative and qualitative terms. And progress will continue to be made in this area.

Table 7 gives figures for per capita household energy consumption in 1975 in major industrialised nations. The United States, of course, is the largest energy-consuming country in this respect, followed by West Germany, the United Kingdom, France, Italy and Japan in this order. Japan is close behind Italy but far behind the other nations.

But Japan's energy use in this sector is expected to continue growing, considering that consumer life in this country is headed for further improvement, in terms of both quantity and quality, thus bringing us closer to the levels in other major industrialised nations. If the Japanese Government tries to curb consumption through prices policy it will have to raise energy prices to unrealistically high levels. In other words, it will be impossible to hold down energy demand if the rise in energy prices remains within reasonable bounds. My feeling is that we cannot arrest the rising trend of energy demand in this sector unless prices are raised several times or even dozens of times.

In general, it is difficult to lower one's living standard. In economic jargon say that it has "downward rigidity". There is a substitutive relationship between energy and labour. For example, a housewife can save her energy bill if she does the laundry by hand instead of by washing machine, or if she sweeps the floor by hand rather than by vacuum cleaner. By the same token, if you do away with your air conditioner and use a fan

Table 7

PER CAPITA HOUSEHOLD ENERGY CONSUMPTION IN MAJOR COUNTRIES (1975)

(data compiled by the Institute of Energy Economics)

United States	23,304 10^3 Kcal	France	11,253 10^3 Kcal
West Germany	14,774	Italy	7,408
United Kingdom	13,420	Japan	7,173

instead you can also cut back on energy use. But given the kind of free economy which we maintain, it will be next to impossible to expect the consuming public to forego such amenities of life and revert to such simple lifestyles.

Now, let me summarize what I have said. The first point I made is that the energy problem is not so simple as to be resolvable through supply and demand adjustments induced by changes in energy prices. Another point is that, although energy conservation is a very important question and therefore a very important policy objective there is no magic formula for it. It is a very difficult job, to say the least, but it is something that must be done.

TWO APPROACHES

The energy problem, taken as a whole, is extremely complex and involves many difficult questions. We have to consider it in the broad context of the world economy. We must take into account, for example, trends in the world economy, the intricate power relationship there, and economic policies of foreign countries. We must also consider political and military problems, the problem of natural resources, the problem of technology development, and social or psychological tendencies in various countries.

Paradoxical though it may sound, the more we explain the energy problem the more difficult to understand it becomes. It would be much easier to understand the energy problem if it were explained in simplistic terms − such as the argument that we will have no oil at all 30 years from now, or that we will have major power failures if we do not step up efforts to develop alternative sources of energy. Actually, the energy problem is a vastly complex problem. And if we try to explain all the complexities involved theproblem gets only more difficult to understand.

The complexity of the energy problem makes it all the more difficult to make reasonable forecasts concerning energy supply and demand. But this is a problem we must resolve in order to make sure that the Japanese economy is placed solidly on the stable growth path. This means that we must take the necessary measures now to ensure the hoped-for economic growth.

However, if the Japanese economy remains in the doldrums and expands at low rates of 4-5 per cent in, say, the next ten years, perhaps no serious problems will develop, at least outwardly, in the area of energy supply and demand. But the sustained, deep economic stagnation would cause various distortions in such areas as employment and corporate management. In these economic circumstances the energy problem would be "contained", as it were, because of the low rate of economic growth. So at least in appearance, the energy problem would be resolved. In other words, we would achieve a semblance of stability in this potentially unstable area.

So we face two choices or approaches. In my view, the right approach is to get the Japanese economy back on the course of stable growth. This should be our primary objective. And in order to achieve this goal the Japanese Government should think seriously what needs to be done to secure the necessary energy supplies or restrain the growth of energy demand.

Then, what is the meaning of "stable growth"? In terms of growth rate it is generally understood to mean that the Japanese economy should continue to expand in years ahead at the annual rate of about 6 per cent.

Table 8

AVERAGE CAR MILEAGES IN JAPAN (PRIVATE AND COMMERCIAL CARS)

(data compiled by the energy conservation division)

F.Y. 1973	8.78 km/l
F.Y. 1974	8.75 "
F.Y. 1975	8.78 "

It is unrealistic to anticipate, however, that the Japanese economy will grow by 6 per cent every year. So the growth rate is expressed in terms of an average value. An average annual growth rate of 6 per cent, for example, means that the economy probably will have to expand more than 7 per cent, perhaps close to 8 per cent, when business cycle is uptrend. Perhaps this much growth is needed in order to make up for slower growth in recessionary periods. But on average the economy is expected to achieve 6 per cent growth over a period of time.

Table 9

ENERGY CONSUMPTION ELASTICITY IN JAPAN (TO GNP)

(1) by year and energy

F.Y.	Elasticity to GNP		
	Petroleum	Electric power	Total energy
1960-1965	2.28	1.10	1.20
1965-1970	1.55	1.12	1.22
1970-1973	0.95	1.11	0.99
1970-1976	0.69	1.17	0.76

Notes: 1) Data compiled by the Institute of Energy Economics.

2) Figures under "petroleum" do not include consumption in the area of gas.

(2) by year and sector

F.Y.	SECTOR					
	Final demand sectoral total	Mining and manufac-turing	Energy	Trans-port	Agricul-tural indus-tries	Household-commerce
1960-1965	1.20	1.19	1.07	1.11	1.04	1.32
1965-1970	1.22	1.29	1.37	1.03	1.22	1.14
1970-1973	0.99	0.82	1.27	0.99	1.82	1.38
1970-1976(*)	0.76	0.44	1.28	1.00	0.62	1.37

*) Figures for 1976 are estimated.

Notes: 1) Energy consumption elasticity values in the "tentative fore-cast". 1975-85 0.89 (accelerated policy case), 1985-90, 0.74.

2) Data compiled by the Institute of Energy Economics.

Now, let me turn briefly to the question of energy conservation ratio. As I have already said, the ratio is likely to reach 5.5 per cent at the most, as estimated in the reference case. So energy demand is bound to increase when the Japanese economy enters a period of stable growth. The question here is how we should secure the necessary volume of energy to meet the increased demand. Of course, petroleum will be the first thing we have to consider in this respect, because oil remains the primary source of energy. So we must secure adequate oil supplies first. But it is ex-tremely difficult to estimate with any reasonable degree of accuracy how much oil we must import. However, 7.4 MB/D is probably what we can take as a good rule of thumb.

Table 10 gives some of the figures published by the OECD in the World Energy Outlook. The basic assumption is that the total productive capacity of OPEC will reach 45 billion bbl. per day in 1985. On this assumption,

Table 10

WORLD OIL SUPPLY AND DEMAND FORECAST (OECD)

(Unit: million bbl./day)

	1985	
	Standard case	Accelerated policy case
Imports into Japan	8.7	7.6
Imports into United States	9.7	4.3
Imports into OECD, Europe	14.7	11.0
Imports to OECD, other regions	1.9	1.5
(OECD total)	(35.0)	(24.4)
Imports into Communist bloc	0.8	0.8
Imports into non-oil LDCs	4.2	4.2
Imports into non-OPEC oil-producing	3.8	3.8
Countries inventory buildup	0.5	0.5
(subtotal)	(0.1)	(0.1)
Total import demand	35.1	24.5
Consumption in OPEC countries	4.2	4.1
Required output in OPEC	39.3	28.6
Productive capacity Saudi Arabia	15.0	15.0
Productive capacity Iraq	7.0	7.0
Productive capacity Iran	6.0	6.0
Productive capacity, other OPEC nations	17.0	17.0
(OPEC capacity)	(45.0)	(45.0)

Notes: 1) World Energy Outlook.

2) The standard case is where energy conservation and alternative energy development will not be positively promoted, while in the accelerated policy case countries will positively carry out their energy policies.

3) Targets set by the IEA ministerial meeting in October 1977: IEA total 26 million B/D (28 million B/D including France) United States 6, Japan 7.4, EC 10 (12 including France), other countries 2.6 (figures by country are estimated).

the OECD estimates that production in the OPEC area will have to be kept at the level of 28 to 29 MB/D if oil-consuming countries are to maintain a degree of surplus – that is, if we want to see the world oil market change from seller's market to buyer's market. For this purpose, however, energy demand will have to be limited, as projected in the OECD's accelerated policy case. On this score, the IEA adopted a similar approach at its ministerial meeting in October 1977. Specifically, the council agreed to

set total oil import ceiling for its member countries in 1985 at 26 MB/D.
Although this target is not binding on the member countries, the understanding is that they will make efforts to limit their oil imports within
this framework. No official information has been published about a breakdown of this 26 MB/D figure. But estimates made on the basis of the
available information produce the following breakdown: 6 MB/D for the
United States, provided President Jimmy Carter's energy programme proves
effective; 7.4 MB/D for Japan in the accelerated policy case; 10 MB/D for
the EC (12 MB/D including France, which is not an IEA member) and 2.6 MB/D
for the other member countries.

In Japan's case, there is a possibility that actual imports might exceed 7.4 MB/D. Nor is it likely that United States imports will be limited
to 6 MB/D. Probably the figure will exceed 10 MB/D. So it seems extremely
difficult to attain the IEA target.

The next question is what needs to be done to improve the supply structure of energy other than oil. This can be done in a number of ways, by
using a combination of different types of energy at different levels of
consumption. For reference, chronological changes in Japan's energy supply structure are presented in Table 11.

Table 11

SUPPLY STRUCTURE OF PRIMARY ENERGY (%)

(data compiled by the Institute of Energy Economics)

(1) Chronological changes	Petroleum	Coal	Hydropower	Other	Total
1933 (1/10 of 1975 consumption)	8	69	12	11	100
(1948 ")	5	60	21	14	100
1966 (1/2 of 1975 consumption)	60	26	11	3	100
1975	72	18	6	4	100

Three different years - 1933, 1948 and 1966 - were selected, in part
because energy consumption in the first two years happened to be one-tenth
that in 1975. If we were to consume now the same amount of energy that we
used in 1933 and 1948, then we would not have to import any energy. This
is because Japan currently produces about 10 per cent of its total energy
needs. In those two years, petroleum made us 5-8 per cent of the total

energy supply, coal more than 60 per cent, hydropower 12-21 per cent, and other sources of energy such as firewood and charcoal more than 10 per cent.

In the last of the three years selected, energy consumption was half the level in 1975. In 1966, we depended on petroleum for as much as 60 per cent of our total energy needs. Coal made up 26 per cent of the total, hydropower 11 per cent, and natural gas and other types of energy about 3 per cent. I should think that this is a relatively stable supply structure comparing to present situation.

In other words, if our energy consumption is about half the present level the Japanese economy will be able to maintain a rather stabilized energy supply structure comparable to those in other major industrialised nations. Actually, Japan's present energy supply structure is highly vulnerable, because we maintain a very high level of energy consumption in a small country with a very large population.

Table 12 indicates differences in the energy supply structures of major industrialised countries including Japan. To take petroleum as an example, the rate of dependence in Japan in 1975 was much higher than those in West Germany and France, to say nothing of the United States.

The Table shows clearly that Japan has an unstable supply structure as compared with other major industrialised countries. As we have already seen, this structure has become increasingly unstable over a period of time.

As with Japan, other countries also have energy programmes designed to improve their supply structures. Table 13 shows how energy consumption is estimated to change in Japan, Germany and France in 1985. In Japan's case, figures for both 1985 and 1990 given in the accelerated policy case of the "tentative forecast" are presented. These are compared with 1985 estimates for Germany and France. Though the Japanese pattern in 1985 is much better than now and is expected to become more stable in 1990, in the case of petroleum, the rate of dependence in Japan both in 1985 and 1990 is still considerably higher than in Germany and France. So the vulnerability of Japan's energy supply structure will stay at a relatively high level.

It is essential, therefore, that we do our best to achieve at least the targets set in the accelerated policy case. But we will have to overcome great difficulties in order to attain these goals. It is regrettable in this connection that not much positive effort has been made in this direction since the "tentative forecast" was prepared in June 1977. Current moves suggest that our dependence on petroleum will either stay at the present level or drop only slightly below it in the future. This

is why it is difficult to achieve the accelerated policy targets, despite
the strong need to attain them.

Table 12

ENERGY SUPPLY STRUCTURES IN MAJOR COUNTRIES (1975)

(data compiled by the Institute of Energy Economics)

	Petroleum	Coal	Hydropower	Natural gas	Nuclear	Total
United States	44	20	4	30	2	100
United Kingdom	45	35	1	16	3	100
West Germany	53	29	2	14	2	100
France	65	15	8	10	2	100
Italy	69	8	7	15	1	100
Japan	72	18	6	2	2	100

Table 13

ENERGY PROGRAMMES IN MAJOR COUNTRIES

(data supplied by the Institute of Energy Economics)

	Petro-leum	Coal	Hydro-power	Natural gas	Nuclear	Other	Total
Japan, accelerated case, 1985	66	15	4	7	7	1	100
Japan, accelerated case, 1990	57	16	4	8	11	4 (incl.	100
West Germany, 1985	45	22	–	18	13	2 hydro-power)	100
France, 1985	42	11	6	16	24	1	100

Table 14 was prepared from a more realistic point of view, with these
difficulties in mind. That is, the Table indicates a probable pattern of
Japan's energy supply in 1985, on the assumption that the energy situation
will remain largely unchanged. The overall projection is that the primary
energy supply will reach 660 million kiloliters as calculated in terms of
oil. What this means is that the economic growth rate may drop to less
than 5 per cent, depending on what kind of energy supply system we will
have.

Even in the probable case cited in Table 14, the rate of dependence
on imported oil will stay at nearly 70 per cent. So vulnerability in
energy supply will remain high. Consequently, in order to achieve

Table 14

PROBABLE ENERGY SUPPLY PATTERN IN 1985

		(reference case)	(reference) (accelerated case)
Hydropower	2,000 million kW (3.3%)	1,950	2,250
Geothermal	20 million kW (0.1%)	50	100
Domestic oil and gas	800 million kl (1.2%)	800	1,100
Domestic coal	1,700 million t (2%)	2,000	2,000
Nuclear	2,600 million kW (6.2%)	2,600	3,300
LNG	2,400 million t (5.5%)	2,400	3,000
Imported steam coal	800 million t (0.8%)	600	1,600
Imported oil	4.3 billion kl (69.4%)	5.05	4.32
(Imported coking coal)	(8,600 million t)	(8,700)	(8,600)
Total	6.2 billion kl (oil equivalent)	7	6.6
Conservation	0.4 billion kl (5.5%)	0.4 (5.5%)	0.8 (10.8%)
Grand total	6.6 billion kl	7.4	7.4

○740 million kl 660 million kl (down 11 per cent), corresponding to growth rate of about
○4.5 per cent.

Vulnerability in energy supply will remain high.

210

annual average economic growth of 6 per cent we must make much more effort to secure adequate energy supplies. Otherwise, the Japanese economy could, in the worst event, enter a long period of stagnation. And in this process of economic slowdown, ironically enough, would not reach serious proportions but stay largely in the background.

Table 15 shows what needs to be done in order to reduce our dependence on foreign oil to at least about 60 per cent. For the time being only three types of energy - imported steam coal, nuclear energy and LNG - can be substituted for petroleum. There is a view in Japan as well as in the United States that there is a "soft path" to oil substitution - namely, that we can use solar energy, wind power, and geothermal power, for example. Energy conservation is another "soft path", according to this view. Those who think this way point out the difficulties involved in the "hard path" such as nuclear energy. My own view is that these "soft" types of energy may have some particular areas but are far from sufficient to meet the growing energy demand in Japan. Therefore, the "soft path" approach is unrealistic.

Table 15

TWO CASES IN WHICH IMPORTED OIL DEPENDENCE CAN BE REDUCED TO 60 PER CENT

(1) Where one of the following types of energy is substituted for oil		
Imported steam coal	8 million tons	102 million tons
Nuclear energy	26 million kW	66.4 million kW
LNG	24 million tons	66.4 million tons
(2) Where the following types of energy are substituted equally for oil		
Imported steam coal	8 million tons	39 million tons
Nuclear energy	26 million kW	39 million kW
LNG	24 million tons	38 million tons

The realistic approach is to rely on the three types of alternative energy I just mentioned. This is the only way to reduce our heavy dependence on imported oil. But if we were to substitute only one type of energy for petroleum in order to cut the rate of dependence to 60 per cent or so, the supply of that particular energy would have to be expanded inordinately. For instance, we would need more than 100 million tons of imported fuel coal, or 66 million kW of nuclear energy or 66 million tons of LNG. What if we split the burden equally among the three types of energy? In this case, we would have to use 40 million tons of imported fuel coal, 40 million kW of nuclear energy and 40 million tons of LNG. These are also unrealistically large figures. All of this shows that petroleum carries overwhelmingly heavy weight in Japan's energy supply structure. To use a popular analogy, oil is to alternative energy what a sneeze is to a cold. If oil "sneezes", alternative energy will "catch cold".

So it is extremely difficult, I must say, to stabilize Japan's present economic structure, particularly the energy supply structure. This is all the more reason why we must attack this problem. The right approach we should take in our energy strategy is to reduce the vulnerability in our energy supply in ways compatible with stable economic growth.

CONCLUSIONS - FIVE PROPOSALS

What are the key points in future energy strategy?

The first is the implementation of large-scale expansionary economic measures centering around energy-related investment and the stabilization of the Japanese economy. Of course, various measures are being carried out to promote economic recovery, and further measures are scheduled for implementation in the immediate future. However, I do not think that these measures will be sufficient to put the economy on the stable growth path.

Table 16

FIVE PROPOSALS TO MAKE JAPAN INDISPENSABLE TO THE WORLD

---

1. Implementation of large-scale expansionary economic measures center-ing around energy-related investment and stabilization of the Japanese economy.

2. Reducing the instability in oil import:
    a) Expanding Government stockpiles of oil and establishment of emergency supply system.
    b) Reducing the dependence on foreign oil on a gradual basis.

3. Deployment of comprehensive alternative energy policy:
    a) Drawing a "borderline" between oil and alternative energy.
    b) Improving the receiving system for imported coal, imported LNG and LPG.
    c) Promotion of nuclear energy policy, particularly strict compliance with the policy of peaceful use and drastic improvement of the safety surveillance system.

4. Strengthening the energy industry in response to changes in the energy supply structure.

5. Building Japan into the world as an integral part of it.

My view is that greater measures should be taken to stimulate internal demand. In this case, priority should be given to energy-related investments as an essential means of ensuring a steadier economic recovery.

The second key point in future energy strategy is the reduction of the instability in the oil import situation. An isolated economic giant like Japan must be prepared at all time for any possible change in the oil supply. In this respect it is particularly important for the Japanese Government to expand its oil stockpiles. Beyond this, the Japanese public should be educated on the importance of oil stockpiles as a means of ensuring the economic security of Japan. On the strength of such a favourable public consensus, Government stockpiles should be drastically expanded.

It is also necessary to establish as soon as possible an emergency oil supply system designed to avoid any dislocations in the domestic economy in cases of emergency. This is because member countries of the IEA, including Japan, will be required to restrain domestic oil demand - and this will include rationing - if and when oil supply in the world is restricted. The IEA is now preparing an international energy programme including oil sharing in times of emergency.

And, of course, it is basically important to reduce our dependence on foreign oil on a gradual basis. For this purpose we must step up efforts to promote the development and use of alternative energy. The main thrust of such efforts should be directed to the independent development of oil and natural gas.

The third point to be emphasized with respect to energy strategy is that Japan should pursue a comprehensive alternative energy policy. There is a great deal of uncertainty over the future availability of oil in the world. There is indeed a fear that if demand in oil-consuming countries exceeds a certain level they will start scrambling for this essential energy. If this happens, major industrialised nations in the Free World will be split in two, and the Western world will be thrown into confusion, politically and economically. So, in order to avoid such a calamity these countries should stay together. In fact, this is the basic reason why the IEA is calling for co-operation among its member countries. According to some energy experts, country-by-country oil rationing will become necessary when an international oil scramble develops.

This makes it imperative that Japan reduce its heavy dependence on foreign oil. However, Japanese policy regarding alternative energy is such that there is no clear-cut relationship between oil and its substitutes. For instance, if nuclear power output falls short of the target, the shortage will be covered by increasing oil imports. In fact, this is what has happened. Oil imports also have been increased to make up for fuel shortages at electric power plants, because not much progress has been made in the use of coal. By the same token, delays in LNG imports have been covered also by increasing oil consumption.

What is needed here is a clearly defined substitutive relationship between oil and alternative energy. An energy shortage resulting from delays in nuclear power plant construction should be covered not by increasing oil imports but by using more alternative energy, such as coal and LNG. In other words, we must establish something like a borderline between oil and alternative energy so that oil will not be used easily to substitute for other types of energy. Otherwise, Japan will not be able to deal effectively with the changing oil situation in the world. This is an important point.

This points up also the importance of improving the system for receiving imported coal or imported LNG and LPG. But more important is the role of nuclear energy. Any energy programme in Japan cannot be effective unless nuclear energy is incorporated. What we need at this point is bold policy initiatives so that nuclear power development in this country will get really under way. Nuclear energy policy, of course, involves more complex factors than in the case of other types of energy. In my view, existing policies should be re-examined in order to expand nuclear power output while promoting the public acceptance of nuclear energy development.

Two points must be made in this respect. The first is that Japan should commit itself completely to the peaceful use of nuclear energy. This is, however, easier said than done. It is difficult to convince other countries that Japan will never use nuclear power for military purposes. What Japan should do under the present circumstances is to participate more actively in the International Nuclear Fuel Cycle Evaluation (INFCE). Japan should promote nuclear energy development and utilisation by dealing flexibly with various problems that have been presented in connection with this programme. We should try to develop a sense of balance internationally so that we can promote co-operation with other countries.

The other point that must be made is that we should establish a better safety surveillance system. Opponents of nuclear energy claim that the present system is unreliable. But I do not agree. The system as it is operated now is already sufficiently reliable to maintain its authority. But, considering that nuclear energy is the most important type of alternative energy, and one which needs to be developed by winning the public acceptance of it, it would be essential that we have a still better surveillance system.

Regarding the question of strengthening the energy industry in response to changes in the energy supply structure, one thing that is likely is that such changes may weaken the financial health of individual enterprises that comprise the energy industry. As a result, it may become necessary to raise energy prices on a continuing basis. What is at stake is not only short-term problems such as the one involving foreign exchange gains, a matter which has invited criticism from some quarters. Also at stake is the longer-term question of improving the business foundations of the energy industry and its member enterprises.

This is particularly true of the oil industry, in which the capital cost is expected to rise as a result of increased capital investment. The cost of maintaining oil stocks is also rising. Moreover, if the energy supply structure is stabilized in the future, oil demand will either level off or decline. And this will happen at a time when depreciation reaches a peak. Thus, the oil industry must make huge investments even though the oil market is expected to stop expanding or shrink at some future date. Certainly this does not make economic sense, at least under normal circumstances.

Yet, in the situation described above, Japan's energy industry, particularly the oil industry, could become a "second JNR (Japan National Railway"). In the early part of the decade beginning in 1965, the JNR's financial position was expected to deteriorate, but the corporation made large-scale investments, such as the project to build Shinkansen (super express) bullet-train lines. Now the JNR's financial house is in a shambles. In order to avoid the mistake made by the national railways it is absolutely necessary to formulate a long-range programme designed to improve the financial health of the oil industry and component enterprises. Of course, it is essential that such a programme be carried out effectively.

Finally, let me say that Japan's vulnerability in the area of energy supply is something we cannot get over with completely, no matter what energy policy is adopted. But isn't there something we can do? Yes, there is. I think the best thing to do, after all has been said, is to build a kind of interlocked system that would make the Japanese economy indispensable to the world economy. Under such a system of interdependence, dislocations in the Japanese economy due to major energy shortages would have a great impact on the world economy. To put it the other way round, it would serve the interests of the world economy to keep the Japanese economy stable rather than drive it into a corner.

I think that establishing such a structure of interdependence is the only way by which Japan can ensure her economic survival while depending on imported oil. In order to make such high energy vulnerability compatible with its stable economic growth, Japan should step up efforts to internationalise the Japanese economy – namely, to integrate its economy more closely into the world economy. It is my hope that if we look at the energy problem from this angle we will be able to find new ways to solve it.

ENERGY DEMAND

AUTHORS AND PARTICIPANTS IN WORKSHOP ON ENERGY DEMAND

Mr. A. Bain
  Imperial Oil
  Toronto, Canada

Dr. O. Bernardini
  Montedison
  Milan, Italy

Prof. E. Berndt
  M.I.T.
  Cambridge, Mass.

Mr. B.J. Choe
  World Bank
  Washington, D.C.

Mr. A. Clarke &
  Mr. du Moulin
  Shell Oil
  London

Mr. L. Da Silva
  Chief, Industrial Economics
  and Infrastructure Section
  Inter-American Development Bank
  Washington, D.C.

Dr. R. Eden
  Head, Energy Research Group
  Cavendish Laboratory
  University of Cambridge
  England

Mr. J-R Frisch
  Attaché à la Direction Générale
  Electricité de France
  Paris

Mr. Friese
  Swiss Aluminium
  Switzerland

Prof. J. Griffin
  University of Houston
  Texas

Mr. P. Hesse
  Manager of Purchases
  Union Carbide Europe S.A.
  Geneva, Switzerland

Dr. H. Kraft
  MVW
  Hamburg, Germany

Mr. Matthews
  Forecasting Division
  BP Trading, London

Dr. D. Schmitt
  Cologne University
  Germany

Mr. D. Sternlight
  Atlantic Richfield Co.
  Los Angeles, Calif.

Prof. J. Sweeney
  Stanford University
  Palo Alto, Calif.

Prof. L. Waverman
  University of Toronto
  Institute of Policy Analysis
  Toronto, Canada

Mr. C. Waeterloos
  EEC - Directorate-General for
  Energy
  Brussels

THEORETICAL APPROACHES TO ENERGY DEMAND

FACTOR MARKETS IN GENERAL DISEQUILIBRIUM: DYNAMIC MODELS OF THE INDUSTRIAL DEMAND FOR ENERGY

by

E.R. Berndt, M.A. Fuss and L. Waverman(*)

INTRODUCTION

Existing models of firms' demand for inputs can be viewed as either based on static optimisation with instantaneous adjustment or as myopic optimisation with constant marginal costs of adjustment. Such models suffer from several serious drawbacks. First, it is quite unlikely that adjustment of firms to desired input levels takes place instantaneously or that marginal costs of adjustment are constant (i.e. that supply curves of inputs are perfectly elastic). Second, if one views the world as always being in long-run equilibrium, one will ignore the adjustment process itself. Clearly the adjustment process is of prime importance, not only for the planning of an individual firm, but also for the conduct of public policy. For example, if energy prices triple, the long-run equilibrium analysis is unable to address itself to the following crucial issues: How long will it take firms to adjust to the new higher energy prices? How could the implementation of certain investment incentives affect the speed of adjustment? Will the short-run response to the higher energy prices be one which increases unemployment, while the long-run solution would involve a more labour-intensive economy? Clearly, to address itself to these issues, one must model the nature of the disequilibrium process and the costs of adjustment to the new set of optimal factor demands. Since short-run responses are likely to differ from long-run responses, forecasts of factor demands are likely to be misleading unless they explicitly incorporate the adjustment process.

*) This is an abridged version of a report funded by the Electric Power Research Institute, Palo Alto, California (EPRI EA-580, Project 683-1, Interim Report, November 1977). All errors or omissions are the fault of the authors. The views expressed are not necessarily the views of EPRI.

We take two approaches in developing dynamic disequilibrium models of factor demand. First, we incorporate ad hoc Koyck adjustment matrices in an essentially static system of interrelated factor demand equations, while imposing appropriate sets of constraints on both the final equilibrium and the adjustment path. There are a number of severe theoretical and econometric problems with this generalisation of single equation ad hoc adjustment models, as will become clear in subsequent analysis.

We then proceed to derive an entirely different theoretical model which incorporates costs of adjustment into the long-run optimisation process rather than appending them in constant form. The firm is envisioned as operating with a set of variable factors and a set of quasi-fixed factors. Quasi-fixed factors are subject to increasing marginal costs of adjustment. These rising factor supply curves exist because of costs, which to the firm, are either external (delivery delays, etc.) or internal (competition between production and investment for use of the firm's resources). Utilising restricted profit functions, we derived a set of simultaneous factor demand equations. The speed of adjustment for quasi-fixed factors is endogenous, non-constant, and in particular is affected by the interest rate, the user cost of the quasi-fixed factor, technological parameters, and investment tax policy.

CHARACTERISTICS OF THE ADJUSTMENT PROCESS

Under the assumption of <u>instantaneous adjustment</u>, it is assumed that firms and industries respond immediately and fully to changes in input prices and output levels. Since almost all time-series empirical implementations of factor demand models use either quarterly or annual data, the instantaneous adjustment assumption in effect is equivalent to assuming that, within the time span of one quarter or one year, respectively, firms adjust fully and completely to their desired or equilibrium factor demand levels.

There are a number of reasons why this instantaneous adjustment assumption is likely to be violated. First, firms often find themselves with long-lived equipment or buildings, and typically find it suboptimal in the short run to dispose of such long-lived assets, or to purchase immediately additional physical capital. For example, delivery lags or borrowing constraints are a not uncommon experience. Second, firms are often under contractual agreements with salaried employees, other workers, suppliers of materials, etc. and thus downward adjustment might be costly and likely to involve a considerable period of time. With respect to expansive adjustment, firms typically need time to draw up plans and to take bids for new construction, to recruit and train additional personnel, and to obtain commitments on additional supplies of purchased materials. No doubt other reasons could be given. An implication of these few remarks, however, is that the assumption of instantaneous adjustment is not likely to be realistic, and that an assumption of lagged adjustment might be more reasonable and reliable.

The lagged adjustment process can be specified in a number of ways. Before doing so, however, we believe it useful to develop criteria for evaluating alternative specifications of the lagged adjustment process. First, the lagged adjustment process specification should incorporate the optimising behaviour of economic theory. In particular, we suggest that it is important to recognise explicitly in model specifications that firms take into account costs of adjustment to "equilibrium" values, and that speeds of adjustment are endogenous choice variables rather than exogenous and predetermined.

Second, the lagged adjustment process specification should allow for the possibility of "general disequilibrium". For example, suppose there suddenly occurs a sharp increase in the price of a particular type of energy - say, natural gas - while other input prices remain fixed. The competitive firm would of course prefer to reduce its demand for natural gas and perhaps substitute towards electricity (i.e. move instantaneously to its new equilibrium position), but because its equipment is "tied" to natural gas, it cannot do so immediately. Thus the lagged adjustment of the firm to reduced natural gas consumption implies a disequilibrium in the electricity demand market (most likely electricity consumption would initially be less than optimal) and in the demand for capital equipment market (a shift in demand from natural gas-using to either electricity-using or dual capacity equipment). The specification of lagged adjustment in the electricity demand equation must therefore take account of disequilibrium in the natural gas and capital equipment demand equations, i.e. lagged adjustment specifications must not be isolated, but must take into account "general disequilibrium". It is generally not possible for only one factor market to be out of equilibrium.

Third, the specification of lagged adjustment should be consistent with the basic Le Chatelier principle, which in this context states that short-run own price elasticities should be smaller in absolute value than long-run own price elasticities. Intuitively, the longer the period of time allowed the firm for adjustment, the greater the extent of adjustment. In particular, the Le Chatelier principle is inconsistent with specifications that permit short-run own price elasticities to be larger in absolute value than long-run own price elasticities.

Finally, the specification of lagged adjustments should ensure that output feasibility constraints are satisfied throughout the disequilibrium process. For example, suppose that observed capital, labour, and electricity demands are all different from their equilibrium values. The specification of the lagged adjustment mechanism should reflect the constraint that the observed output could be produced by the observed levels of capital, labour, and electricity. An extreme counter-example is the case when all observed input levels are less than their equilibrium levels for the given output; clearly these input levels are not sufficient to produce the observed output. Thus it is important that the specification of the lagged adjustment mechanism explicitly incorporate the production feasibility constraints.

AD HOC ADJUSTMENT AND THE DYNAMIC TRANSLOG

In our first approach, we will incorporate an ad hoc adjustment process within an essentially static model. Only a summary of the conceptual development and empirical results will be given here. For a detailed exposition see Berndt, Fuss and Waverman (1977). We assume that the cost function can be approximated by a translog cost function so that the optimal long-run input demand functions are given by the cost share equations:

$$M^*_{it} = \frac{\hat{w}_{it} V_{it}}{C_t} = \delta_i + \sum_j \nu_{ij} \ln \hat{w}_{jt} + \nu_{iY} \ln Y_t \, . \quad \begin{array}{l} i = 1, \ldots, m \\ t = 1, \ldots, T \end{array}$$

where $\hat{w}_{it}$ is an input price, Y_t is output and δ_i, ν_{ij} and ν_{iY} are parameters of the cost function.

Following the Nadiri and Rosen (1973) generalisation of the single equation Koyck adjustment model we specify the ad hoc adjustment process as:

$$M_t - M_{t-1} = B(M^*_t - M_{t-1})$$

where M_t, M^*_t are m x 1 vectors of actual and long-run optimal shares respectively, and B is an mxm constant coefficient adjustment matrix. If $B = I_m$, then $M_t = M^*_t$, i.e. adjustment of cost shares to their equilibrium levels is instantaneous. Since the M_{it} are cost shares, it is required that they add to unity for each time period t. As shown by Berndt-Savin (1975) in the context of autoregressive models, this "adding up condition" implies that the column sum of B must be the same constant for all columns in B. When B is assumed to be diagonal, this condition requires that the diagonal elements of B must all be equal, i.e. all cost shares must adjust at the same rate.

The basic problem with the cost share adjustment specification is that it does not necessarily imply that input _levels_ are adjusting to their desired values. For example, cost shares could be in equilibrium, but expenditures on all inputs could be too large by a factor of λ. This would imply (with exogenous input prices) that levels of all input demands would be λ times as large as desired, even though cost shares would be in equilibrium. Nevertheless, the static translog model has recently been used widely[1] and therefore we believe it is of interest to examine the theoretical and empirical implications of a dynamic ad hoc generalisation of the static model.

EMPIRICAL ESTIMATION OF THE DYNAMIC TRANSLOG

To illustrate the dynamic translog specification, we employ data on input services of capital (K), labour (L), aggregate energy (E), and other non-energy intermediate materials (M) for United States manufacturing, 1947-71. /E.R. Berndt and David O. Wood (1975)7. Since the partial adjustment model includes as regressors lagged values of the dependent variables, it is necessary to drop the first observation; hence all estimates are based on 1948-71 data. Our estimation method is maximum likelihood, based on the algorithm outlined in E.R. Berndt, B.H. Hall, R.E. Hall, and J.A. Hausman (1974). These parameter estimates are asymptotically consistent and efficient.

Assuming that the u_t^m disturbance vector is independently and identically normally distributed with mean vector zero and constant singular covariance matrix Ω, we estimate parameters in (10) under three alternative adjustment specifications: (deleting one equation due to singularity of Ω)

   i) adjustment is instantaneous;
  ii) the adjustment matrix is diagonal which implies that the diagonal elements must be equal;
 iii) B = a full matrix with no restrictions on the adjustment parameters. As noted in Berndt et al. (1977) it is not possible to identify the individual parameters in B unless prior restrictions are imposed on B. In this model such identifying restrictions are not imposed. However, characteristics of production such as short- and long-run price responsiveness are identified.

---

1) See, for example, Hudson and Jorgenson (1974), and Fuss (1977).

For these three specifications we report the implied short- and long-run price elasticities. In the first column of Table 1, we present own price elasticity estimates of the translog function assuming instantaneous adjustment. These elasticity estimates are quite stable over the 1948-71 sample period, and therefore we report only the 1971 estimates. The 1971 figures reported in the first column of Table 1 correspond very closely with those reported by Berndt-Wood $\underline{/}$(1975), Table 5, p. 265$\underline{/}$.

Table 1

ESTIMATED OWN PRICE ELASTICITIES WITH ALTERNATIVE AD HOC
DYNAMIC ADJUSTMENT SPECIFICATIONS AND THE TRANSLOG COST FUNCTION
UNITED STATES MANUFACTURING DATA, 1948-71
(Elasticity Estimates for 1971)

Estimated Elasticity	Instantaneous Adjustment Long Run = Short Run	Diagonal Adjustment Matrix		Full Adjustment Matrix	
		Long Run	Short Run	Long Run	Short Run
ϵ_{KK}	-.25	-.19	-.27	-.76	-.41
ϵ_{LL}	-.46	-.46	-.48	-.46	-.43
ϵ_{EE}	-.49	-.47	-.51	-.70	-.69
ϵ_{MM}	-.24	-.24	-.26	-.24	-.15

In the second and third columns of Table 1 we report estimated short- and long-run price elasticities based on the diagonal adjustment matrix. The column figures illustrate that our translog model with a diagonal adjustment matrix can produce results inconsistent with the Le Chatelier principle. For example, the short-run own price elasticity estimates are all larger in absolute value than the long-run estimate.(1)

---

1) Suppose the constant adjustment parameter is k. Comparing price elasticities for the short run ($0 < k < 1$) and the long run ($k = 1$), after cancelling out M_{it} (which is assumed to be approximately equal in the two cases) we obtain:

$$\epsilon_{ii,t} \big| k=1 - \epsilon_{ii,t} \big| 0 < k < 1 = \gamma_{ii} - k \gamma_{ii} = (1 - k)\gamma_{ii}.$$

If $\gamma_{ii} > 0$, then the long-run own price elasticity would be smaller in absolute value than the short-run own price elasticity.

But, $\gamma_{ii} = \dfrac{\delta \ln M^*_{it}}{\delta \ln \hat{w}_{it}}$

Where $\gamma_{ii} > 0$, the demand for the ith input is <u>share inelastic</u> and the Le Chatelier principle is likely to be violated.

We now generalise our adjustment matrix and permit non-zero off-diagonal elements, i.e. we now allow for "general disequilibrium". We first test the null hypothesis of a diagonal adjustment matrix against the alternative hypothesis of a full adjustment matrix. We obtain a likelihood ratio test statistic of 28.128; the chi-square critical value for the eight restrictions at the .01 level of significance is 20.090. Thus we must reject the restrictions of the diagonal model. Intuitively, our results suggest that the cost in terms of goodness of fit of imposing a diagonal adjustment matrix is very high. However, once one imposes the restrictive diagonal specification, a further test indicates that instantaneous adjustment cannot be rejected. The additional cost of imposing an instantaneous adjustment model is negligible.

Unfortunately, the characteristic roots of the estimated full adjustment matrix do not satisfy the conditions for stability; the real root is .2687 but two complex roots occur: 1.1075 + .12326i and its complex conjugate 1.1075-.12326i. Thus, point estimates of the characteristic roots lie outside the unit circle. It is not possible to test whether they are "significantly" outside the unit circle, for the distribution of the roots under the alternative hypothesis is not known.

Even though the estimated full adjustment model does not appear to be stable, for reasons of completeness we compute the implied short- and long-run price elasticities. These elasticities are presented in the final column of Table 1. All long-run own price elasticities are larger in absolute value than the short-run elasticities.

In summary, results based on the generalised ad hoc adjustment model suggest that (i) the diagonal constraint is too restrictive, but that if one insists on a diagonal adjustment matrix there is little additional cost in assuming instantaneous adjustment; (ii) with the diagonal specification estimated own price elasticities fail to satisfy the Le Chatelier principle; and (iii) the estimated full adjustment model provides a considerably better fit, but the estimated adjustment matrix fails to satisfy conditions for stability and convergence of the adjustment process. Based on these results, the generalised ad hoc specification does not appear to be a viable dynamic specification.

DYNAMIC FACTOR DEMAND SYSTEMS WITH EXPLICIT COSTS OF ADJUSTMENT

The ad hoc dynamic models analysed in the previous sections can be interpreted as approximations to cost of adjustment models with increasing marginal costs of adjustment. The limitations of the ad hoc model with its constant adjustment matrix, /I.e. where the path to the steady state is independent of the exogenous variables (prices)/, are extremely stringent /Treadway (1974)/ and are unlikely to be met in practice. While this problem has been examined in the theoretical literature; to our knowledge, there exists no empirical analysis which relaxes these restrictions.

In this section, using recent advances in the duality literature /Lau (1976)/ we are able, at least in the case of one quasi-fixed factor, to implement Lucas' (1967) and Treadway's (1971) theoretical cost of adjustment models with variable adjustment paths. The procedure is first to generate from normalised restricted profit functions the short-run demand for variable factors. Since these short-run demands are derived from optimising behaviour restricted by the quasi-fixed factor, they will be on the efficient frontier (as long as the restricted profit function satisfies the usual regularity conditions). The feasibility problem discussed earlier which is associated with ad hoc partial adjustment models thus is not present in this analysis. Second, the demand for the quasi-fixed factor can be obtained by explicitly solving for the adjustment path as is done in Lucas. The resulting system can then be estimated using systems of equations estimation procedures.

The model presented in this chapter satisfies the four criteria established for a valid dynamic system. As we will show below, the adjustment process arises explicitly from the application of optimisation theory and speeds of adjustment are endogenous choice variables dependent on the parameters of the production and cost of adjustment functions, and on the exogenous price variables. Second, the lagged adjustment process allows for the possibility of "general disequilibrium" with respect to deviations from long-run equilibrium. While the fully variable factors are in short-run equilibrium at any point in time, their values will differ from those taken in long-run equilibrium (unless the quasi-fixed factors are themselves in long-run equilibrium). The "disequilibrium" adjustment paths of the quasi-fixed factors are specified explicitly. The net result is that any disequilibriating shock which results in all factors being out of long-run equilibrium is consistent with the model. Third, as will be demonstrated below, the basic Le Chatelier principle that short-run own price elasticities should be smaller than long-run own price elasticities is satisfied in this model. Intuitively this must be the case, since the short-run/long-run dichotomy in our model corresponds to Alfred Marshall's distinction which provided the first application in economics of the Le Chatelier principle. Finally, as mentioned above, output feasibility constraints are satisfied. The normalised restricted profit function, which is the basic building block of our model, incorporates output feasibility (and input efficiency) since it is dual to (and thus indirectly represents) the short-run production frontier.

We will begin by presenting Lucas' model of dynamic factor demands with external costs of adjustment. We then derive a form of the model suitable for econometric estimation.

A THEORETICAL MODEL WITH EXTERNAL COSTS OF ADJUSTMENT

Lucas' (1967) Model

In this section we summarise and adapt where necessary the main features of Lucas' model. Assume factor markets for variable inputs $v = (v_j)$ $i = 1, \ldots M$ are perfectly competitive with prices $\hat{w} = (\hat{w}_j)$ $i = 1, \ldots M$ which are assumed to be known with certainty and to remain stationary over time. Assume quasi-fixed inputs $x = (x_i)$ $i = 1, \ldots N$ can be varied at a cost $C_i(\dot{x}_i)$ where $\dot{x}_i = \frac{dx_i}{dt}$ is the rate of change of the lth quasi-fixed factor and where:

$$C_i(0) = 0 \ , \ C_i'(\dot{x}_i) > 0 \ , \ C_i''(\dot{x}_i) > 0 \ , \ i = 1, \ldots N \qquad (1)$$

Constraints (1) imply the existence of positive, increasing marginal costs of adjustment. If p denotes product price, which we assume is given to the firm by the market and is assumed by the firm to be stationary and known with certainty, net receipts (cash flow) at time t are:

$$R(t) = p \ F\left[x(t), \ v(t)\right] - \sum_{i=1}^{N} C_i\left[\dot{x}_i(t)\right] - \sum_{j=1}^{M} \hat{w}_j v_j(t) \qquad (2)$$

where $F(\cdot)$ is a concave production function. Costs of adjustment are external to production activity since F is not a function of $\dot{x}$. Therefore current production is unaffected by accumulation or decumulation of the quasi-fixed factors, but of course, future production is affected by these changes. Further, since $C_i(0) = 0$, zero depreciation is assumed. These assumptions will be relaxed in subsequent sections. Now suppose the firm finances entirely by borrowing at a fixed, constant rate r. The present value of net receipts (at time 0) is then:

$$V(0) = \int_0^\infty e^{-rt} R(t) dt$$

The firm's problem is to choose time paths v(t), x(t) so as to maximise V(0), given any initial x(0) and v(t), x(t) 0. The optimal solution can be characterised by the calculus of variations (or the Pontryagin maximum principle) as the solution to the first order conditions:

$$p \, F_{v_j} - \hat{w}_j = 0$$

$$\text{or} \quad F_{v_j} - w_j = 0 \text{ where } w_j = \hat{w}_j/p \qquad j = 1, \ldots M \qquad (4)$$

$$p \, F_{x_i} - r \, C_i'(\dot{x}_i) + C_i''(\dot{x}_i) \, \ddot{x}_i = 0 \quad i = 1, \ldots N \qquad (5)$$

and the endpoint or transversality conditions

$$\lim_{t \to \infty} e^{-rt} \, C_i'(\dot{x}_i) = 0 \qquad\qquad i = 1, \ldots N \qquad (6)$$

We assume that all functions have continuous first and second order derivatives. Given strict concavity of the production function in the variable factors, equations (4) can be solved for the unique profit maximising short-run demand functions

$$\bar{v}(t) = v[w(t), x(t)] \qquad (7)$$

Suppose we now form the function

$$H(t) = F[\bar{v}(t), x(t)] - \sum_j w_j(t) \cdot \bar{v}_j(t) = H[w(t), x(t)] \qquad (8)$$

$p(t)H(t)$ is the <u>maximum</u> variable profit attainable at time t , <u>conditional</u> on the level of the quasi-fixed inputs x(t). Therefore, H(t) is a restricted profit function normalised by dividing through by p; i.e., it is a <u>normalised restricted profit function</u>. H(t) possesses five important properties /Lau (1976)/ which will be utilised below:

i) H is increasing in x and decreasing in w   (8a)

ii) H is convex in w   (9)

iii) H is concave in x   (10)

iv) $\dfrac{\partial H}{\partial w_j} = -\bar{v}_j$ the conditional profit maximising input level   (11)

v) $\dfrac{\partial H}{\partial x_i}$ = the normalised shadow price of the service flow from the ith quasi-fixed input.   (12)

Substituting from (8) into (12) we obtain:

$$\bar{R}(t) = p \, H\left[x(t), w(t)\right] - \sum_{i=1}^{N} C_i\left[\dot{x}_i(t)\right] \qquad (13)$$

and

$$\bar{V}(0) = \int_0^\infty e^{-rt} \, \bar{R}(t) \, dt \qquad (14)$$

Maximising (2) with respect to v(t), x(t) is equivalent to maximising (14) w.r.t. x(t) since (14) incorporates the optimal v(t), conditional on x(t). The first order conditions can be obtained by changing (5) to:

$$p \, H_{x_i} - r \, C_i'(\dot{x}_i) + C_i''(\dot{x}_i) \, \ddot{x}_i = 0 \qquad\qquad i = 1, \ldots N \qquad (15)$$

Equations (15) have a stationary solution $x^*(p, w, r)$ which is obtained by setting $\dot{x}_i = \ddot{x}_i = 0$ and therefore satisfies

$$p \, H_{x_i}(x^*, \bar{v}(x^*)) - r \, C_i'(0) = 0 \qquad\qquad (16)$$

Since $p \, H_{x_i}(x^*, \bar{v}) = p \, F_{x_i}(x^*, \bar{v})$, equations (16) specify that

in the steady state equilibrium, the marginal value product of the quasi-fixed factor equals its marginal accumulation cost (at $\dot{x}_i = 0$). $x^*(t)$ is the steady state or long-run profit maximising demand for the vector of quasi-fixed factors obtained by solving (16). The steady state demand for the variable factors can be obtained as $v^*(t)$ by substituting $x(t) = x^*(t)$ in equations (17).

Lucas links his model to the ad hoc partial adjustment or flexible accelerator literature by showing that the short-run demand for the quasi-fixed factors can be generated from equations (15) and (16) as an approximate solution (in the neighbourhood of $x^*(t)$) to the linear differential equation system:

$$\dot{x} = B^*[x^*(t) - x(t)] \qquad\qquad (17)$$

where B^* is a matrix of adjustment parameters. These parameters are not constants as in the ad hoc literature but depend on the exogenous variables, the production technology, and the cost of adjustment functions. In particular Lucas shows that if N=1 (i.e. there is only one quasi-fixed factor) equations (17) reduce to:

$$\dot{x}_1 = B^*[x_1^*(t) - x_1(t)] \qquad\qquad (18)$$

where $B^* = -\frac{1}{2}\left[r - \left(r^2 - 4\frac{H''(x_1^*)}{C''(0)}\right)^{\frac{1}{2}}\right]$ \qquad\qquad (19)

and $H''(x_1^*) = F''(x_1^*) < 0$ for uniqueness of x_1^*. Equation (19) illustrates clearly the variable and endogenous nature of the adjustment coefficient. First, since $H''(x_1^*) < 0$ and $C''(0) > 0$, $0 < B < 1$ so that the actual stock is always moving towards the stationary stock. Second, as marginal adjustment costs increase sharply ($C''(0) \to \infty$), $B^* \to 0$ so that no adjustment occurs. Third, as marginal adjustment costs approach a constant ($C''(0) \to 0$), equations (15) → equations (16) and $B^* \to 1$. Fourth, the curvature of the restricted profit function (and the production function) with respect to the quasi-fixed factor affects the speed of adjustment since $\frac{\partial B^*}{\partial |H''|} > 0$. Finally, $\frac{\partial B^*}{\partial r} < 0$, so that reducing the interest rate increases the speed of adjustment. Thus, the speed of adjustment is endogenously determined by the basic parameters and variables of the economic model.

AN ECONOMETRIC ADAPTION OF LUCAS' MODEL

Lucas' model as presented in section 3.2 needs to be modified in several ways before it can be implemented empirically. First, the zero depreciation assumption should be relaxed. Second, the cost units of the cost of adjustment functions need to be determined. Finally, functional forms for the normalised restricted profit function and cost of adjustment functions must be specified. We will discuss these three problems in turn.

Assume that the stock of each quasi-fixed factor x_i depreciates exponentially at rate δ_i. Let z_i be the gross addition to the stock of factor i. Then:

$$\dot{x}_i = z_i - \delta_i x_i \tag{20}$$

Suppose the cost of adjustment functions $C_i(\dot{x}_i)$ are specified as:

$$C_i(\dot{x}_i) = \hat{q}_i z_i + \hat{q}_i D_i(\dot{x}_i) \tag{21}$$

where $\hat{q}_i$ is the asset purchase price of quasi-fixed factor i. Then it can be shown $\underline{/}$Berndt et al (1977)$\underline{/}$ that equation (15) becomes:

$$H_{x_i} - u_i - rq_i D_i'(\dot{x}_i) + q_i D_i''(\dot{x}_i)\ddot{x}_i = 0 \qquad i=1, \ldots N \tag{22}$$

where $u_i = (\hat{q}_i/p)(r + \delta_i)$ is the normalised user cost associated with the service flow from quasi-fixed factor i when marginal costs of adjustment are constant, and $q_i = \hat{q}_i/p$ is the normalised asset price. Equations (16), (18), and (19) can be adjusted accordingly.

We are now in a position to specify the functional forms to be used for the normalised restricted profit function and the cost of adjustment functions. We have chosen quadratic approximations to those functions for our econometric specification. The quadratic normalised restricted profit function has been advocated by Lau (1976) since the Hessian of second order partial derivatives is a matrix of constants thus facilitating the linking of short- and long-run responses. We make use of this convenient property in the next section. A further advantage of the quadratic specification is the fact that the characterisation of the optimal paths for the quasi-fixed factors given by (15) and (16) is globally as well as locally valid since the underlying differential equations are linear. $\underline{/}$Treadway (1974)$\underline{.}$$\underline{/}$

AN ECONOMETRIC MODEL OF THE DYNAMIC DEMAND FOR ENERGY WITH CAPITAL
AS A QUASI-FIXED FACTOR

In this section we apply the preceding analysis to the demand for
aggregate energy. Assume there exists a production function:

$$Q = F(E, L, M, K) \tag{23}$$

where Q = gross output
   E = aggregate energy input
   L = labour input
   K = capital input
   M = materials input

Suppose factor and product prices are exogenously determined, and
capital is fixed in the short run (i.e., quasi-fixed). The theory of
duality between profit and production implies that, given short-run profit maximising behaviour, the characteristics of production implied by
(23) can be uniquely represented by a normalised restricted profit function of the form:

$$H = H(p_E, p_L, p_M, K) \tag{24}$$

where p_i, i = E, L, M are _normalised_ (by output price) factor prices.
A quadratic approximation to H is given by:

$$H = \alpha_0 + \sum_i \alpha_i p_i + \alpha_K K + \frac{1}{2} \sum_i \sum_j \gamma_{ij} p_i p_j + \sum_i \gamma_{iK} p_i K + \frac{1}{2} \gamma_{KK} K^2 \tag{25}$$

$$i, j = E, L, M$$

Utilising the property that $\frac{\partial H}{\partial p_i} = -x_i$, the short-run profit maximising
quantity demand of the ith input, we obtain:

$$-E = \alpha_E + \gamma_{EE} p_L + \gamma_{EL} p_L + \gamma_{EM} p_M + \gamma_{EK} K$$

$$-L = \alpha_L + \gamma_{LE} p_L + \gamma_{LL} p_L + \gamma_{LM} p_M + \gamma_{LK} K$$

$$-M = \alpha_M + \gamma_{ME} p_L + \gamma_{ML} p_L + \gamma_{MM} p_M + \gamma_{MK} K \tag{26}$$

The properties of neoclassical production theory imply symmetry of
the matrix of second order partial derivatives $\frac{\partial X_i}{\partial p_j}$, which leads to the
following parameter restrictions:

$$\gamma_{EL} = \gamma_{LE} \; ; \; \gamma_{EM} = \gamma_{ME} \; ; \; \gamma_{LM} = \gamma_{ML} \tag{27}$$

The system (25), (26) subject to the constraints (27) constitute the
means of estimating the short-run demand for the variable factors. We
now turn to a specification of the optimal path of the quasi-fixed factor,
capital.

A quadratic approximation to the cost of adjustment function $D(K)$ is given by $D(\dot{K}) = d_0 + d_K\dot{K} + \frac{1}{2} d_{KK}\dot{K}^2$ (28)

Utilising (25) and (28), the first order condition (22) for the optimal path of the quasi-fixed factor becomes:

$$\alpha_K + \sum_i \gamma_{iK}p_i + \gamma_{KK}K - u_k - rq_K(d_K + d_{KK}\dot{K}) + q_K d_{KK}\ddot{K} = 0 \qquad (29)$$

The steady state solution can now be written as:

$$\alpha_K + \sum_i \gamma_{iK}p_i + \gamma_{KK}K^* - u_K - rq_K d_K = 0 \qquad (30)$$

The steady-state or long-run demand for capital can be solved from (30) as:

$$K^* = -\frac{1}{\gamma_{KK}}\left[\alpha_K + \sum_i \gamma_{iK}p_i - u_K - rq_K d_K\right] \qquad (31)$$

where $\gamma_{KK} = H''(x_i^*) < 0$. Note that the steady-state demand for capital corresponds to the static demand (with u_K as the "price" variable) only if $d_K = 0$, i.e., only if there is no linear term in (28). Utilising our previous results we can characterise the optimal path of K by the flexible accelerator formulation:

$$\dot{K} = B^*(K^*-K) \quad \text{where } B^* = -\frac{1}{2}\left[r - (r^2 - \frac{4\gamma_{KK}}{q_K d_{KK}})^{\frac{1}{2}}\right] \qquad (32)$$

Two final assumptions allow us to complete the deterministic part of our econometric model. First, we assume that short-run production activity is conditional on the capital stock at the beginning of the period, (i.e. K_{t-1}) so that any capital stock adjustment during the period does not affect production (or short-run variable profit) until the following period. ⌐This assumption is relaxed in Berndt et al (1977), where we consider Treadway's non-separable internal cost of adjustment model.⌐ Second, we assume that (32) can be replaced by the discrete approximation:

$$\Delta K_t = B^* (K_t^* - K_{t-1}) \qquad (33)$$

We now collect our system of equations to be estimated:

Normalised Variable Profit $= Q - p_E E - p_L L - p_M M$

$$\begin{aligned}
&= \alpha_0 + \alpha_E p_E + \alpha_L p_L + \alpha_M p_M + \alpha_K K_{-1} \\
&\quad + \frac{1}{2}(\gamma_{EE}p^2 + \gamma_{LL}p_L^2 + \gamma_{MM}p_M^2 + \gamma_{KK}K^2_{-1}) \\
&\quad + \gamma_{EL}p_E p_L + \gamma_{EM}p_E p_M + \gamma_{LM}p_L p_M \\
&\quad + \gamma_{EK}p_E K_{-1} + \gamma_{LK}p_L K_{-1} + \gamma_{MK}p_M K_{-1} \qquad (34)
\end{aligned}$$

$$- E = \alpha_E + \gamma_{EE}p_E + \gamma_{EL}p_L + \gamma_{EM}p_M + \gamma_{EK}K_{-1} \tag{35}$$

$$- L = \alpha_L + \gamma_{EL}p_E + \gamma_{LL}p_L + \gamma_{LM}p_M + \gamma_{LK}K_{-1} \tag{36}$$

$$- M = \alpha_M + \gamma_{EM}p_E + \gamma_{LM}p_L + \gamma_{MM}p_M + \gamma_{MK}K_{-1} \tag{37}$$

$$K - K_{-1} = - \frac{1}{2} \left[r - (r^2 - \frac{4\gamma_{KK}}{q_K d_{KK}})^{1/2} \right] . \left[- \frac{1}{\gamma_{KK}} (\alpha_K + \gamma_{EK}p_E + \gamma_{LK}p_L \right.$$
$$\left. + \gamma_{MK}p_M - u_K - rq_K d_K) - K_{-1} \right] \tag{38}$$

Equations (34) – (38) form a system of 5 equations, the parameters of which completely specify the dynamic demand functions. In order to demonstrate this fact we will compute the short- and long-run price elasticities of the demand for energy with respect to the normalised price of energy p_E. The short-run elasticity is given by:

$$\varepsilon_{EE}^s = \frac{d \log E}{d \log p_E} \bigg|_{K=K_{-1}} = \left[\frac{p_E}{E} \right] \cdot \left[\frac{\partial E}{\partial p_E} \right] = \left[\frac{p_E}{E} \right] \cdot (-\gamma_{EE}) \tag{39}$$

(where $\gamma_{EE} > 0$)
and the long-run elasticity by:

$$\varepsilon_{EE}^L = \frac{d \log E}{d \log p_E} \bigg|_{K=K^*} = \left[\frac{p_E}{E} \right] \left[-\gamma_{EE} + \frac{\gamma_{EK}^2}{\gamma_{KK}} \right] \tag{40}$$

Since $\gamma_{KK} < 0$, $\left| \varepsilon_{EE}^s \right| < \left| \varepsilon_{EE}^L \right|$ and the Le Chatelier principle is satisfied. The various own and cross-price elasticities can be computed in an analogous manner to provide a complete summary description of the dynamic behaviour of the factor demands. To illustrate further the richness of the description, we consider the effect on the demand for energy of a change in fiscal policy. We note first that changes in q_K induce changes in B*. This fact allows a particularly detailed analysis of the effects of fiscal policy incentives. The three main fiscal policy instruments which affect demand for factors of production are: (1) the business (corporate) income tax; (2) the investment tax credit; and (3) the provision for capital cost allowance deductions. In order to determine the effects of these instruments, we note that each can be viewed as changing the net asset price of the quasi-fixed factor as calculated by the user firm. Let the net (implicit) asset price of the capital investment good be q_K. Then Hall and Jorgenson (1967, page 393) have shown that $\tilde{q}_K$ is related to $\tilde{q}_K$ through the formula:

$$\tilde{q}_K = q_K \left[\frac{(1-k)(1-tZ)}{1-t} \right] \tag{41}$$

where k = tax credit rate allowed on investment expenditure

 t = (constant) business income tax rate

 Z = present value of the capital cost allowance on
       1 dollar's investment,

and where it is assumed that the depreciation base is reduced by the amount of the tax credit. The firm's response to fiscal policy incentives can be analysed by replacing q_K by $\tilde{q}_K$ from equation (41) in the user cost of capital services variable $u_K = q_K(r + \delta)$ and the cost of adjustment function $q_K D(\dot{K})$ (which implies $q_K d_{KK}$ is replaced by $\tilde{q}_K \cdot d_{KK}$ in equation (32) for B*). Now suppose there is an increase in the investment tax credit rate k. The implicit asset price $\tilde{q}_K$ declines so that u_K declines and B* increases. What is the effect on the demand for energy? In the short-run, there is no effect. In the long-run there will be an increase in the equilibrium demand for capital and, therefore, an increase in the demand for energy if capital and energy are long-run complements and a decrease if they are long-run substitutes. In the intermediate run, there is a compounding of two effects: The change in the long-run equilibrium demand discussed above and an increase in the rate of convergence (B*) to the long-run equilibrium. Hence, in our model fiscal policy affects not only long-run equilibrium demand for factors of production, but also the endogenous adjustment lags. In particular, we can compute the effect of a change in k on the intermediate run demand for energy as:

$$\frac{\partial E}{\partial k} = \gamma_{EK} \frac{\partial K}{\partial k} = \gamma_{EK} \left[B^* \frac{\partial K^*}{\partial k} + K^* \frac{\partial B^*}{\partial k} \right]$$

$$= \gamma_{EK} \left[\Sigma(r + \delta) \, q_K \frac{(1-tZ)}{(1-t)} \left\{ \frac{B^*}{(-\gamma_{KK})} + K^* \frac{\partial B^*}{\partial q_K} \right\} \right] \qquad (42)$$

where

$$\frac{\partial B^*}{\partial q_K} = \frac{1}{2} \left\{ r^2 - \frac{4\gamma_{KK}}{q_K d_{KK}} \right\}^{-1/2} \cdot \left\{ \frac{1}{q_K^2 \, d_{KK}} \right\} > 0 \qquad (43)$$

Both terms in the second set of $\{\}$ brackets in equation (42) are positive. Hence any effect of an increase in the investment tax credit on energy demand consists of a long-run effect and a path adjustment effect which work in the same direction. Whether there is an increase or decrease in the demand for energy depends on the sign of γ_{EK}. The effects of the other fiscal policy variables can be analysed in a similar manner. This analysis is left to the interested reader.

AGENDA FOR FUTURE RESEARCH

We are currently estimating the dynamic model introduced in Section 3 using data drawn from United States manufacturing. In addition, we are extending the econometric model to include the case of two or more quasi-fixed factors, so that technical change can be incorporated as the endogenous adjustment of the stock of knowledge. A preliminary version of this model of technical change can be found in Berndt et al. (1977).

BIBLIOGRAPHY

Berndt, E.R., Bronwyn H. Hall, Robert E. Hall, and Jerry A. Hausman (1974), "Estimation and Inference in Nonlinear Structural Models", Annals of Economic and Social Measurement, October 1974, pp. 653-665.

Berndt, E.R. and N. Eugene Savin (1975), "Estimation and Hypothesis Testing in Singular Equation Systems with Autoregressive Errors", Econometrica, September-November, pp. 937-957.

Berndt, E.R. and David O. Wood (1975), "Technology, Prices, and the Derived Demand for Energy", Review of Economics and Statistics, August 1975, pp. 259-268.

Berndt, E.R., M. Fuss and L. Waverman (1977), "Dynamic Models of the Industrial Demand for Energy", Electric Power Research Institute Research Report, Palo Alto, California, August.

Fuss, M.A., (1977), "The Demand for Energy in Canadian Manufacturing: An Example of the Estimation of Production Structures with Many Inputs", Journal of Econometrics, January, pp. 89-116.

Hall, R.E. and D.W. Jorgenson, (1967), "Tax Policy and Investment Behaviour", American Economic Review, June 1967, pp. 391-414.

Hudson, E.A., and D.W. Jorgenson, (1974), "U.S. Energy Policy and Economic Growth, 1975-2000", Bell Journal of Economics and Management Science, Autumn.

Lau, L.J., (1976), "A Characterisation of the Normalised Restricted Profit Function", Journal of Economic Theory, February 1976, pp. 131-163.

Lucas, R., (1967), "Optimal Investment Policy and the Flexible Accelerator", International Economic Review, February 1967, pp. 78-85.

Nadiri, M.I. and Sherwin Rosen (1973), A Disequilibrium Model of Demand for Factors of Production, New York: National Bureau of Economic Research, General Series No. 99, 1973.

Treadway, A.B., (1974), "The Globally Optimal Flexible Accelerator", Journal of Economic Theory, 7, pp. 17-39.

Treadway, A.B., (1971), "The Rational Multivariate Flexible Accelerator", Econometrica, 39, pp. 845-56.

THE DEMAND FOR GASOLINE IN THE UNITED STATES: A VINTAGE CAPITAL MODEL

by

J.L. Sweeney

ACKNOWLEDGMENTS

The author is an associate professor of Engineering-Economic Systems at Stanford University. The Original version of the model described here was developed while the author was employed by the United States Federal Energy Administration (FEA). Descriptions of the basic framework and previously estimated equations appear in the 1976 National Energy Outlook,(4) in the FEA draft document "Passenger Car Use of Gasoline: An Analysis of Policy Options", by James Sweeney,(11) and in "The Capital Stock Adjustment Process and the Demand for Gasoline: A Market Share Approach", by Cato Rodekohr, and Sweeney.(2) The author would like to thank, without implicating, Mark Rodekohr and Derriel Cato for their assistance in developing earlier drafts of this paper and for writing the computer software to run the model; Lewis Rubin for beginning the process of updating the data and econometrics from the previous version; and Anjum Mir for research assistance in developing the version reported here.

I. INTRODUCTION

Numerous automobile gasoline consumption models have been developed in response to current energy conditions, as have many analyses of options aimed at reducing gasoline consumption. Most gasoline demand studies - such as those conducted by Chamberlain;(3) Verleger;(17) Ramsey, Rasche, and Allen;(10) Houthakker and Kennedy;(5) Houthakker, Verleger, and Sheehan;(7) and McGillivray(8) - have estimated gasoline demand using various forms of a flow adjustment model. The basic flow adjustment model expresses gasoline consumption as a function of the real price of gasoline, disposable income, and gasoline consumption in the previous period. While these models have produced interesting, but often seriously conflicting, results they suffer from several defects. First, these models typically lack a strong theoretical structure and rely upon reduced form econometrics. Second, they can be used to examine only one potential policy, an increase in the price of gasoline. Finally, the econometric specifications impose severe restrictions on adjustment dynamics, since they do not explicitly allow for differential adjustment speeds associated with changes in the character of the capital stock and changes in its utilisation. Other approaches to the estimation of gasoline demand such as the RAND studies(1, 19) and the Verleger and Osten study(18) used a multi-equation approach. While the RAND studies explicitly model the behavioural relationships involved in gasoline demand, they do not examine the choice of new car efficiency or the vintaging process that takes place over time. The Verleger/Osten model exogenously specifies the new car efficiency and thus ignores the most significant influence of changing prices and incentives.

This paper presents a vintage capital approach to estimating the demand for gasoline, an approach in which explicit attention is paid to the processes of capital stock adjustment and stock utilisation. Equations are developed and econometrically estimated in order to describe the critical determinants of capital stock adjustment and utilisation. These equations allow endogenous determination of the trajectories of new car sales, size of the automobile fleet, new car efficiency, fleet efficiency, vehicle miles, and gasoline consumption, as functions of macro-economic variables, gasoline prices, automobile technology, and policy measures directed at automobile efficiency or utilisation.

The model varies from typical econometric demand studies in that the structural detail is sufficient to examine adjustment dynamics, to examine policies acting through traditional economic variables or policies directed specifically at the capital equipment and its characteristics, and to examine influences of technical changes. It varies from typical engineering models in that careful attention has been paid to behavioural relationships. The micro-economic theory has been developed and econometric estimations have been conducted based upon the theory.

The remainder of the paper includes four sections. Section II develops the theory underlying the model. Attention is paid to the determinants of vintage efficiency of new car sales, of vehicle miles for the fleet and for individual vintages, and to the capital stock adjustment processes. Section III presents the econometric estimations of the relationships discussed in Section II. Results of alternative specifications deriving from the theory are presented and a preferred equation is selected for each relationship. Section IV examines operation of the overall

model. The results of a base case run and several sensitivity studies
are presented. Finally, Section V presents a summary and conclusion.

II. THEORY

Basic Concepts

The aggregate demand for automobile use of gasoline will be viewed
as a demand derived from consumer desire for mobility, as translated into
vehicle miles, and consumer preferences over the capital stock of automobiles by which those vehicle miles are obtained. Thus, the demand for
vehicle miles and the demands for automobiles and their characteristics
will be explicitly examined along with the relationships between these
two demands.

For estimating gasoline consumption, the significant summary variable
describing the capital stock of automobiles is the average (or, more precisely, the harmonic mean) efficiency of the fleet, with weights for the
averaging corresponding to relative miles driven by automobiles of different efficiencies. One efficiency, as measured by miles per gallon, of
the fleet is known, gasoline consumption is simply vehicle miles divided
by average fleet efficiency.

For this analysis, individual model years, or vintages, will be explicitly examined, while any heterogeneity within a vintage will be
averaged into a summary statistic for that vintage. Within this framework,
then, the aggregate demand for gasoline by the fleet of automobiles can
be expressed as the sum over all vintages of the demands by automobiles
from the given vintage. Thus, gasoline demand can be expressed by the
following tautology:

$$(1) \qquad GAS = \sum_i \frac{VMPC_i * N_i}{mpg_i} ,$$

where GAS represents the aggregate gasoline demand, $VMPC_i$ represents vehicle miles per car obtained by the ith vintage, N_i represents the number of
cars in existence from vintage i, and mpg_i is the efficiency (in terms of
miles-per-gallon) of automobiles from the ith vintage.

Equation (1) can be decomposed into two factors by defining a second
tautology:

$$(2) \qquad VM = \overline{VMPC} * N ,$$

where VM represents the total vehicle miles obtained by the fleet of automobiles; N, the number of cars in existence; and $\overline{VMPC}$, the average over

all autos of the vehicle miles per car. Using (2), equation (1) can be
rewritten:

$$(3) \qquad GAS = \left\{ \sum_i \left(\frac{1}{mpg_i}\right)\left(\frac{VMPC_i}{\overline{VMPC}}\right)\left(\frac{N_i}{N}\right) \right\} VM ,$$

or

$$(4) \qquad GAS = \frac{VM}{\overline{mpg}}$$

where $\overline{mpg}$ can be evaluated by equation (5).

$$(5) \qquad \overline{mpg}_t = 1/ \left\{ \sum_{i=-\infty}^{t} \left(\frac{1}{mpg_i}\right) \left(\frac{VMPC_{it}}{\overline{VMPC}_t}\right) \left(\frac{N_{it}}{N_t}\right) \right\} .$$

Here t subscripts have been added to denote the time-dependent terms.

The term $\overline{mpg}$ is the weighted harmonic mean efficiency of the various
vintages, and will be referred to more simply as the average efficiency
of the fleet. The weights used in calculating this mean correspond to the
fraction of cars in existence from the given vintage times a measure of
the relative intensity of utilisation of automobiles from the vintage.
The measure of the relative intensity of utilisation is the ratio of vehicle miles per car from the ith vintage to the average over all automobiles
of vehicle miles per car.

Equation (4) highlights the two fundamental factors underlying gasoline demand: the efficiency of the capital stock of automobiles and the
intensity of utilisation of that stock. More simply, consumers make two
critical choices that determine gasoline consumption. The first is the
number of vehicle miles or the amount of travel chosen. The second is
the efficiency with which these miles are travelled.

The two factors of equation (4) can be expected to exhibit very different dynamics. Vehicle miles is the most volitive component of consumer
choice and can adjust very rapidly to changes in the economic variables
influencing travel behaviour. On the other hand, average stock efficiency
changes very slowly, since in any year the new cars are numerically dominated by the existing fleet of automobiles. Thus, in the short run, the
effect on gasoline demand of changing market conditions is felt almost
exclusively through the intensity of utilisation of the capital stock.
Only in the long run is the adjustment of that stock important to gasoline demand.

As suggested in equation (4), the demand for gasoline for automobiles
can be estimated by evaluating separately the aggregate vehicle miles

travelled by all cars and the average efficiency of the stock. Vehicle miles can be estimated as a function of income, driving cost, and other economic variables. This term is allowed to vary rapidly as the independent variables change. Average efficiency can be estimated by evaluating (a) the efficiency of autos from each vintage, (b) the relative miles driven by autos from different vintages, and (c) the numbers of cars in existence from each vintage. Thus, average efficiency is allowed to evolve slowly as new cars are purchased and old cars are scrapped or utilised less intensively over time.

These ideas form the basis of a model used to examine the demand for gasoline for automobiles. The basic framework is illustrated in Figure I. Subsequent sections of this paper describe the theory underlying various model components, the econometrically estimated equations, and, finally the simulations performed using the model.

Capital Stock Adjustment Processes

As indicated in equation (5), the average efficiency of the fleet of automobiles can be examined through an analysis of the adjustments in the capital stock of automobiles occurring over time. Subsequent subsections describe the theory underlying the various factors from equation (5). In sequence are discussed: Vintage Efficiency, Intensities of Utilisation, Capital Stock Composition. Subsequent sections discuss stock adjustments and new car sales.

Vintage efficiency

The capital stock at any given moment is composed of automobiles from different vintages, which, once produced, have relatively fixed characteristics. In particular, it will be assumed that the average fuel efficiency from each vintage remains constant over time.

Two factors contribute to a determination of average efficiency of automobiles of a given vintage: the technical efficiencies of specific models and the market shares captured by each model. Either an increase in the efficiency of each car or a shift in market shares from less efficient to more efficient autos will increase average share efficiency. These two factors - market shares and technical efficiencies - are not independent. This section will examine the relationships between market shares, technical efficiencies and average efficiencies.

It will be assumed that in purchasing new cars, consumers consider capital costs and projected operating costs of the various models as well as other characteristics of the available array of autos. Furthermore,

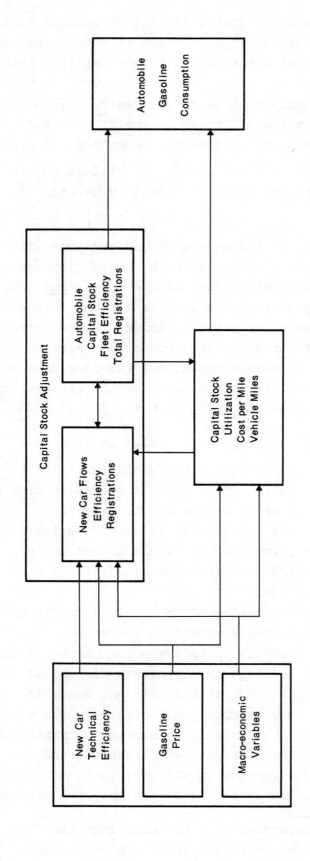

Figure 1

BLOCK DIAGRAM OF MODEL

Capital Stock Adjustment

Automobile
Capital Stock
Fleet Efficiency
Total Registrations

New Car Flows
Efficiency
Registrations

Capital Stock
Utilization
Cost per Mile
Vehicle Miles

Automobile
Gasoline
Consumption

New Car
Technical
Efficiency

Gasoline
Price

Macro-economic
Variables

in estimating operating costs, it will be assumed that the consumer uses current gasoline prices as a forecast of future prices; in particular it will be assumed that the expected future gasoline price is equal to the current prices.

Let S^i be a vector whose k^{th} elements, S^i_k, indicates the market shares captured by the kth model in the ith vintage. Let G be a vector whose k^{th} component, G^i_k, indicates the gallons of gasoline consumed per mile driven for the kth car in the ith vintage (G_k is the inverse of fuel efficiency). Then the following identity relates G^i, S^i, and mpg_i:[1]

$$(6) \qquad mpg_i = 1/\left[G^i \cdot S^i \right]$$

The vector S^i is assumed to depend upon operating costs of the various autos. The operating costs of the kth model depends upon maintenance costs, physical characteristics of the model, and fuel costs. Gasoline price enters only through the last factor and then only in so far as it influences the fuel cost per mile drive. Hence operating costs of the kth model are influenced by the product of gasoline price times the G^i_k.

Holding all factors constant but gasoline cost and the model efficiencies, the above discussion leads to the mathematical relationships:

$$(7) \qquad mpg_i = 1/\left[G^i \cdot S^i (P^i_G \, G^i) \right] \ ,$$

where P^i_G is the (scalar) gasoline price in time i.

By equation (7) the average efficiency of the ith vintage can be influenced by the gasoline price or by the efficiencies of the various models. A gasoline price change can be expected to alter the market shares of the various models and thus to change the weights used for averaging efficiencies. In particular one would expect an increase in gasoline price to shift buyers away from fuel-inefficient cars so as to increase the mpg of the vintage.

Changes in efficiencies of the various models have two effects. First, for market shares constant an increase (or decrease) in efficiency of a given model leads to an increase (or decrease) in efficiency of the vintage. Second, a change in efficiency of a model leads to changes in market shares of that and other models and hence to a change in the vintage efficiency. For example, an increase in the efficiency of one model, all else unchanged, leads to increases in the market share of that model and decreases in the market shares of other models.

A 10 per cent increase in the efficiency of each car would have two effects. With market shares held constant, the vintage efficiency would

---

1) Two vectors separated by a dot will indicate the inner product of the two vectors. For example $G^i \cdot S^i$ represents $\sum_k G^i_k * S^i_k$.

increase 10 per cent. However, the efficiency increase would induce a market shift toward the less fuel efficient cars, therefore leading to a net efficiency increase of less than 10 per cent.

The properties of equation (7) generally depend upon the matrix of own and cross elasticities of demand for the various models. Thus little can be said in general about the response to gasoline price changes or to changes in model efficiencies. However for one special case the elasticity of vintage efficiency with respect to model efficiencies can be related to the elasticity with respect to gasoline price. The special case is one in which the efficiency of each model changes by the same percentage.

Let efficiencies of each vintage be allowed to vary such that $G^i = G^0/EFF$ where EFF is a scalar equal to unity when there has been no change in model efficiencies. EFF will be referred to as the "technical efficiency" of the vintage. Then if technical efficiency - EFF - increases by 10 per cent, all model efficiencies (the inverses of the G^i_k's) must increase by 10 per cent. Equation (7) can be written:

$$(8) \qquad mpg_i = 1 / \left[\frac{G^0}{EFF_i} \cdot S^i \left(\frac{P^i_G}{EFF_i} \, G^0 \right) \right]$$

The elasticity of mpg_i with respect to P^i_G evaluated with EFF_i equal to unity will be denoted as η^i_p. Then[1]

$$n_p = - \, P^i_G \; mpg_i \; \left[G^0 \cdot \nabla S \cdot G^0 \right] \quad \geq \quad 0,$$

where ∇S is the gradient of S. The elasticity of mpg_i with respect to EFF evaluated with EFF equal to unity will denoted as η_{EFF}. Then

$$\eta_{EFF} = 1 + P^i_G \; mpg_i \; \left[G^0 \cdot \nabla S \cdot G^0 \right].$$

Combining the two equations above gives the following relationship between the two elasticities:

$$(9) \qquad \eta_{EFF} = 1 - \eta_p$$

Equation (9) can be intuitively understood by associating the two terms on the right hand side with the two separate effects of changes in technical efficiency. The first term expresses the elasticity if market shares do not change. In this case a constant proportional change in each model efficiency is met by the same proportional change in the vintage efficiency. The second term expresses the effect of efficiency improvements inducing a shift toward the less efficient cars. This term is just

1) The inequality derives from the characteristics of ∇S. ∇S is a negative semi-definite matrix since each diagonal element is negative, each off-diagonal element is positive, and each column sum is zero.

the negative of the price elasticity of vintage efficiency. Intuitively this occurs because a given percentage increase in all efficiencies has the same impact on all operating costs as would the same percentage decrease in gasoline price.

Equation (9) establishes that the greater is the elasticity of vintage efficiency with respect to gasoline price, the lower will be the elasticity with respect to technical efficiency. In fact, for large price elasticities in excess of unity, the elasticity with respect to technical efficiency would be negative. In such a case, an increase in each efficiency would alter market shares enough to give a net decrease in the efficiency of the vintage!

Intensities of Utilisation

A second factor contributing to the determination of the fleet-wide mean efficiency of travel is the relative intensity of usage of cars from each vintage. In particular it has been observed that as a general rule, older cars are driven fewer miles per year than are newer cars. For example, the 1972 United States Department of Transportation "Nationwide Personal Transportation Study" developed data relating car age to vehicle miles driven. These data are reproduced in Appendix 1.

Theoretically, gasoline price, new car price, repair costs, and incomes should influence how intensively older cars are driven in comparison to newer cars. However, for the model described here a particularly simple relationship was chosen - the relative miles driven of automobiles was assumed to be a geometrically decreasing of the automobile age:

$$(10) \qquad \frac{VMPC_{it}}{VMPC_{jt}} = \gamma^{(j-i)} ,$$

for j; i $\leq$ t and for $\gamma < 1$.

While the form of equation (10) is too simple for some applications, it leads to computational simplification. In particular, it provides a convenient expression for evaluating the factor ($VMPC_{it}/\overline{VMPC}_t$) from equation (5). Using equation (10) one obtains:

$$(11) \qquad \frac{VMPC_{it}}{\overline{VMPC}_t} = 1 / \left\{ \sum_{j=-\infty}^{t} (N_{jt}/N_t)^{(i-j)} \right\} .$$

Equation (11) shows that the ratio of vehicle miles per car for the ith vintage to the average vehicle miles per car can be simply calculated as an average of the geometrically declining utilisation factor $\alpha^{(i-j)}$. The value of this simplification will become apparent later.

Capital Stock Composition

The third factor required to calculate the fleet-wide mean efficiency of travel, through equation (5), is the relative number of automobiles in existence from each vintage, i.e., the capital stock composition. Therefore, it is necessary to estimate both sales of new cars and scrappage rates of existing cars.

Theoretically, scrappage rates should depend upon the previous utilisation of a given vintage, upon new car prices and repair costs. However, since the major focus of this analysis is upon the demand for gasoline, rather than upon scrapping per se, a simple assumption of exponential scrapping is used. In particular, it is assumed that a fraction $(1 - \delta)$ of the automobiles remaining from a given vintage are scrapped during the year. Under this assumption, the number of cars from vintage i in existence in year t can be related to new cars sold from vintage i as follows:

$$(12) \qquad N_{it} = NPCR_i\ \delta^{(t-i)}, \text{ where}$$

$NPCR_i$ is the number of new passenger cars registered from vintage i in year i. In equation (12) the term $\delta^{(t-i)}$ represents the fraction of cars from the ith vintage remaining in year t.

Exponential scrappage was chosen for computation and econometric convenience, as will be seen at a later point. However, this assumption is not crucial for the model. Further work will be directed toward developing a better representation of scrapping.

Stock adjustment

Equation (5), describing the evaluation of the automobile stock, can now be simplified. Using equations (11) and (12), and some simple algebraic manipulation, equation (5) becomes:

$$(13) \qquad \overline{mpg_t} = \frac{\displaystyle\sum_{i=-\infty}^{t} NPCR_i (\delta\ \gamma)^{(t-1)}}{\displaystyle\sum_{i=-\infty}^{t} NPCR_i (\delta\ \gamma)^{(t-i)}\ \frac{1}{mpg_i}}$$

In equation (13) the factor $\gamma^{(t-i)}$ can be interpreted as the intensity of usage of automobiles from the ith vintage in comparison with autos from the tth generation. This factor equals unity when t equals i and declines geometrically for i decreasing from t. Thus, the factor $\gamma^{(t-i)}$ can be interpreted as translating the actual number of existing autos of the ith generation to an _effective_ number of new cars. The notion is that if a used car were used only one-half as much as a new car, it would provide the effective services of only one-half a new car. The factor $\{NPCR_i \ \delta^{(t-i)}\}$ is the actual number of autos from vintage i existing at time t. Thus, the term $\{NPCR_i(\delta \gamma)^{(t-i)}\}$ can be interpreted as the effective number of autos from vintage i at time t. The numerator of equation (13) can simply be interpreted as the effective number of automobiles existing at time t. This effective stock will be denoted as ES_t.

Equation (13) can be slightly rewritten:

$$(13') \qquad \frac{1}{\overline{mpg_t}} = \sum_{i=-\infty}^{t} \frac{NPCR_i(\delta \gamma)^{(t-i)}}{ES_t} \left(\frac{1}{mpg_i}\right)$$

The term within brackets of equation (13') is simply the fraction of the effective stock represented by vintage i. The second term is the inverse of efficiency of the ith vintage. Thus, the fleet efficiency is simply the geometric mean of vintage efficiencies, with weights for averaging corresponding to the effective stocks from each vintage, rather than the more normal actual stocks.

Equation (13) can be evaluated by two simple recursive relationships. By definition of ES_t:

$$ES_t = \sum_{i=-\infty}^{t} NPCR_i(\delta \gamma)^{(t-1)}$$

$$\frac{ES_t}{\overline{mpg_t}} = \sum_{i=-\infty}^{t} NPCR_t(\delta \gamma)^{(t-i)} (1/mpg_i) \ .$$

These lead to the following two recursive equations:

$$(14) \qquad ES_t = ES_{t-1}(\delta \gamma) + NPCR_t$$

$$(15) \qquad \frac{ES_t}{\overline{mpg_t}} = \frac{ES_{t-1}}{mpg_{t-1}}(\delta \gamma) + \frac{NPCR_t}{mpg_t}$$

The recursive nature of these two equations is made possible by the assumption of exponential scrapping and exponential decline of utilisation. Note that although the vintage capital stock model requires summation over all past vintages, the exponential assumptions allow $\overline{mpg_t}$ to be estimated for all $t > j$ if ES_j and $\overline{mpg_j}$ are known and $NPCR_i$ and mpg_i are known for every i between j and t. The entire history need not be explicitly considered.

New car sales

The final element describing the capital stock adjustment processes is the sales and subsequent registration of new cars. Under the notion that consumers desire transportation services, consumers can be envisioned as desiring an effective stock of automobiles with the size of this stock related to the amount of driving that is anticipated. The effective stock desired (ES_t^*) then can be expressed as a function of vehicle miles (VM_t) and other factors (X_t):

$$ES_t^* = f(VM_t, X_t).$$

New car sales is envisioned as a process of moving the effective stock from its current level toward its desired level. Hence, new passenger car registrations should be an increasing function of desired effective stock and a decreasing function of the effective stock remaining from the previous year:

$$(16) \qquad NPCR_t = g(f(VM_t, X_t), ES_{t-1}).$$

It should be noted here that vehicle miles is assumed to influence desired stock of autos rather than vice versa. This assumption is consistent with the notion that the demand for automobiles is derived from the demand for transportation services.

The effective stock variable is used here rather than a more conventional measure of the number of automobiles (e.g., passenger car registrations) because the more conventional measures do not account for the relative intensity of utilisation of automobiles of various vintages. The use of the effective stock variable implies that an older automobile is less effective in adding to the existing stock of automobiles than is a new car.

Vehicle Miles

While the capital stock adjustment processes for the automobile fleet are very slow, the utilisation of that capital stock can change rapidly as conditions change. Hence, adjustment processes are less important in analysing utilisation of the capital stock than in analysing the composition of that stock.

For this model, it has been assumed that vehicle miles obtained is a function of cost of that travel as well as other variables:

$$(17) \qquad VM_t = h(CPM_t, X_t),$$

where CPM_t is the cost per mile of automobile travel in time t.

The first component of travel cost is the fuel cost and this cost depends upon the capital stock characteristics as well as upon fuel price. In particular, increasing efficiency of the fleet reduces fuel cost and leads to increases in vehicle miles desired by the population. Since P_G^t represents the gasoline cost per gallon and $\overline{mpg_t}$ represents the mileage obtained per gallon consumed, the per mile fuel cost of driving is simply the ratio:

$$P_G^t \,/\, \overline{mpg_t} \,.$$

A second component is the time cost of travel, here treated as a fraction of the value of the time if used to earn an income. For a car travelling at an average speed equal to $AVSPEED_t$, each rider bears a travel cost of $WF \cdot WR_t/AVSPEED_t$, where WR_t is the wage rate at time t and WF is the factor relating value of time at work to value of time in driving.[1] In order to make time and cost and fuel cost per mile equivalent, one of the factors must be scaled by number of passengers per car; either gasoline cost must be divided by the number of passengers per car to put all costs on a passenger-mile basis, or per person time cost must be multiplied by number of riders per car to put all costs on a vehicle mile basis. The latter procedure was used to give the time cost component equal to

$$WF \cdot WR_t \cdot RPC/AVSPEED_t,$$

where RPC is average number of riders per car.

Other driving costs theoretically should be included whenever they are variable costs of driving. Automobile depreciation depends upon the auto age and upon the cumulative vehicle miles driven. That component of depreciation associated with miles driven is a cost which should influence vehicle miles, but that component associated with auto age is not. Similarly, insurance costs should not in general be included.

While costs other than time cost and fuel cost should be included as components of cost per mile of driving - CPM_t - data limitations lead to their exclusion. Thus, cost per mile was specified as follows:

(18) $$CPM_t = P_G^t/\,\overline{mpg_t} + WF \cdot WR_t \cdot RPC/AVSPEED_t.$$

III. ECONOMETRIC ESTIMATION

The theoretical relationships described in Section II were tested using national data, and parameters of equations derived from Section II

---

1) Studies of model choice suggest that WF lies between .3 and .7. For the purppse of this study WF is chosen to equal .5.

were estimated. The first econometric estimates were developed in 1974 and the resultant equations were incorporated into the Federal Energy Administration's Project Independence Evaluation System (PIES), a large-scale energy supply and demand model (see 1976 National Energy Outlook).(4) Subsequently in 1976, with new data available, the equations were re-estimated, generally using the same functional forms, as well as alternative specifications. All equations estimated using the most recent data are presented here, as are the specific equations incorporated in the 1974 version of the model.

Vintage Efficiency

Following equation (8), vintage efficiency would be estimated as a function of gasoline price and of a technical measure of automobile efficiency:

$$(19) \qquad \qquad mpg_i = EFF_i \ g(P_G^i/EFF_i).$$

The variable EFF_i is a normalization(1) of the factor "C" as defined by the United States Environmental Protection Agency.(2) This factor describes the average fuel economy for cars of a given year, <u>standardized</u> for the weight of the car. This factor is defined such that the efficiency of a given car from a given year is approximately equal to the factor "C" for that year divided by the automobile weight. The term P_G^i is the average price of regular grade gasoline in year i, deflated by the CPI.

Data on average efficiency of newly purchased cars from a given model year are published by the Environmental Protection Agency based upon their tests of automobiles using the Federal Driving Cycle.(3) Unfortunately, these efficiency measures overestimate actual on-the-road performance. To account for this difference, versions of equation (19) were estimated in log-linear form based upon EPA test data. Then the constant term was adjusted to best fit Federal Highway Administration data on fleetwise efficiency. This procedure then simply provided a multiplicative scaling of the relationship based upon EPA data to become consistent with FHWA data. In the subsequent results the constants are those calculated using EPA data.

---

1) EFF is proportional to "C" and set equal to unity for 1974.

2) United States Environmental Protection Agency, Office of Air and Water Programs, Mobile Source Pollution Control Program, <u>Fuel Economy and Emission Control</u>, November 1974.

3) The specific data are from "A Report of Automotive Fuel Economy", U.S. Environmental Protection Agency, Office of Air and Water Programs, Office of Mobile Source and Pollution Control, October 1973.

Several equations $\lfloor$related to equation (8)$\rceil$ were estimated, all in log-linear specifications. First were log-linear specifications of equation (8). The independent variables were logarithms of P_G^t and EFF_t, and the coefficients on the two variables were constrained to sum to unity. Results from this specification, for two different sample periods, is presented as equations (19a) and (19b) in Table 1. Similar equations were also estimated in which the sum of elasticities was not constrained to equal unity. These are presented as equations (19e).

Variations on these equations were estimated in which gasoline price was lagged by one year. There are several theoretical reasons why this specification may be preferable to those with current gasoline price. First, the data on vintage mpg and EFF are for the model year, not the calendar year, while gasoline price is defined as an average over a calendar year. However, model years are generally introduced in late summer or early autumn of the preceding calendar year. Thus, the lagged price observations may help to compensate for the temporal mismatch.

Second, the mix of automobiles of different sizes is determined partially by manufacturer decisions as to the models to offer for sale and the characteristics of each model. These decisions must be made in advance of the introduction and sales of the model-year automobiles. These decisions are guided by manufacturer perceptions of market conditions. To the extent that manufacturer decisions and manufacturer market perceptions depend upon current gasoline prices, a lagged gasoline price is the relevant variable. Equations (19c) and (19d) constrain the elasticities to sum to unity and hence are parallel to equations (19a) and (19b). Equations (19g) and (19h) do not constrain the sum of elasticities and hence parallel equations (19e) and (19f).

Equations (19i) through (19l) each add per capita income as an explanatory variable, both defined in the current year and lagged one year. Equations (19i) and (19j) parallel equation (19a); similarly, equations (19k) and (19l) parallel equation (19c). Income is added, since one may expect consumers to buy larger, more luxurious cars as their incomes increase. On the other hand, many models of high efficiency imports may be purchased as incomes rise. Thus, theoretically, the sign of the income coefficient is indeterminate.

In general, the equations that include lagged gasoline price perform well, with R^2 values equal to about .85 and Durbin-Watson statistics between 2.0 and 2.2. Numerical values of the parameters vary only slightly with specification and with estimation interval. On the other hand, those equations including cotemporaneous price behave less well. The estimated parameters vary radically with estimation interval, the values of R^2 are

Table 1

NEW CAR EFFICIENCY (MPG) EQUATIONS

Eq	Constant	P_G	$P_G(-1)$	EFF	YD/N	Sample Period	R^2	D.W.
19a	3.348 (22.2)	.716 (5.5)		.284(a)		57-73	.75	2.3
19b	2.93 (14.2)	.357 (2.0)		.643(a)		57-74	.45	1.5
19c	3.308 (32.0)		.689 (7.6)	.311(a)		57-73	.85	2.2
19d	3.344 (32.0)		.721 (7.9)	.279(a)		57-74	.86	2.1
19e	3.272 (20.5)	.645 (4.6)		.194 (1.3)		57.73	.78	2.3
19f	2.836 (12.8)	.271 (1.4)		.510 (2.3)		57-74	.49	1.2
19g	3.264 (28.2)		.648 (6.3)	.258 (2.4)		57-73	.86	2.2
19h	3.325 (28.9)		.703 (6.9)	.249 (2.2)		57-74	.86	2.0
19i	3.349 (21.4)	.728 (4.9)		.272(a)	.011 (0.2)	57-73	.75	2.3
19j	3.02 (14.8)	.320 (1.9)		.680(a)	-.119 (-1.6)	57-74	.54	1.9
19k	3.307 (31.0)		.704 (6.9)	.296(a)	.016 (0.4)	57-73	.85	2.2
19l	3.34 (31.1)		.719 (6.9)	.281(a)	-.003 (-.06)	57-74	.86	2.1

All variables appear in logarithmic form.
a) Constrained to equal unity minus the price elasticity.

reduced from the values in corresponding equations including lagged price, and many Durbin-Watson statistics are low. Hence, a general evaluation of summary statistics in Table 1 suggests that the theoretically more desirable lagged gasoline price is also econometrically more desirable.

The gasoline price variable is a significant determinant of the average efficiency of automobiles purchased even when controlled for technical efficiency of automobiles offered for sale. The gasoline price variable is significant at the 99 per cent level in nine of the twelve equations, and lagged gasoline price performs better than does current price. Current gasoline price is significant in three of six equations (those estimated over the shorter interval), while lagged price is significant in all six equations. The estimated elasticities with respect to lagged gasoline price range between .648 and .721, a maximum variation of only 11 per cent, while elasticities with respect to current gasoline price vary radically with specification and sample period.

The income coefficient is insignificant in each regression. Its inclusion slightly increases the estimated price elasticities, but the change is not significant. The inclusion of per capita income also slightly lowers the values of $\bar{R}^2$. Given these results, the income variable was dropped from subsequent consideration.

The elasticity with respect to the technical efficiency variable - EFF - is significantly less than unity in all cases. If changes in technical efficiency did not induce an offsetting shift in market shares, then this elasticity would be unity. Hence, it can be concluded that shifts in market shares induced by changes in technical efficiency offset most (in these cases about 75 per cent) of the efficiency effects of changes in technical efficiency.

The theory of Section II implied that the sum of price elasticity and technical efficiency elasticity should equal unity. If there were no influence of efficiency on market shares of automobiles, then the two elasticities would sum to greater than unity. In equations (19e) through (19h) the sums were not constrained to unity. Adding the elasticities gives the sums of .84, .78, .91, and .95 for equations (19e), (19f), (19g), and (19h) respectively, with no sums statistically different from the theoretical prediction of unity. The closest values to unity - .91 and .95 - are derived from equations incorporating the preferable lagged gasoline price, while the other two values are derived from equations containing contemporaneous prices. The results therefore strongly confirm the theory presented in Section II.

In constructing the 1974 version of the simulation model, equation (19c) was chosen. This equation incorporates the preferable lagged gasoline price and constrains the sum of elasticities to equal unity, as suggested by the theory. Hence, this equation was judged preferable to equations (19a), (19e), and (19g). In the current version, equation (19d) is incorporated.

For the automobile simulation model, the constants in equations (19c) and 19d) are scaled up to better reflect on-the-road performance than do the EPA tests(1), which consistently underestimated on-the-road performance. The constant of equation (19c) was chosen to best fit Federal

---

1) Subsequently, EPA has changed their testing procedure so that current procedures overestimate average on-the-road performance by 14 per cent (averaged over the automobile lifetime).

Highway Administration (FHWA) Data(1) on average fleet efficiency. This was accomplished by using equations (15) and (14) to solve for the vintage efficiencies which would just make fleet efficiencies correspond to FHWA data. These calculated vintage efficiencies were used as dependent variables and the constant of equation (19c) was recalculated. The result was to increase the constant from 3.344 to 3.396, a net increase in calculated vintage efficiency of 5.3 per cent.

New car sales

Following equation (16), new car sales would be estimated as a function of vehicle miles, stock remaining from previous years, and other variables. Other variables of interest include population, per capita income, new car and used car prices, and unemployment.

Data were obtained from several sources. Data on new car sales - NPCR - were obtained from the Motor Vehicle Manufacturer's Association.(2) Data on the macro-economic variables were obtained through standard Department of Commerce and Department of Labor publications. Vehicle miles was published by the FHWA,(3) and new car price was the new car consumer price index.(4) The effective stock of automobiles remaining from previous years was calculated based upon estimated factors and in equation (14).

The new car sales equation was fitted in log-linear form, with per capita new car registrations as the dependent variable. Independent variables included effective stock remaining from the previous period, per member of the current population $\sqrt{ES(-1)/N}$, per capita vehicle miles (VM/N), per capita disposable income in constant dollars (YD/N), new car price index divided by the CPI (PCAR), two measures of cost per mile of driving, CPM, the total of gasoline and time cost per mile of driving, the gasoline cost per mile (GCPM), and the unemployment rate (Ru). The latter variable - unemployment - appeared in linear form. All others were in logarithmic form.

---

1) U.S. Dept. of Transportation, Federal Highway Administration, Highway Statistics, 1950 through 1974 issues. Table VM-1, Average miles travelled per gallon of fuel consumed - Passenger Cars.

2) Motor Vehicle Manufacturer's Association, 1973/1974 Automobile Facts and Figures.

3) U.S. Dept. of Transportation, Federal Highway Administration, Highway Statistics, 1950 through 1974 issues. Table VM-1, Total Travel - Passenger Cars.

4) Consumer price index - new cars, U.S. Dept. of Labor, Bureau of Labor Statistics.

Table 2 presents estimated parameters obtained using a sample period of 1957-1974. Also included is the estimated equation based upon data from 1954-1972 and included in the original version of the model.

In general, the equations for new car sales using 1957-1974 data fit well, with the lowest R^2 equal to .86. There is no evidence of serial correlation of the residuals in equations including vehicle miles per capita as an explanatory variable. In these equations, the Durbin-Watson statistics all lie close to 2.0. When vehicle miles is excluded, Durbin-Watson statistics still generally are near to 2.0, but fall as low as 1.5. In the equation using earlier data, the Durbin-Watson statistic of 1.6 is low enough that the hypothesis of serial correlation of the error terms cannot be rejected.

The coefficient on effective stock remaining from the previous year is large, negative, and statistically significant at the 99 per cent level in each equation. The elasticity varies from -3.5 to -4.5 among the equations and hence short-run new passenger car registration elasticities with respect to other variables range between 4.5 and 5.5 times as great as the corresponding long-run elasticities. Thus, changes in income, unemployment, or vehicle miles have a far larger initial impact than long-run impact.

The rate of unemployment has a negative coefficient which is statistically significant at the 99 per cent level in each equation. The coefficient ranges between -0.049 and -0.078, with smaller absolute values occurring when vehicle miles is excluded. When vehicle miles is included, the coefficient ranges between -0.068 and -0.072 for all equations generated using the more recent data. Thus, for this set of equations, a one-percentage point increase in the unemployment rate leads to a 7 per cent decrease in new passenger car registrations in the short run and a 1.4 per cent decrease in the long-run equilibrium rate of registrations.

The income elasticities are all positive and statistically significant, with all but two being significant at the 99 per cent level. However, the estimated short-run elasticities vary considerably in magnitude. When the equations are controlled for unemployment rate, short-run income elasticities vary between 3.1 and 3.4 if vehicle miles is included in the regression and between 3.7 and 4.9 if this variable is excluded. When unemployment is not controlled for, the short-run income elasticity varies between 6.0 and 6.4 if vehicle miles is included and between 4.9 and 5.6 if vehicle miles is excluded from the regressions.

Income has an indirect effect on new passenger car registrations as well as the direct effect measured by the per capita income coefficient. This indirect effect derives through the influence of income on vehicle

Table 2

NEW CAR SALES PER CAPITA (NPCR/N) EQUATIONS*

	Const.	ES(-1)/N	VM/N	YD/N	PCAR	CPM	GCPM	Ru	R^2	D.W.
16a	22.542 (5.38)	-4.53 (-4.7)	-1.10 (-1.3)	6.38 (5.2)					.87	1.9
16b	20.994 (7.0)	-4.11 (-6.0)	1.27 (1.5)	3.31 (2.8)				-0.072 (-3.8)	.94	1.9
16c	19.607 (4.9)	-3.98 (-5.2)	1.17 (1.3)	3.36 (2.7)		-0.286 (-.5)		-0.069 (-3.4)	.94	1.9
16d	18.30 (4.3)	-3.73 (-4.6)	0.918 (1.0)	3.30 (2.8)			-0.291 (-.9)	-0.068	.94	2.0
16e	17.42 (2.9)	-3.60 (-3.0)	-0.94 (-1.1)	6.20 (5.1)	1.03 (1.2)				.89	2.3
16f	19.19 (4.5)	-3.93 (-4.3)	1.23 (1.4)	3.38 (2.7)	.21 (.3)			-0.070 (-3.4)	.94	2.0
16g	19.65 (4.1)	-3.99 (-4.1)	1.17 (1.2)	3.35 (2.6)	-0.28 (.03)	-0.31 (-.3)		-.067 (-3.2)	.94	1.9
16h	19.09 (4.2)	-3.94 (-4.3)	0.76 (.76)	3.10 (2.4)	-0.61 (-.6)		-0.519 (-1.0)	-0.072 (-3.4)	.95	2.0
16i	17.91 (3.1)	-4.06 (-3.9)	-1.15 (-1.4)	6.16 (5.0)		-0.92 (-1.1)			.89	2.2
16j	17.52 (3.0)	-3.80 (-3.4)	-1.52 (-1.7)	6.07 (4.9)			-0.528 (-1.2)		.89	2.4
16k	16.99 (2.7)	-3.68 (-2.9)	-1.01 (-1.1)	6.16 (4.8)	0.72 (.5)	-0.39 (-.3)			.89	2.3
16l	16.84 (2.7)	-3.60 (-2.9)	-1.26 (-1.1)	6.10 (4.8)	.557 (.4)		-0.31 (-.4)		.89	2.4
16m	22.65 (5.3)	-4.58 (-4.7)		5.26 (5.7)					.86	1.5
16n	21.38 (6.9)	-4.20 (-5.9)		4.79 (7.0)				0.052 (-3.8)	.93	2.0
16o	18.93 (4.3)	-3.96 (-5.0)		4.66 (6.6)		-0.494 (-.8)		-0.050 (-3.6)	.93	2.1
16p	17.32 (4.2)	-3.61 (-4.5)		4.17 (5.3)			-0.42 (-1.4)	-0.055 (-4.2)	.94	2.2
16q	16.73 (2.82)	-3.50 (-2.9)		5.25 (5.8)	1.186 (1.4)				.88	2.0
16r	19.65 (4.24)	-3.89 (-4.1)		4.82 (6.9)	.359 (.5)			-0.049 (-3.3)	.93	2.1
16s	19.06 (3.9)	-4.01 (-4.0)		4.64 (5.9)	-0.93 (-.1)	-0.56 (-.5)		-0.050 (-3.3)	.93	2.1
16t	18.61 (4.2)	-3.93 (-4.3)		3.71 (3.8)	-0.83 (-.8)		-0.70 (-1.6)	-0.063 (-3.8)	.94	2.2
16u	18.36 (3.0)	-4.15 (-3.9)		5.01 (5.2)		-0.85 (-1.0)			.87	1.7
16v	20.46 (3.4)	-4.27 (-3.7)		4.93 (4.4)			-0.23 (-.5)		.86	1.6
16w	16.74 (2.6)	-3.50 (-2.7)		5.25 (5.2)	1.9 (0.9)	0.008 (.0)			.88	2.0
16x	17.31 (2.7)	-3.53 (-2.8)		5.55 (4.6)	1.48 (1.3)		0.21 (.4)		.87	2.0
16y**	4.08 (0.7)	-3.75 (-6.3)	2.32 (2.5)	1.78 (2.0)				-.078 (-3.6)	.90	1.6

* Sample period 1957-1974, except equation 16y, with sample period 1954-1972. All variables expressed in logarithmic form except Ru, which was not transformed.

** Sample period 1954-1972. The numerical values of VM/N used in equation 16y differ by a factor of 1000 from those in equations 16a through 16x. Changing units to regain consistency would change the constant in equation 16y to 20.11. In 16y the dependent variable is new passenger car sales per capita, while new passenger car registrations per capita were used in all other equations.

miles. In subsequent paragraphs it will be argued that the elasticity of vehicle miles with respect to per capita income is around unity. Thus, the direct plus indirect elasticity of new passenger car registrations with respect to income is approximately equal to the sum of vehicle mile and income elasticities. This total elasticity of new car registrations with respect to income in the long run centres in the range of .90 through 1.0, with 15 of the equations having total long-run elasticities exceeding 1.0, with four of these in the 1.1 to 1.2 range. Only those equations from which unemployment was excluded exhibited total income elasticities exceeding 1.0 in the long run. Since unemployment is such a robust variable, its exclusion seems clearly inappropriate. Of the 13 equations which control for unemployment, nine have total income elasticities in the 0.9 - 1.0 range and eleven have total elasticities in the 0.85 - 1.0 interval.

The estimated elasticities of new car registrations with respect to vehicle miles, for income held constant, fall around two variables. When unemployment is included in the equation, short-run elasticity with respect to vehicle miles averages 1.1, varying from a high of 1.27 to a low of .76 (when using the more recent data). The long-run elasticities in these cases averages .22, varying between .25 and .15. When unemployment is excluded from the estimated equations, short-run elasticities average -1.2, varying between -1.52 and -.94, while long-run elasticities average -.24.

The negativity of the elasticity with respect to vehicle miles provides another reason for rejecting equations which exclude unemployment. When unemployment is controlled for, the vehicle mile elasticity is always positive but at best marginally significant. Part of the difficulty lies in the high degree of correlation between income and vehicle miles in the sample. It is shown in a later part of the paper (see Table 3) that income per capita is a highly significant explanatory variable for vehicle miles per capita. Hence, it is particularly difficult to separate the effects of these two variables on new car sales. While the set of equations taken together suggests a new car elasticity with respect to vehicle miles of around unity, the evidence is far from compelling.

The results relating new car prices to sales were rather surprising. While one would expect the coefficient of PCAR, the new car price index divided by the CPI, to be consistently negative, this result was not obtained. The sign of this coefficient varies among equations following no clearly explicable pattern. And in general, each negative coefficient is insignificant.

The insignificance of new car price is surprising, but this may have several causes. First, used car prices are excluded from the equation, even though these prices should influence and be influenced by the new car sales. The exclusion of this variable could conceivably bias the co-efficient on new car price. However, since the existing stock of used cars is captured by the factor ES(-1)/N, and since this stock helps to determine used car prices, the role of the used car market in explaining new car demand presumably should be captured to a significant extent. Second, there may be a simultaneous equation bias introduced into the estimation. Manufacturers typically make production plans and contract for inputs significantly before sale, based upon estimated sales. In a situation of demand reduction below planned sales, manufacturers engage in promotions (such as price rebates) having the effect of reducing new car price so as to increase demand. What would be observed in such a situation would be a net demand decrease correlated with a price decrease, giving a positive relationship between price and demand. This simultaneous equation bias may be partially responsible for the insignificance of the new car price variable, although the precise cause awaits further research.

Finally, variables related to the operating costs of automobiles have been included. Gasoline cost per mile (GCPM) is the real gasoline price divided by the average efficiency of the stock of cars, while cost per mile (CPM) includes gasoline cost plus time cost of operating automobiles. These variables were included to test whether operating costs of auto-mobiles may have an influence on new car purchase which is independent of its influence through vehicle miles. In general, these variables are in-significant, especially when vehicle miles is included as an independent variable. However, the consistently negative sign in all but two equa-tions suggests that these variables may have an independent influence.

Operating costs could have an independent effect on new car sales if consumers seeing higher operating costs expect to cut down their use of automobiles over time. In such a case, the purchase of automobiles may be reduced before vehicle miles are in fact reduced. This would give an in-dependent relationship between operating costs and vehicle miles.

Whether a relationship between new car sales and operating costs which is independent of vehicle miles exists is still open to question. Given the non-conclusive tests, these variables were left out of the equa-tion finally used in the model.

Equation (16y) was used in the original version of the automobile simulation model, and equation (16b) is used in the current version. Equation (16b) was chosen since it included vehicle miles, disposable in-come, and unemployment as explanatory variables, while excluding those variables that were not consistently significant. These two equations have very similar long-run and dynamic behaviour, except that the elasticity of new car registrations with respect to gasoline price (operating through the influence on vehicle miles) is twice as large for the former equation as for the latter.

It is felt that further research could improve the new car registration equation; however, the research has yet to be conducted. Since gasoline consumption is relatively insensitive to new car sales in the model, improvements to this equation are of relatively low priority for examining gasoline demand. Hence, improvements will await later research.

Vehicle miles

To estimate vehicle miles, all variables were converted to per capita form so that total vehicle miles equals the population times an estimated per capita vehicle mile. The basic equations used derived directly from equations (17) and (18). For vehicle miles, equation (20) was estimated:

$$(20) \qquad\qquad VM/N = v(CPM, YD/N, Ru) \ ,$$

In some equations, in place of cost per mile, gasoline cost per mile was used, a variable that excludes time cost from cost per mile.

Data sources for each variable of equation (20) were described previously. For equation (18), WF was taken as a parameter equal to .5, and RPC was taken as a parameter equal to 1.5. These values were estimates based upon modal choice studies for WF and Department of Transportation studies for RPC. Wage rate was obtained from the Bureau of Labor Statistics,(1) and, as with all price variables, was converted to constant dollars by using the consumer price index. The average speed variable - AVSPEED - was an average of the average speed obtained on main rural highways(2) and an average speed on urban roads, assumed to equal 20 mph. Weights for the averaging were each .5. This split of urban vs rural lies within the range of the data and was approximately correct for 1964.

Several other variables were tested in the vehicle miles equation. A measure of free time - weekly hours of production workers (HPEA)(3) - was incorporated into several equations and a measure of the size of the automobile fleet was also included: PCRPN, the passenger car registrations per capita. In a set of equations not reported here per capita value of

---

1) Hourly Earnings of Production Workers - Total private non-agriculture. "Employment and Earnings Statistics for the United States". BLS Bulletin Series 1312.

2) Reported by the FHWA in Highway Statistics.

3) Weekly hours of production workers - Total private non-agricultural hours per week. U.S. Department of Labor, Bureau of Labor Statistics, BLS Bulletin Series 1312.

effective stock of cars $\underline{/}$ES(-1)/N(-1)$\underline{/}$ was used rather than the variable PCRPN in order to reduce the simultaneous equation problem (Vehicle miles is used to explain new passenger car registrations which in turn is a component of effective stock), and to improve the measure of automobile stock.

In estimating the equations, all variables were used in logarithmic form, except for unemployment, which was not transformed. Table 3 presents the results. Equations (20a) through (20p) were based upon a sample period of 1957 through 1974, while equation (20q) is based upon the sample period 1951-1972. The latter equation was that used in the early version of the simulation model. Equation (20q) was fitted using a first-order autocorrelative transformation, while all others were fitted using ordinary least squares.

All equations fit the historical data closely, as measured by very high values of R^2, but some showed signs of serial correlation of the residuals, as evidence by low values of the Durbin-Watson statistics.

In all of the equations, the variables describing driving cost were negative and were significant in the majority of the cases. However, the variable GCPM, the gasoline cost per mile, generally performed better than did the theoretically preferable variable CPM, the time cost plus gasoline cost per mile. This former variable was significant in all but one equation specification. On the other hand, CPM was significant in only one-half of the equations. The elasticity with respect to GCPM was generally small but stable, varying over a roughly two-to-one range from a low of -0.12 to a high of -0.27. Elasticities on CPM varied over a wider range. Thus, the higher quality of equations including GCPM leads to a rejection of those equations containing CPM rather than GCPM.

The measures of free time were generally successful. Whenever HPEA, the weekly hours of production workers, was included in the equations, its coefficient was negative, the expected sign, and generally significant. Only when the second measure of free time, unemployment, was also included did HPEA ever become insignificant. The sign pattern of the unemployment coefficients is difficult to explain. A positive coefficient would be expected, in general, since increases in unemployment with income held constant imply two conditions. First, more unemployment implies more free time and hence more travel. Second, for current income constant, more unemployment implies more permanent income and hence more travel. Both effects suggest a positive coefficient on unemployment. The reason for the significant negative coefficients in equations (20g) and (20h) is unclear.

Table 3

VEHICLE MILES PER CAPITA (VM/N) EQUATION*

Eq #	C	CPM	GCPM	YD/N	Ru	HPEA	$\frac{PCR(-1)}{N(-1)}$	R^2	D.W.
20a	.0119 (.02)	-.0337 (.14)		1.068 (19.1)				.98	.6
20b	-.389 (-1.0)		.139 (-1.3)	1.040 (28.0)				.99	.5
20c	-.505 (-1.1)	-.173 (-1.0)		1.13 (27.5)	.017 (4.2)			.99	1.7
20d	-.433 (-1.7)		-.119 (-1.6)	1.076 (39.8)	.015 (4.2)			.99	1.7
20e	5.27 (2.9)	-.235 (-1.2)		.912 (13.5)		-1.539 (-3.1)		.99	1.1
20f	5.60 (3.8)		-0.219 (-2.9)	.821 (13.7)		-1.657 (-4.1)		.99	1.4
20g	2.017 (1.0)	-.232 (-1.4)		1.04 (13.9)	-.0127 (2.6)	-.703 (-1.3)		.99	2.0
20h	3.106 (1.7)		-0.17 (2.5)	0.933 (12.3)	-0.0095 (2.1)	-.975 (-2.0)		.995	2.2
20i	-5.95 (-2.8)	-.532 (-2.0)		.28 (1.0)			.936 (2.9)	.99	.9
20j	-4.69 (-3.5)		-0.262 (-2.9)	0.274 (1.2)			.792 (3.3)	.99	.7
20k	-2.54 (-1.1)	-.325 (-1.4)		.834 (2.6)	.014 (2.6)		.334 (0.9)	.99	1.7
20l	-2.20 (-1.4)		-.175 (-2.0)	.750 (2.5)	.012 (2.3)		.328 (1.1)	.99	1.5
20m	.183 (-.06)	-.508 (-2.2)		.430 (1.6)		-1.10 (-2.1)	.621 (1.9)	.99	1.6
20n	1.48 (.6)		-0.272 (-3.8)	.419 (2.1)		-1.229 (-3.0)	.474 (2.1)	.995	1.9
20o	.0949 (0.03)	-.360 (-1.6)		.794 (2.5)	.0103 (1.8)	-.658 (-1.2)	.289 (0.8)	.99	2.1
20p	1.372 (0.6)		-.225 (-2.7)	.632 (2.3)	.0059 (1.1)	-.952 (-2.0)	.306 (1.1)	.996	2.2
20q**	6.518 (12.1)	-.358 (-1.8)		.976 (11.1)	.0026 (0.9)			.996	1.1

* Sample period 1957-1974, except (20q); sample period for (20q) was 1951-1972. All variables expressed in logarithmic form except Ru, which is not transformed.

** Equation (20q) is specified using a first-order autocorrelative adjustment. The parameter rho equals .810 with a t-statistic of (12.7). The numerical values of VM used in equation (20q) differ by a factor of 1000 from those used in equations (20a) through (20p). Changing units to regain consistency would reduce the constant of equation (20q) to -0.390.

The per capita income elasticity is always positive and is generally significant. However, the magnitude of this coefficient and the associated t-statistic depends upon whether or not the stock of automobiles is included as an independent variable. When stock of automobiles is excluded, income elasticities are each close to unity and t-statistics all exceed 12. However, when auto stock is included, the magnitude of the income coefficient is reduced, now varying between .28 and .83. The t-statistics are drastically reduced.

As shown by Table 3 the driving cost and income elasticities are altered when PCR is included as an independent variable. However, PCR itself depends upon income and vehicle miles in the long run. Thus, it is useful to examine the elasticity of vehicle miles with respect to gasoline cost or income while including the indirect effect through PCR as well as the direct effect. The resulting elasticities will be referred to as the system elasticities.

Long-run system elasticities of vehicle miles with respect to income, cost per mile, and gasoline cost per mile can be calculated for each equation that includes passenger car registrations as an independent variable. This is accomplished by calculating the long-run elasticity of passenger car registrations with respect to vehicle miles and income from equation (16_). These long-run elasticities (.25 and .65 respectively) can be incorporated into the various vehicle mile equations through the passenger car registrations factor in order to calculate system elasticities. For equations not including PCR, the system elasticities are simply the elasticities appearing in Table 3. Table 4 presents these system elasticities for the various equations. Equations are displayed in groups of four. Each equation in a group has identical independent variables except that two equations in each group have CPM and two have GCPM as independent variables.

Table 4 shows that the long-run system elasticities with respect to income all are approximately equal to unity. Comparing system income elasticities for those equations including passenger car registrations with income elasticities of those equations that are identical except for the exclusion of passenger car registrations, the system income elasticity is always within 10 per cent of the income elasticity for the corresponding equation from the latter set. Hence, including of passenger car registrations as an exogenous variable in fact has little effect on the system income elasticity estimates.

Long-run system elasticities with respect to driving costs, on the other hand, do depend upon whether PCR is included as an explanatory variable. Incorporation of the PCR term increases the system elasticity of vehicle miles with respect to driving costs. This occurs because increased driving costs reduce vehicle miles, which reduces passenger car registrations in the long run, which in turn further reduces vehicle miles.

Table 4

SYSTEM ELASTICITIES OF VM/N

Indep. Var. Eq #	CPM	GCPM	YD
20a	-.034	[-.012]	1.1
20i	-.70	[-.25]	1.2
20b		-.14	1.0
20j		-.33	1.0
20c	-.17	[-.06]	1.1
20k	-.36	[-.13]	1.2
20d		-.12	1.1
20l		-.19	1.1
20e	-.24	[-.08]	.9
20m	-.60	[-.21]	1.0
20f		-.22	.8
20n		-.31	.8
20g	-.23	[-.08]	1.0
20o	-.39	[-.14]	1.1
20h		-.17	.9
20p		-.24	.9

The cost per mile variable includes gasoline cost per mile and time cost. Hence, the elasticity of vehicle miles with respect to gasoline cost can be calculated for those equations that include cost per mile as the independent variable. These calculated elasticities appear in brackets in Table 4. When PCR is excluded from the equation, the implicit elasticities with respect to gasoline cost vary from a low of -.012 to a high of -.22. When PCR is included, these implicit system elasticities vary over the smaller range of from -.14 to -.33.

While the long-run system elasticity of vehicle miles with respect to income seems to equal or slightly exceed unity, the relative direct influence of income vs the influence of passenger car registrations cannot be sorted out based upon these equations. The high degree of colinearity between these two variables makes such a sorting difficult. However, for those equations desirable on other grounds - i.e., including GCPM and Ru or PHEA as independent variables - the income elasticity varies between .42 and .75 while the passenger car registration elasticities vary between .30 and .47. More precise and more dependable delineation of these two effects requires further research.

For the current version of the model, equation (20p) has been chosen based upon its inclusion of the variables shown to be most desirable from the entire set of equations, based upon coefficients that are representative of those throughout the set of estimated equations, and based upon desirable values of summary statistics.

Other equations

In order to complete estimation of the model, two parameters are needed: γ and δ. The first of these - γ - is the parameter describing the relative intensity of utilisation of automobiles from successive vintage, and is defined in equation (10). The parameter δ describes the fraction of cars from one year surviving to the next year. This parameter is used in equation (12).

To estimate γ, data from the United States Department of Transportation "Nationwide Personal Transportation Study" were used. These data presented average annual vehicle miles for autos of different ages. The following equation was fitted:

(21) $$VMMY = V_1 \cdot \gamma^{AGE},$$

where VMMY was the average vehicle miles for the model year and AGE was the age of automobiles from the given model year. Estimated parameters are presented in Table 5:

Table 5

PARAMETERS FROM EQUATION (21)

Equation #	V_1	γ	AGE Range	$\bar{R}^2$	D.W.
21	16.56 (20.4)	.921 (85.1)	1-10	.84	1.9

To estimate δ, data from the FHWA was used to estimate total auto-mobile registrations - PCR.(1) Equation (22) was fitted for δ:

(22)   PCR = δPCR(-1) + NPCR .

Table 6 presents the estimated parameter:

Table 6

ESTIMATION OF EQUATION (22)

Equation #	δ	Sample Period	$\bar{R}^2$	D.W.
22	.9351 (461)	57-74	.998	.6

The implication of the estimated values of δ and γ on survival and utilisation of automobiles from a given vintage are presented in Table 7. Presented are relative vehicle miles per car remaining, the fraction of the vintage surviving, and the relative vehicle miles for the entire vintage. This last figure is the product of the fraction surviving times the relative vehicle miles per car.

Table 7

ASSUMED AUTOMOBILE SURVIVAL AND
RELATIVE VEHICLE MILES OVER TIME

Vintage age	Relative vehicle miles per car	Fraction surviving	Relative vehicle miles for entire vintage
0	1.0	1	1
5	.66	.71	.47
10	.44	.51	.22
15	.29	.37	.11

1) U.S. Dept. of Transportation, Federal Highway Administration, Highway Statistics, Table VM-1, Number of Vehicles Registered - Passenger Cars.

IV. MODEL OPERATION

The theory of Section II and the econometric estimations of Section III have been combined into an automobile simulation model designed to provide conditional forecasts of gasoline demand and to aid as a tool in examining implications of various policy actions. The overall model has been diagrammed in Figure 1. Table 8 describes the specific equations used for each variable and the independent variables used in each equation. The Appendix includes a Fortran listing of the model.

Three runs of the model will be discussed here in some detail. The base case assumes no policy initiatives and excludes the efficiency standards of the Energy Policy and Conservation Act (EPCA). The second case incorporates EPCA efficiency standards on newly purchased cars and assumes that the standards will be met on the aggregate. The third case again excludes the efficiency standards but assumes the imposition of a gasoline tax which increases gasoline price by 50 per cent above the levels of the base case. The tax revenues are rebated in a way so as to leave real income constant.

The exogenous macro-economic and technological variables are assumed to be identical in each of the three cases. Table 9 presents the assumed growth rate of each exogenous variable for three forecast intervals: 1974-1980, 1980-1985, and 1985-1990. These are compared with the actual growth rates for the 1957-1974 interval. Tables 10 and 11 present outputs of the model, with Table 10 providing growth rates and Table 11, level of each of the relevant endogenous variables. For comparison, historical growth rates and levels are also presented.

The base case assumes an absence of all policies directed at automobile use of gasoline and thus provides a benchmark for analysis of policy choices. The basic conclusions for the base case are:

1. The automobile consumption of gasoline would grow significantly more slowly over the next fifteen years than it did over the previous seven.
2. Vehicle miles would grow slightly more slowly than previously.
3. Vintage efficiency and fleet efficiency would reverse their downward trends. Vintage efficiency would grow at a significantly slower rate than technical efficiency. This growth rate would be reduced in each successive five-year interval but would remain positive. Fleet efficiency would increase at a slower rate than would vintage efficiency over the first six years, but will continue to grow after growth in vintage efficiency declines.

Table 8

EQUATIONS USED FOR AUTOMOBILE SIMULATION MODEL

Independent variable	Equation #	Independent variables*
GCPM		P_G, $\overline{mpg}$
VM	20p	GCPM, YD, Ru, N, HPEA, PCR(-1)
NPCR	16b	ES(-1), VM, YD, Ru, N
PCR	22	NPCR, PCR(-1)
ES	14	ES(-1), NPCR
mpg	19d	P_G(-1), EFF
$\overline{mpg}$	15	ES, ES(-1), $\overline{mpg}$(-1), NPCR, mpg
GAS	(4)	VM, $\overline{mpg}$

*) Variables followed by (-1) are lagged values of the variable, with one-year lags.

Table 9

GROWTH RATES FOR EXOGENOUS VARIABLES

	1974-1980	1980-1985	1985-1990	Actual 1957-1974
YD/N	2.2	2.0	1.2	2.4
N	0.9	1.0	1.0	1.23
EFF	4.6	2.9	2.5	-0.67
P_G	0.9	0.0	0.0	-0.06
Ru	0	-3.9	0.9	1.59
HPEA	-0.4	-0.4	-0.3	-0.34

4. New car sales would grow rapidly as the automobile industry re-
covers from its slump. Subsequently, new car sales growth rates
would more nearly approximate historical trends.

The efficiency standard case explores the impacts on gasoline consumption of a mandatory efficiency standard on newly purchased cars, under
the assumption that the standard is just met in the aggregate. The time
phasing of the standard corresppnds to that of the EPCA, with a 1980 efficiency of 17.6 and a 1985 and 1990 efficiency of 24.1 mpg. These on-the-
road average efficiencies for the vintages would correspond to EPA-
measured efficiencies of 20 mpg and 27.5 mpg respectively. In this run,
it is simply assumed that the efficiency standards would be met, even

Table 10

GROWTH RATES FOR ENDOGENOUS VARIABLES

	GAS	VM	MPG	MPG	NPCR	PCR
Actual: 57-74	4.22	3.92	-1.02	-0.29	1.92	4.00
Base Case						
74-80	2.47	3.53	4.40	1.04	6.68	3.00
80-85						
85-90	1.81	2.81	0.61	0.98	1.07	2.59
Efficiency standard case						
74-80	2.00	3.67	6.30	1.63	6.75	3.01
80-85	0.11	4.26	6.49	4.14	2.90	3.38
85-90	0.35	3.29	0.00	2.93	1.13	2.71
Gasoline tax case						
74-80	-0.93	2.42	9.61	3.38	6.52	2.77
80-85	0.81	4.03	1.19	3.20	2.68	3.33
85-90	1.22	3.00	0.61	1.76	1.08	2.62

though it currently seems unlikely that the existing legislation will motivate a sufficient response to justify an assumption of compliance to the standards.

Basic conclusions about the effect of the efficiency standard are as follows:

1. The standard would greatly reduce the growth of gasoline consumption by automobiles. The rates subsequent to 1980 would be reduced to below 0.5 per cent per year. Consumption would be 12 per cent lower than in the base case in 1985 and 18 per cent lower in 1990.

2. The standard would reduce the per mile cost of driving and would lead to increases in vehicle miles over the base case. This effect would be more than offset by the increase in fleet efficiency.

3. The standard would have only a very slight effect on new car registrations or total registrations.

The third case imposes a tax which increases the gasoline price by 50 per cent beginning in 1976. No efficiency standard is included. This case can be used both to simulate the effects of a tax and to calculate the dynamic demand elasticities of the model. The basic conclusions are:

1. The gasoline tax would motivate an immediate and lasting impact on gasoline consumption growth rates. Growth rates through 1980

Table 11

PROJECTED VALUES FOR ENDOGENOUS VARIABLES

	GAS (Billion gallons)	VM (Billion miles)	MPG*	$\overline{\text{MPG}}$	NPCR (Million cars)	PCR (Million cars)
Base Case						
1974	74.3	1018	12.2	13.7	8.7	110
1980	85.9	1253	15.8	14.6	12.8	131
1985	95.5	1497	16.8	15.7	14.6	154
1990	104.5	1720	17.3	16.5	15.4	175
Efficiency standard case						
1974	74.3	1018	12.2	13.7	8.7	110
1980	83.6	1263	17.6	15.1	12.9	131
1985	84.1	1556	24.1	18.5	14.9	155
1990	85.6	1830	24.1	21.4	15.7	177
Gasoline tax case						
1974	74.3	1018	12.2	13.7	8.7	110
1980	70.2	1175	21.2	16.7	12.7	129
1985	73.1	1432	22.4	19.6	14.5	152
1990	77.6	1660	23.1	21.4	15.3	174

*) Corresponds to average on-the-road performance. To obtain projections cor-
responding to the EPA tests, multiply by 1.14.

would be negative, but would become positive in the later periods. Through 1985 consumption levels would remain below 1974 consumption levels.

2. The gasoline tax would motivate immediate reductions in vehicle miles, in comparison to the base case values, reductions which would decline in percentage terms over time. The 1976 reduction of 8.7 per cent would decline to 6.2 per cent in 1980 and 3.5 per cent by 1990.

3. The 50 per cent gasoline tax would motivate a more rapid increase in new car efficiency than that determined by the efficiency standard. In all years after 1982, the vintage efficiency motivated by the gasoline tax would approximate the EPCA efficiency standard, with the gasoline tax leading to slightly lower efficiency.

4. The fleetwide average efficiency under the gasoline tax would exceed that obtained under the efficiency standard for all years before 1990.

5. The gasoline tax would slightly reduce the number of new cars registered each year.

Using the data of Table 11 one can examine the dynamic price elasticities of demand for gasoline. These equal -.22 in the first year, -.50 for four years, -.66 for nine years, and -.73 for 14 years. These elasticities can be broken into two components: the vehicle mile elasticity and the fleet efficiency elasticity. Vehicle mile elasticities are -.22, -.16, -.11, and -.09, while fleet efficiency elasticities are .00, .34, .55, and .64 for these same four years, respectively. The vehicle mile elasticity declines over time while the fleet efficiency elasticity increases. Over time the latter effect dominates the former. These elasticities converge in the long run to -.60 and .72, with a long-run equilibrium gasoline demand elasticity equal to -.78. The monotonically declining vehicle mile elasticity occurs because driving cost initially increases greatly as gasoline price increases, but then slowly decreases as the fleet of automobiles becomes more efficient.

In the short run, then, an increase in gasoline price has a small impact on gasoline consumption and the entire impact is felt through changing vehicle miles. In the long run, however, the effect on vehicle miles is reduced by a factor of three, while the impact on fleet efficiency grows large. In the long run, the influence on gasoline consumption felt through changes in the capital stock efficiency dominates the effect felt through vehicle miles by an order of magnitude.

The model has also been run to examine dynamic income elasticities of demand. The one-year elasticity is .60, while the fifteen-year elasticity is .86. The short-run elasticity is slightly less than the income elasticity of vehicle miles with the stock of cars held constant, while the 15-year elasticity includes the effects of increasing automobile stock as incomes increase. The difference between gasoline demand elasticity and vehicle miles elasticity comes about because increases in income lead to more rapid evolution of the automobile stock and hence speed the rate at which more efficient cars are introduced into the stock. Increases in income have pronounced impacts on new car registrations, with one-year and fifteen-year elasticities equal to 4.1 and 0.9 respectively.

Finally, the effects of a 10 per cent increase in technical efficiency - EFF - have been examined. Since increases in technical efficiency induce shifts toward large cars, a net increase in vintage efficiency of about 2.7 per cent results from a 10 per cent increase in technical efficiency. Fleet efficiency increases by 1.7 per cent in five years, by 2.3 per cent in ten years, and by 2.5 per cent in 15 years. Vehicle miles increases by about 0.5 per cent in each of these years. Thus, the net impact of a 10 per cent technical efficiency improvement would be only a 1.3 per cent reduction in gasoline demand in five years, 1.7 per cent in 10 years, and 1.9 per cent in 15 years. Hence, increases in technical efficiency may have far smaller impacts than is often supposed.

V. SUMMARY AND CONCLUSIONS

This paper has presented a structural, dynamic model of the automobile use of gasoline. Within a vintage capital framework, it has been possible to model the capital stock of energy-using equipment - in this case automobiles - and the utilisation of that stock. The theoretical model has been implemented into a simulation model whose critical relationships have been estimated econometrically.

A central relationship concerns the effect of fuel price and technical characteristics of automobiles on the average efficiency of a vintage. It has been estimated that a 10 per cent increase in gasoline price leads to a 6.5 per cent - 7.5 per cent increase in new car average efficiency, while a 10 per cent increase in the efficiency of each model of automobile leads to only a 2.5 per cent - 3.5 per cent increase in average efficiency. The effect of these changes on the fleet efficiency is felt progressively more over time, with about 60 per cent of the ultimate effect felt over five years and 85 per cent over 10 years.

The utilisation of the capital stock can be described in terms of the relative per car vehicle miles of old vs new cars, and in terms of the vehicle miles obtained by the fleet. The quantitative impact of changes in gasoline price on vehicle miles has not been conclusively demonstrated, but the evidence presented suggests a small elasticity of vehicle miles with respect to gasoline price.

This model represents a departure from conventional econometric demand studies in that dynamic processes are examined by explicit representation of capital stock adjustments rather than through a simple distributed log formulation. It is felt (and hoped) that such explicit examination of capital stock adjustments may prove fruitful in other econometric studies, as well as casting light onto automobile demands for gasoline.

REFERENCES

1. Burright, Burke K. and Enns, John H., Econometric Models of the Demand for Motor Fuel, Report R-1561-NSF/FEA, The Rand Corporation, Santa Monica, April 1975.

2. Cato, Derriel, Rodekohr, Mark, and Sweeney, James L., "The Capital Stock Adjustment Process and the Demand for Gasoline: A Market Share Approach", in Econometric Dimensions of Energy Demand and Supply, John Kraft (ed.), Lexington Books, 1976.

3. Chamberlain, Charlotte, "Models of Gasoline Demand", unpublished working paper, U.S. Department of Transportation, Transportation System Center/SP-21, Cambridge, Massachusetts, October 1973.

4. Federal Energy Administration, National Energy Outlook, FEA-N-75/713, Washington, D.C., February 1976.

5. Houthakker, H.S. and Kennedy, Michael, "The World Demand for Petroleum Model", unpublished paper, Data Resources Inc., Lexington, Massachusetts.

6. Houthakker, H.S., and Verleger, Phillip, "The Demand for Gasoline: A Mixed Cross-Sectional and Time Series Analysis", unpublished paper, Data Resources, Inc., Lexington, Massachusetts, May 1, 1973.

7. Houthakker, H.S., Verleger, P.K., and Sheehan, D.P., "Dynamic Demand Analyses for Gasoline and Residential Electricity", American Journal of Agricultural Economics, May 1974.

8. McGillivray, Robert G., "Gasoline Use by Automobiles", Working Paper 1216-2, The Urban Institute, Washington, D.C., December 1974.

9. Motor Vehicle Manufacturer's Association, 1975 Automobile Facts and Figures, Detroit, Michigan, 1975.

10. Ramsey, J., Rasche, A., and Allen, B., "An Analysis of the Private and Commercial Demand for Gasoline", unpublished paper, Department of Economics, Michigan State University, February 18, 1974.

11. Sweeney, James L., "Passenger Car Use of Gasoline: An Analysis of Policy Options", Federal Energy Administration, Office of Energy Systems, Working Paper, February 7, 1975.

12. U.S. Bureau of Labor Statistics, Employment and Earnings Statistics for the U.S., BLS Bulletin Series 1312, U.S. Govt. Printing Office.

13. U.S. Department of Transportation, Nationwide Personal Transportation Study (1972), U.S. Government Printing Office, Washington, D.C.

14. U.S. Department of Transportation, Federal Highway Administration, <u>Highway Statistics</u>, 1950-1974, U.S. Government Printing Office.

15. U.S. Environmental Protection Agency, Office of Air and Water Programs, Mobile Source Pollution Control Program, <u>A Report of Automotive Fuel Economy</u>, October 1973.

16. U.S. Environmental Protection Agency, Office of Air and Water Programs, Mobile Source Pollution Control Program, <u>Fuel Economy and Emission Control</u>, November 1974.

17. Verleger, Phillip, "A Study of the Quarterly Demand for Gasoline and Impacts of Alternative Gasoline Taxes", Data Resources, Inc., Lexington, Massachusetts. Report submitted to the Executive Office of the President, Council on Environmental Quality, December 1973.

18. Verleger, Phillip and Osten, James, "An Econometric Model of Consumer Demand for New and Replacement Automobiles", unpublished.

19. Wildhorn, W., Burright, B., Enns, J., and Kirkwood, T.F., <u>How to Save Gasoline: Public Policy Alternatives for the Automobile</u>, Report R-1560-NSF, The Rand Corporation, Santa Monica, California, 1974.

ENGINEERING AND ECONOMETRIC APPROACHES TO INDUSTRIAL ENERGY CONSERVATION AND CAPITAL FORMATION: A RECONCILIATION

by

Ernst R. Berndt and David O. Wood(1)

1) This paper is a considerably shortened version of another manuscript with the same title issued as Working Paper MIT-EL-77-40WP at the MIT Energy Laboratory and Working Paper No. 77-14 at the UBC Programme in Natural Resource Economics. Our interest in this topic was piqued by discussions with colleagues on the National Academy of Sciences – National Research Council Committee on Nuclear and Alternative Energy Systems (CONAES). We are indebted to Cathy White, M.S. Khaled and Supriya Lahiri for exceptional research assistance, and to Charles Blackorby, Erwin Diewert, William Hogan, Dale Jorgenson, Laurence H. Linden, Robert Pindyck and members of the UBC National Resource Economics Workshop for helpful comments.

"Diuerse pathes leden diuerse folke the rihte wey to Roome."
- Geoffrey Chaucer, A Treatise on the Astrolabe, 1391.

I. INTRODUCTION

One of the dominant themes of economic history and development after the Industrial Revolution is the reduction in human toil and labour brought about by the more efficient harnessing of energy to provide power for industrial equipment. Ever since late 1973 however, when OPEC announced a series of dramatic price increases in crude oil, a great deal of attention has been focussed on how industrialised economies might conserve on energy and still maintain reasonable economic progress. In this paper we consider one aspect of this issue, namely, the relation between energy conservation and capital formation.

The energy-capital relationship has been analysed extensively in the last few years by, among others, mechanical and industrial engineers. One set of engineering studies has compared actual energy efficiency of existing plant and equipment to the maximum efficiency possible based on the Second Law of Thermodynamics and has shown, not unexpectedly, that the current average efficiency of about 5-7 per cent leaves considerable

room for improvement.(1) Another set of engineering studies has been
more applied and has shown in considerable detail how either through
retrofitting or completely new blueprint designs, equipment can be made
more energy efficient but in almost all cases at the cost of a larger
initial capital outlay than would be required if the equipment were less
energy efficient.(2) In brief, the salient finding of this engineering
literature is the trade-off between energy consumption and initial capital cost outlay.

Another approach to the energy-capital relationship utilises economic
theory and econometric procedures. This approach begins with the specification of a production function or dual cost function, then obtains derived demand equations for energy and non-energy inputs, and finally
yields estimated price elasticities among energy and non-energy inputs.(3)
A somewhat counter-intuitive empirical finding of our earlier published
econometric paper /Berndt and Wood (1975)7 was that in United States manufacturing, 1947-71, energy and capital were complementary inputs rather
than substitutes. This empirical finding appears to contradict the basic
"hard" intuition of engineering energy-capital substitutability. Skepti-
cism about the reliability and credibility of our energy-capital complementarity finding is not, however, confined to engineers. A recent econometric study by James M. Griffin and P.R. Gregory (1976) reports energy-
capital substitutability and argues for the plausibility of this finding
using engineering examples.

The above remarks suggest that it would be desirable to develop an
analytical and empirical framework that relates engineering energy-capital
substitutability to economic energy-capital complementarity, and that also
reconciles the seemingly disparate econometric evidence. That is the purpose of this paper. In Section II we present a summary of the underlying

---

1) See, for example, E.P. Gyftopoulos et. al. (1974) and the reference
   cited therein.

2) It is not possible for us to provide an exhaustive set of literature
   references. In addition to references cited in the much publicised
   studies of the Ford Foundation Energy Policy Project (1974) and the
   U.S. Federal Energy Administration Project Independence Report, see
   U.S. Office of Emergency Preparedness (1972), U.S. National Academy
   of Sciences (1977), National Petroleum Council (1971, pp. 24-31),
   Eric Hirst et. al. (1976), (1977), Lee Schipper and Joel Darmstadter
   (1977), A.H. Rosenfeld (1977), E.P. Gyftopoulos et. al. (1977), and
   R.H. Socolow (1977).

3) A partial listing of relevant studies here includes E.R. Berndt and
   D.O. Wood (1975), E.R. Berndt and D.W. Jorgenson (1973), M.G.S. Denny
   and C. Pinto (1976), M. Fuss (1977), J.M. Griffen and P.R. Gregory
   (1976), Robert S. Pindyck (1977), Barry C. Field and Charles Grebenstein
   (1977), Jan R. Magnus (1977) and Paul Swain and Gerhard Friede (1976).

economic theory of cost and production, functional separability, derived factor demand functions, and various net and gross measures of substitution elasticities among inputs. Readers familiar with this literature may wish to move on directly to Section III, where we focus attention on a weakly separable two input energy-capital model and introduce the notion of "utilised capital" – a composite of energy and capital. In Section IV we imbed this utilised capital subfunction into a four input capital, labour, energy, and intermediate materials production model and demonstrate that although the relationship between energy and capital is one of apparent substitutability, in fact it is one of energy-capital complementarity (in the sense of Hicks-Allen). In Section V we derive implications of our analytical framework for a reconsideration and reconciliation of the seemingly disparate econometric findings noted above. In Section VI we present a model which does not rely on the utilised capital separability specification, yet still is able to reconcile the engineering notion of energy-capital substitutability with the Hicks-Allen concept of energy-capital complementarity. Finally in Section VII we provide concluding remarks and offer suggestions for further research.

II. THEORETICAL FOUNDATIONS

We begin by considering a positive, twice differentiable, strictly quasi-concave production function with a finite number of inputs,

$$(1) \qquad Y = F(x) = F(x_1, x_2, \ldots, x_n), \qquad x_i > 0$$

relating the maximum possible output, Y, obtainable from any given set of inputs. The set of n inputs is denoted $N = (1, \ldots, n)$, and $\partial F / \partial x_i = F_i$, $\partial^2 F / \partial x_i \partial x_j = F_{ij}$. Both engineering and econometric approaches utilise this production function concept.

In the case of engineering design studies, all the inputs are considered since the objective is to develop and ultimately implement a detailed plan for the production process. Subsequent efforts to improve upon this plan will focus upon any input or subset of inputs which appear to provide opportunities for substantial cost reduction. In contrast, engineering energy conservation studies have tended to concentrate on the efficiency of energy use in the production process. Hence the focus is on a small subset of the inputs to the production process. For example, as will be discussed in greater detail later, a number of engineering studies have shown that energy savings of a particular fuel input, say x_i, are possible if certain new types of equipment inputs, say x_j, are employed. Such detailed engineering process analysis studies typically either ignore all other inputs x_k, $k \neq i,j$, in the production function (1) or else implicitly assume that these other input quantities all remain constant. Econometric studies, on the other hand, often aggregate the myriad of inputs into a much smaller number of composite inputs. Both the engineering process analysis and the econometric approaches frequently rely on the notion of functional separability, which we now define.

Let us partition the set of n inputs, $N = (1,..,n)$ into r mutually exclusive and exhaustive subsets $(N_1, N_2, ..N_r)$, a partition we shall call R. The production function $F(x)$ is said to be weakly separable with respect to the partition R if the marginal rate of substitution between any two inputs x_i and x_j from any subject N_s, $s=1,..,r$, is independent of the quantities of inputs outside of N_s,(1) i.e.

(2) $\qquad \dfrac{\partial(F_i/F_j)}{\partial x_k} = 0$, for all $i, j \in N_s$· and $k \notin N_s$·

The production function $F(x)$ is said to be strongly separable with respect to the partition R if the marginal rate of substitution between any two inputs from subsets N_s and N_t does not depend on the quantities of inputs outside of N_s and N_t, i.e.

(3) $\qquad \dfrac{\partial(F_i/F_j)}{\partial x_k} = 0$, for all $i \in N_s$, $j \in N_t$, $k \notin N_s \cup N_t$·

These separability conditions can alternatively be written as

(4) $\qquad F_j F_{ik} - F_i F_{jk} = 0$

where subscripts follow the pattern noted in (2) and (3).

Weak separability, as employed in engineering and econometric studies of the relationships among input factors, has several important implications. First, weak separability with respect to the partition R is necessary and sufficient for the production function $F(x)$ to be of the form $F(X^1, X^2, ..., X^r)$ where X^s is a positive, strictly quasi-concave, homothetic function of only the elements in N_s, i.e.

(5) $\qquad X^s = f_s(x_i)$, $\qquad i \in N_s$, $\qquad s=1,...,r.$(2)

---

1) A more complete discussion of separability is presented in E.R. Berndt and L.R. Christensen (1973).

2) We rule out the unlikely possibility that prices of all x_k $(k \notin N_s)$ are perfectly correlated with output price.

When f_s is linear homogeneous in x_i, X^S is called a consistent aggregate index of the inputs in N_s. Thus a consistent aggregate index of a subset of inputs exists if and only if the subset of inputs is weakly separable from all other inputs. Engineering process analysis studies which focus only on a small subset of the inputs and ignore all other inputs are appropriate if and only if the subset of inputs are weakly separable from all others. Similarly, econometric studies which utilise input aggregates such as labour (or energy) are valid if and only if the components of labour (energy) are weakly separable from all other non-labour (non-energy) inputs.

A second closely related implication of weak separability is that it permits sequential optimisation.(1) More specifically, if F(x) is weakly separable, then in production decisions relative factor intensities can be optimised within each separable subset, and finally overall optimal intensities can be attained by holding fixed the within-subset intensities and optimising the between-set intensities; the corresponding factor intensities will be the same as if the entire production optimisation decision had been made at once. For example, if electricity and refrigerators are weakly separable from all other inputs, then both engineers and economists do not have to worry about other inputs such as labour or natural gas, but can simply choose refrigerators of the optimal electricity-efficient design, knowing that their optimal energy/refrigerator ratio is independent of other input optimisations.

The most important implication of strong separability is that with respect to the partition R it is a necessary and sufficient condition for the production function to be of the form $F(x) = F(X^1 + X^2 + \ldots + X^r)$, where X^S, $s=1,\ldots,r$, is a function of the elements of N_s only. Notice that strong separability implies weak separability but in general the reverse is not true.

Both economists and engineers analyse production processes with the objective of minimising the cost of production. In particular, it is very useful to specify the optimisation problem as that of minimising the costs of producing a given level of output, subject of course to the exogenous input prices p_1, p_2,...,p_n and the technological constraints embodied in F(x). W. Erwin Diewert (1974) has shown that when F(x) is strictly quasi-concave and twice differentiable, the cost minimisation assumption implies the existence of a dual cost function

(6) $C = G(Y, p_1, p_2, \ldots, p_n)$

---

1) For further discussion, see Charles Blackorby, Daniel Primont, and R. Robert Russell (1975).

relating the minimum possible total cost of producing output Y to the positive input prices $p_1, p_2, \ldots, p_n$ and the output level Y. Notice in particular that the cost function (6) reflects the technological parameters in (1). Also, when the production function is weakly separable with respect to the partition R and when each of the f_s functions is linearly homogenous, the corresponding dual cost function has the same partition in input prices, i.e. the dual cost function is

$$(7) \qquad C = G(Y, P^1, P^2, \ldots, P^r)$$

where the input price aggregates P^s are positive, strictly quasi-concave linearly homogenous functions of only the elements in N_s, i.e.,

$$(8) \qquad P^s = g_s(p_i), \qquad i \in N_s, \qquad s = 1, \ldots, r.$$

The usefulness of the cost function in demand theory is that the cost minimising derived demand equations are directly obtainable from the derivatives of G. As before, denote $\partial G / \partial p_i = G_i$, $\partial^2 G / \partial p_i \partial p_j = G_{ij}$. By Shephard's Lemma, [1] the optimal derived demands are

$$(9) \qquad x_i = G_i, \qquad i = 1, \ldots, n.$$

A question of considerable interest, especially to economists, is the sensitivity of the optimal derived demand for x_i to a change in input prices. Such sensitivity measures will vary of course depending on what other variables are held fixed. The most common elasticity measure, due to J.R. Hicks (1933) and R.G.D. Allen (1938) is the demand <u>price elasticity</u> defined as

$$(10) \qquad \varepsilon_{ij} = \frac{\partial x_i}{\partial p_j} \frac{p_j}{x_i} = \frac{\partial \ln x_i}{\partial \ln p_j}$$

where output Y is held constant, only p_j changes, but all factors are allowed to adjust to their optimal levels. Hicks-Allen defined inputs x_i and x_j as substitutes, independent, or complements according to ε_{ij} was positive, zero, or negative, respectively. Furthermore, when there are only two inputs in the production function (say x_1 and x_2), the strict quasi-concavity condition on the underlying production function F will ensure that $\varepsilon_{12} > 0$, i.e. the two inputs must be substitutable.

In general, $\varepsilon_{ij} \neq \varepsilon_{ji}$. The <u>Allen partial elasticities of substitution</u> σ_{ij} are essentially normalized price elasticities,

$$(11) \qquad \sigma_{ij} = \varepsilon_{ij} / M_j$$

where M_j is the cost share of the jth input in total costs, i.e.

---

1) For further discussion, see W.E. Diewert (1974).

$$(12) \qquad M_j = p_j x_j / \sum_{i=1}^{n} p_i x_i.$$

The effect of this normalization is to make the Allen partial elasticities of substitution symmetric, i.e. $\sigma_{ij} = \sigma_{ji}$, even though $\varepsilon_{ij} \neq \varepsilon_{ji}$. H. Uzawa (1962) has shown that the σ_{ij} can be derived conveniently from the cost function (6) as

$$(13) \qquad \sigma_{ij} = \frac{GG_{ij}}{G_i G_j}, \qquad i,j=1,\ldots,n.$$

Two less familiar but nonetheless interesting alternative substitution elasticity measures are the direct and shadow elasticities of substitution. Daniel McFadden (1963) and Ryuzo Sato and Tetsunori Koizumi (1973), among others, have shown that the <u>direct elasticity of substitution</u> d_{ij} can be defined as

$$(14) \qquad d_{ij} = \frac{\partial \ln (x_i/x_j)}{\partial \ln (p_j/p_i)}, \qquad i \neq j$$

Hence the direct elasticity measures the percentage change in x_i/x_j given an exogenous percentage change in p_j/p_i. We would expect that in general d_{ij} would differ from σ_{ij} or ε_{ij}, since with the d_{ij} only the two inputs x_i and x_j move to their optimal levels (output fixed) while the σ_{ij} or ε_{ij} all factor quantities – not just x_i and x_j – are allowed to adjust to their cost-minimising levels. Two implications of this are worth noting. First, the strict quasi-concavity "curvature" restrictions on $F(x)$ require that all $d_{ij} > 0$, $i \neq j$, even though certain ε_{ij} or σ_{ij} may be negative. Hence we can simultaneously have negative Allen partial elasticities of substitution indicating that inputs i and j are complementary and positive direct elasticities of substitution reflecting two-space substitutability. Second, when there are only two inputs, x_1 and x_2, $\sigma_{12} = d_{12}$.

The final elasticity of substitution measure we consider is the <u>shadow elasticity of substitution</u> S_{ij}, which can be shown to be equal to

$$(15) \qquad S_{ij} = \frac{\partial \ln (p_i/p_j)}{\partial \ln (x_j/x_i)}, \qquad i \neq j,$$

where imputed or shadow prices of the remaining factors and imputed total cost are fixed. The S_{ij} therefore measures the percentage change in the ratio of the input prices p_i/p_j in response to an exogenous percentage change in the quantity ratio x_j/x_i, where prices of the remaining factors

$x_k(k \neq i,j)$ and total shadow cost are held fixed. Again, the curvature conditions on $F(x)$ imply that $S_{ij} > 0$, even though ε_{ij} and σ_{ij} may be negative. Also when there are only two factors of production, x_1 and x_2, $d_{12} = \sigma_{12} = S_{12}$. With more than two production factors, however, in general d_{ij}, S_{ij}, and σ_{ij} will differ.[1] Thus considerable care must be taken in discussing "substitution elasticities", for unless the context makes it clear what variables are being held constant, needless confusion can easily result.

In practice both process engineers and econometricians interested in measuring energy conservation potential may choose to analyse only a subset of the inputs or may be unable to obtain sufficiently detailed or reliable data on all n inputs in (1). For this and other reasons researchers may choose to focus attention on estimating substitution elasticities among and within only a subset of the n inputs. As an example, suppose a researcher has data only on the inputs belonging to the N_s subset in the partition R, and that he wishes to obtain estimates of the price elasticities ε_{ij} among the q inputs within the N_s subset. Although the marginal rate of substitution F_i/F_j, $(i,j \in N_s)$ is independent of all $x_k(k \notin N_s)$ when inputs in N_s are weakly separable from all other inputs, it can be shown that unless additional information is available the researcher cannot obtain estimates of the Hicks-Allen elasticities $\varepsilon_{ij}(i,j \in N_s)$ based only on data for inputs belonging to the weakly separable N_s subset.

Let us define the gross price elasticity ε^*_{ij} $(i,j \in N_s)$ as the derivative $\partial \ln x_i / \partial \ln p_j$ where X^S, the aggregate input or "output" of the weakly separable f_s subfunction (5) is held constant, all x_k $(k \in N_s)$ are allowed to adjust to their optimal levels, but all x_m $(m \notin N_s)$ are held fixed. Let us also define the net price elasticity ε_{ij} as $\varepsilon_{ij} \equiv \dfrac{d \ln x_i}{d \ln P_j}$, $i,j \in N_s$, where now the output $\bar{Y}$ is held constant at $Y = Y$ and all other inputs — not just those in N_s — are allowed to move to their new cost minimising positions. Notice that the gross price elasticity ε^*_{ij} treats X^S as fixed (say, $X^S = \bar{X}^S$), whereas the net price elasticity allows X^S to respond to the change in p_j. More specifically,

$$(16) \qquad \left.\frac{d \ln x_i}{d \ln p_j}\right|_{Y = \bar{Y}} = \left.\frac{\partial \ln x_i}{\partial \ln p_j}\right|_{X^S = \bar{X}^S}$$

$$+ \; \frac{\partial \ln x_i}{\partial \ln X^S} \cdot \frac{\partial \ln X^S}{\partial \ln P^S} \cdot \left.\frac{d \ln P^S}{d \ln p_j}\right|_{Y = \bar{Y}}, \qquad i,j \in N_s.$$

1) Relationships among these three alternative elasticity of substitution concepts in the multiple input case have been analysed by R. Sato and T. Koizumi (1973).

The left-hand side of (16) is the net price elasticity and the first term on the right-hand side is the gross price elasticity; we shall call the last expression on the right side of (16) the scale elasticity. The three terms comprising the scale elasticity may be interpreted as follows. First, because f_s is linearly homogeneous, $\partial \ln x_i / \partial \ln X^S = 1$. Next, the derivative $\partial \ln X^S / \partial \ln P^S$ is the price elasticity for the aggregate X^S given a change in the price of the aggregate input P^S, where $Y = \overline{Y}$. Let us denote this own price elasticity as ε_{ss}. Finally, since $P^S = \sum_i p_i x_i / X^S$, $i \in N_s$, and using Shephard's Lemma $\dfrac{\partial P^S}{\partial p_i} = \dfrac{x_i}{X^S}$, we have

$$(17) \qquad \frac{\partial \ln P^S}{\partial \ln p_j} = \frac{p_j}{P^S} \frac{\partial P^S}{\partial p_j} = \frac{p_j x_j}{P^S X^S}$$

Equation (17) indicates that $\partial \ln P^S / \partial \ln p_j$ is simply the cost share of the jth input in the total cost of producing the input aggregate X^S; hereafter we denote this share as N_{js}. Combining these expressions and substituting into (16), we have that the net elasticity ε_{ij} is the sum of the gross elasticity ε^*_{ij} and the scale elasticity $\varepsilon_{ss} N_{js}$, i.e.

$$(18) \qquad \varepsilon_{ij} = \varepsilon^*_{ij} + \varepsilon_{ss} N_{js}, \qquad i,j \in N_s.$$

Thus the net effect on the derived demand for x_i given a change in the price of x_j $(i,j \in N_s)$ is the sum of a gross effect which holds the input aggregate X^S fixed plus the cost share of the jth input in the total cost of input aggregate X^S times the price elasticity of demand for X^S. Since the curvature conditions on $F(x)$ require that $\varepsilon_{ss} < 0$ and since the cost share $N_{js} > 0$, it follows that $\varepsilon_{ss} N_{js} < 0$ and therefore that $\varepsilon_{ij} < \varepsilon^*_{ij}$. Notice that inputs i and j can be gross substitutes ($\varepsilon^*_{ij} > 0$) but net complements ($\varepsilon_{ij} < 0$).

The implication of this for the researcher dealing only with data on inputs in the weakly separable subset N_s is that in general he cannot estimate ε_{ij}, given data for the variables comprising the subset N_s; all that can be done is to estimate the gross price elasticity ε^*_{ij} and of course the cost share N_{js}. In order to estimate the net price elasticity ε_{ij}, additional information on ε_{ss} is necessary.

The implications of this discussion for reconciling engineering and econometric studies of energy conservation can now be stated. Both engineering and econometric approaches employ the concepts of a production function, weak separability of inputs, and similar assumptions as to the optimising behaviour of producers. Engineering energy conservation studies tend to focus upon the changes in capital design, and the manner in which capital and energy are jointly utilised. Their emphasis is on the subset of inputs including capital and energy inputs, with the assumption of weak separability between these inputs and the other factors of production. Thus, engineering energy conservation studies typically provide information on gross elasticities of substitution between the inputs comprising a subset of factor inputs. Econometric studies, on the other hand, have tended to focus on the relations between all inputs, employing the assumption of weak separability and consequent input aggregation to reduce the number of inputs to a manageable number. When all inputs are included, econometric studies can provide information on the gross elasticities of substitution between factors within a subset, as well as scale elasticities between subsets and consequently net elasticities between any pair of inputs.

III. TWO INPUT ENERGY-CAPITAL MODELS

The recent engineering literature contains numerous blueprint examples of how equipment and appliances could be redesigned or retrofitted to consume less energy, but at the cost of a larger initial capital outlay. For example, E.P. Gyftopoulos et al. (1974) have compared actual fuel use in industry with the theoretically most efficient use based on the Second Law of Thermodynamics; in E.P. Gyftopoulos and T.F. Widmer (1977), percentage changes in fuel efficiency are compared with percentage changes in initial capital cost outlay. Other studies /see, for example, U.S. Environmental Protection Agency (1973)7, have attempted to rank alternative capital-energy use combinations using "life cycle" costing. On the basis of such two-input studies, some economists have been led to conclude that energy (E) and capital (K) are substitutes. For example, J.M. Griffen and P.R. Gregory (1976) state that "....in the long run, one might expect K and E to be substitutes since new equipment could be redesigned to achieve higher thermal efficiencies but at greater capital costs."(1)

Engineering examples of the reduced energy consumption-higher initial cost trade-off are numerous.(2) However, very little econometric work on this issue has been published. Makoto Ohta (1975) reports that the initial capital costs of boilers and turbogenerators purchased by United

---

1) J.M. Griffen and P.R. Gregory (1976), p. 856.
2) See the references cited in footnote 2

States steam generating electric utilities have varied positively with energy efficiency characteristics (essentially, the ratio of electrical output to fuel input per time period). In the residential market, econometric analysis based on hedonic approaches suggests that for air conditioners[1] and refrigerators,[2] when cooling capacity is held fixed, list price and energy efficiency are positively correlated; equivalently, with cooling capacity fixed, there is a trade-off between lower electricity consumption and higher initial capital cost. For example, Berndt and Wood (1977) show that the list price of a 16.5 cubic foot self-defrosting, 1976 model refrigerator (say, 12 cubic feet of refrigerator volume and 4.5 cubic feet of freezer volume) is predicted to increase from about \$425 to about \$460 as average monthly electricity required decreases from 150 to 125 kilowatt hours.

These examples do not necessarily imply, however, a trade-off between the flow of capital services and the quantity of energy. Initial capital cost must first be decomposed into a quantity service flow and a rental price per unit of time. In order to construct a capital quantity flow, additional information would be needed on, for example, durability and physical deterioration over time; similarly, capital rental price measurement would require information on economic depreciation, the opportunity cost of capital, and in the industrial sector, effective rates of corporate taxation. Nonetheless, the proposition that there is a trade-off between capital quantity flow and energy consumption per unit of time, other inputs now considered, seems reasonable to us.[3]

It is obvious that a great deal of industrial equipment relies for power on some type of energy. Therefore let us hypothesise a two-input weakly separable function which combines the inputs of aggregate capital and aggregate energy and produces an output called "utilised capital".[4] In the context of refrigerators or air conditioners, utilised capital services could be the number of hours in which a specified amount of space is cooled to a certain temperature; such utilised capital services would be the output of a production process with two inputs - a refrigerator or air conditioner and DWH of electricity.

---

1) See Jerry A. Hausman (1978).

2) See E.R. Berndt and David O. Wood (1977).

3) Thomas G. Cowing (1974), Table III, p. 149, provides a measure of this trade-off for steam boilers and turbogenerators, but does not express it in elasticity form.

4) For other two input E-K discussions, see Paul W. Zarembka (1974) and Cowing (1974). Our approach differs from that which uses electricity consumptionsas a "proxy" measure for actual capital services, e.g. see Carlisle E. Moody, Jr. (1974) and the references cited therein.

More formally, assume that gross output Y is produced by a cost-minimising competitive firm according to the positive, twice differentiable, strictly quasi-concave production function

(19) $Y = F(K,L,E,M)$,

where K is an input aggregate of capital services, L of labour services, E of energy, and M of non-energy intermediate materials. Within the production function (19), we assume there exists a weakly separable linearly homogeneous utilised capital subfunction with only two inputs, aggregate capital and aggregate energy.

(20) $K^* = f(K,E)$

which together produce the utilised capital output K*. With output quantity and input prices exogenous, the regularity conditions on (19) and (20) imply the existence of a gross output dual cost function relating total cost $C = P_K K + P_L L + P_E E + P_M M$ to gross output Y and to the input prices P_K, P_L, P_E, and P_M,

(21) $C = G(Y,P_K,P_L,P_E,P_M)$;

the existence of the separable subfunction (20) in the production function (19) implies the existence of a dual separable utilised capital cost subfunction in the same input partition,

(22) $C_{K*} = g(K^*,P_K,P_E)$

where $C_{K*} = P_K K + P_E E$; It might be noted here that the assumption of a linearly homogeneous weakly separable utilised capital subfunction implies that the optimal E/K ratios within the utilised capital subfunction depend solely on P_K, P_E, and not on the other input prices P_L, P_M or the level of gross output Y. Hence, under the above separable utilised capital specification, engineers interested in energy conservation issues are able to choose K/E ratios so as to minimise the unit cost C_{K*}/K^* of producing utilised capital services, without having to consider prices of other inputs such as P_L or P_M. In turn, the firm will then determine the cost-minimising total amount of utilised capital K* it demands - a decision which will depend of course on the price of utilised capital services (P_{K*}), and on P_L, P_M, and Y.

Based on the utilised capital cost subfunction (22), we can define gross price elasticities as follows:

(23) $\varepsilon^*_{ij} = \dfrac{\partial \ln x_i}{\partial \ln p_j}$, $i,j = K,E$

where utilised capital output K* is fixed. These gross price elasticities must be interpreted carefully. For example, ε_{KE}^{*} measures the percentage change in the quantity of capital demanded in response to a percentage change in the price of energy, assuming P_K and utilised capital output K* is fixed. Hence these gross price elasticities do not allow for the scale effect mentioned in the previous section, wherein the energy price change affects the unit cost of producing utilised capital services, which in turn brings about a change in the amount of K* demanded and thus results in a change in the derived demand for K.

Since the utilised capital subfunction has only two inputs, the regularity conditions on (22) require E and K to be substitutes; hence ε_{KE}^{*} and ε_{EK}^{*} must be positive. Equivalently, in this two-input model E and K must be substitutable along a strictly convex utilised capital isoquant. We construe much of the recent literature dealing with possibilities for energy conservation as focussing on the real world possibilities for movement along (or a shift in) the utilised capital isoquant. As noted earlier, either through equipment retrofitting or through new equipment design, the engineering-economic energy conservation literature illustrates the fact that a given amount of utilised capital services can be produced with less E but more K. Since this literature typically deals with only two inputs, E-K complementarity is ruled out; for E-K complementarity to be present, it is necessary that the production function have more than two inputs.

IV. AN INTERPRETATION OF ENERGY-CAPITAL COMPLEMENTARITY

Although economics students typically learn about complementary inputs or commodities in their principles courses, the underlying intuitive basis for complementarity remains surprisingly elusive. Paul A. Samuelson (1974) illustrates the possible confusion with a classic coffee, tea, and lemon example:

> "..... we 'know' that coffee and tea are 'substitutes' because we can drink one or the other; in the same way, we know that tea and lemon are 'complements', because tea with lemon makes up our desired brew. And probably we feel that tea and salt are somewhere between being substitutes and being complements: relatively speaking, tea and salt are in the nature of 'independents'.
> "Beyond these simple classifications the plain man may hesitate to go. Thus, sometimes I like tea and lemon; sometimes I like tea and cream. What would you say is the relation between lemon and cream

for me? Probably substitutes. I also sometimes take cream with my coffee. Before you agree that cream is therefore a complement to both tea and coffee, I should mention that I take much less cream in my cup of coffee than I do in my cup of tea. Therefore a reduction in the price of coffee may reduce my demand for cream, which is an odd thing to happen between so-called complements; at least this is in contrast to the case of the tea-and-lemon complements where we should expect a reduction in the price of either to increase the demand for both (as I am induced to consume more cups of lemoned tea).

"Things are not so plain sailing after all, We ..... are not so sure what it is that we know."(1)

It is fitting, therefore, that we attempt to develop a more precise intuition for E-K complementarity consistent with the Hicks-Allen demand framework, yet also simultaneously consistent with the engineering-economic notion of gross E-K substitutability.

Recall that we have specified a weakly separable linearly homogeneous utilised capital production subfunction $K^* = f(K,E)$ imbedded within a "master" production function $Y = F(K,L,E,M) = F(K^*,L,M)$ and the corresponding dual unit cost subfunction for utilised capital $C_{K*}/K^* = g(P_K,P_E)$ nested within a master unit cost function $C/Y = G(P_K,P_L,P_E,P_M)$. Gross price elasticities between E and K were defined in (23).

The net price elasticities for the general case have been derived and defined in (16) and (18) of Section II. In the present context, the net price elasticities for E and K are

$$(24) \qquad \varepsilon_{KE} = \varepsilon^*_{KE} + \varepsilon_{uu}N_E, \qquad \varepsilon_{EK} = \varepsilon^*_{EK} + \varepsilon_{uu}N_K$$

$$(25) \qquad \varepsilon_{KK} = \varepsilon^*_{KK} + \varepsilon_{uu}N_K, \qquad \varepsilon_{EE} = \varepsilon^*_{EE} + \varepsilon_{uu}N_E$$

where ε_{uu} is the price elasticity of demand for utilised capital services K^* (gross output fixed) and N_K and N_E are the cost shares of K and E in the total cost of producing output K^*. Equation (24) indicates, for example, that the net price elasticity ε_{KE} along a gross output isoquant (where $Y = \overline{Y}$) is equal to the positive gross substitution elasticity ε^*_{KE} along a utilised capital isoquant (where $K^* = \overline{K}*$) plus another term which reflects the cost share of energy in the K^* subfunction (N_E) times ε_{uu}, the price elasticity of demand for K^*. Notice in particular that even though the gross substitution term ε^*_{KE} is positive, the sign of

_____

1) P.A. Samuelson (1974), p. 1255.

the net elasticity ε_{KE} is indeterminate <u>a priori</u>. If the negative scale effect $\varepsilon_{uu}N_{E_*}$ is larger in absolute value than the positive gross price elasticity ε_{KE}^*, then energy and capital will be gross substitutes but net complements.

At this point it might be useful to demonstrate this gross substitute-net complement phenomenon geometrically. For simplicity, we first specify another weakly separable lineraly homogeneous production subfunction with two inputs,

$$(26) \qquad L^* = h(L,M)$$

where L^* is the output of the labour-materials production subfunction. Hence the "master" production function can be written as $Y = F(K,L,E,M) = F(K^*,L^*)$ and the dual master cost function as $C = G(Y,P_K,P_L,P_E,P_M) = G(Y,P_K^*,P_L^*)$ where P_K^* and P_L^* are unit costs of the utilised cost subfunctions,

$$(27) \qquad P_K^* = C_K^*/K^* = g^*(P_K,P_E), \qquad P_L^* = C_L^*/L^* = h^*(P_L,P_M)$$

where $C_L^* = P_L L + P_M M$.

We now illustrate (24) and (25) with the following example. Suppose that a cost-minimising competitive firm was in equilibrium producing gross output $Y = Y_1^*$. Given the original input prices P_K^* and P_L^*, the slope of the isocost line AA' in Figure 1 is $-P_L^*/P_K^*$, and the firm minimises costs of production Y_1^* at O_1 using K_1^* units of utilised capital and L_1^* units of the labour-materials composite. Given the original prices P_K and P_E as reflected in the isocost line BB' in Figure 2, the firm produces the K_1^* output at O_2 using K_1 units of capital and E_1 units of energy; similarly, the L_1^* output, given P_L and P_M as reflected in the isocost line CC' in Figure 3, is produced at O_3 using L_1 units of labour and M_1 units of materials.

Now let us assume that the federal government introduces investment incentives that lower P_K. The total effect on the elasticity of demand for capital, ε_{KK}, and on the demand for energy, ε_{EK}, consists of two components - as shown in (24) and (25). First, holding fixed the output of the utilised capital subfunction $K^* = K_1^*$ the steeper isocost line DD' in Figure 2 (due to the lower P_K) indicates that demand for capital would increase from K_1 to K_2, and that demand for energy would fall from E_1 to E_2; in (24) and (25), these gross substitution effects are represented by ε_{KK}^* and ε_{EK}^*, respectively. Second, since the investment incentives decrease P_K, this reduces the cost C_K^* of producing utilised capital services, and by (27) lowers P_K^*. This changes the isocost line in Figure 1 from AA' to a steeper isocost line FF', and results in a new cost-minimising equilibrium at O_5 where derived demand for utilised capital increases from K_1^* and K_2^*, while demand for L^* falls from L_1^* to L_2^*. This results in an outward shift of the K^* isoquant, as shown in Figure 2, increasing derived demand both for capital and for energy; at the new equilibrium O_6, this scale effect increases derived demand for capital from K_2 to K_3 and demand for energy from E_2 to E_3. For capital, the gross substitution effect (K_1 to K_2) and scale effect (K_2 to K_3) reinforce each other, but for energy the two effects work in opposite directions; the gross substitution effect decreases energy demand from E_1 to E_2, whereas the scale effect increases demand for energy from E_2 to E_3. Note that E_3 is larger than E_1. In the particular example of Figure 2, the scale effect $N_k \varepsilon_{uu}$ dominates the gross substitution effect ε_{EK}^*, and thus, although in this example E and K are gross substitutes ($\varepsilon_{EK}^* > 0$), they are net complements ($\varepsilon_{EK} < 0$). Notice that net complementarity implies that the investment incentives contribute to increased (not reduced) energy demand. It might also be noted that the effect of the investment incentives is to lower L^* from L_1^* to L_2^*; as seen in Figure 3, at the new equilibrium O_7 the scale effect results in a

Figure 1
MASTER FUNCTION ISOQUANT

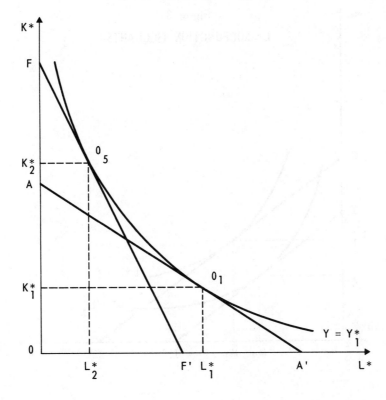

Figure 2
UTILIZED CAPITAL SUBFUNCTION ISOQUANTS

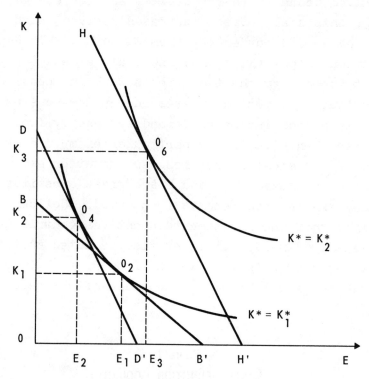

Figure 3
L* SUBFUNCTION ISOQUANTS

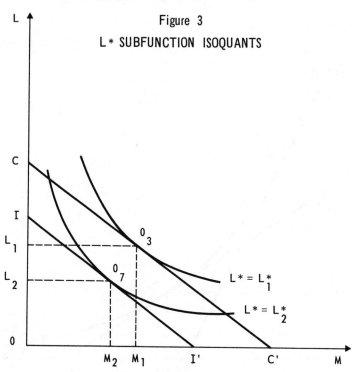

reduction of derived demand both for L and M from L_1 to L_2 and from M_1 to M_2.(1)

Our simple model suggests then that whether net K-E substitutability or net K-E complementarity exists depends on whether the gross substitution effect or the scale effect is dominant.(2) This is, of course, an empirical issue. In order to implement our simple model empirically, we must specify functional forms for the master function and for the K^* and L^* subfunctions. For convenience, we will specify forms for the dual cost functions. In specifying the separable cost subfunctions, we wish to employ a rather flexible or unrestrictive function. For both the K^* and L^* cost subfunctions, we employ the translog form

$$(28) \quad \ln(C_{K^*}/K^*) = \ln \delta_{K^*} + \alpha_K \ln P_K + \alpha_E \ln P_E + \frac{1}{2} \gamma_{KK}(\ln P_K)^2$$

$$+ \gamma_{KE}\ln P_K \ln P_E + \frac{1}{2} \gamma_{EE}(\ln P_E)^2$$

and

$$(29) \quad n(C_{L^*}/L^*) = \ln \delta_{L^*} + \alpha_L \ln P_L + \alpha_M \ln P_M + \frac{1}{2} \gamma_{LL}(\ln P_L)^2$$

$$+ \gamma_{LM}\ln P_L \ln P_M + \frac{1}{2} \gamma_{MM}(\ln P_M)^2$$

where

$$(30) \quad \alpha_K + \alpha_E = \alpha_L + \alpha_M = 1$$

$$\gamma_{KK} + \gamma_{KE} = \gamma_{KE} + \gamma_{EE} = 0$$

$$\gamma_{LL} + \gamma_{LM} = \gamma_{LM} + \gamma_{MM} = 0$$

The translog form is attractive in that it imposes no a priori restrictions on the Allen partial elasticities of substitutions.

To complete our empirical model specification, we assume the master production function is the familiar strongly separable linearly homogeneous Cobb-Douglas function with two inputs, K^* and L^*. We also assume that

---

1) In terms of the refrigerator example mentioned earlier, the investment incentives would induce a shift towards a more capital and less energy intensive refrigerator of a given type (the gross substitution effect), but because these investment incentives would reduce the unit cost (price) of utilised refrigerator services, they would also encourage purchases of a larger, or more sophisticated self-defrosting refrigerator that uses more capital and more energy, but less labour time (the scale effect).

2) This result is analogous to the familiar income and substitution effects of consumer demand theory.

any technical change is constant exponential Hicks-neutral. The dual master unit cost function is then written as

$$(31) \quad \ln(C/Y) = \ln\beta_o + \alpha_t t + \beta_K{}^* + \beta_L{}^* \ln P_L{}^*$$

where $P_K{}^*$ and $P_L{}^*$ are defined as equalling the unit cost of K^* and L^* /see (27)/, t represents time, and where $\beta_K{}^* + \beta_L{}^* = 1$.

Substituting (28) and (29) into (31) and using (27), we can write the master cost function in terms of the separable subfunction prices and parameters:

$$(32) \quad \ln(C/Y) = n\alpha_o + \alpha_t t + \sum_i \beta_i \ln P_i + \frac{1}{2} \sum_i \sum_j \beta_{ij} \ln P_i \ln P_j, \quad i,j = K,L,E,M$$

where $\beta_{ij} = \beta_{ji}$, $\sum_i \beta_{ij} = \sum_j \beta_{ij} = 0$, $\sum_i \beta = 1$, and

$$(33) \quad \ln\alpha_o = \ln\beta_o + \beta_K{}^* \ln\delta_K{}^* + \beta_L{}^* \ln\delta_L{}^*$$

$$\beta_{KL} = \beta_{KM} = \beta_{LE} = \beta_{EM} = 0, \qquad \beta_M = \beta_{L*}\alpha_M$$

$$\beta_{KK} = \beta_K{}^* \gamma_{KK} \qquad \beta_{KE} = \beta_K{}^* \gamma_{KE} \qquad \beta_{EE} = \beta_{K*} \gamma_{EE}$$

$$\beta_{LL} = \beta_L{}^* \gamma_{LL} \qquad \beta_{LM} = \beta_L{}^* \gamma_{LM} \qquad \beta_{MM} = \beta_{L*} \gamma_{MM}$$

$$\beta_K = \beta_K{}^* \alpha_K \qquad \beta_L = \beta_L{}^* \alpha_L \qquad \beta_E = \beta_K{}^* \alpha_E$$

In summary, our simple empirical model includes a strongly separable Cobb-Douglas master cost function of K^* and L^*, where the K^* and the L^* subfunctions are translog with inputs K and E, L and M, respectively. This simple model appears to have substitutable relationships everywhere – between K and E in the two input utilised capital subfunction, between the L and M in the two input L^* subfunction, and between K^* and L^* in the two input master Cobb-Douglas gross output cost function. As we shall now show, however, this simple model is completely consistent with energy-capital complementarity.

Based on the master cost function specification (32)-(33), we utilise (13) and compute Allen partial elasticities of substitution σ_{ij} and price elasticities ε_{ij} along a gross output isoquant as

$$(34) \quad \sigma_{ij} = \frac{\beta_{ij} + M_i M_j}{M_i M_j}, \qquad i,j = K,L,E,M, \quad i \neq j$$

$$\sigma_{ii} = \frac{\beta_{ii} + M_i^2 - M_i}{M_i^2}, \qquad i = K,L,E,M$$

and

$$(35) \qquad \varepsilon_{ij} = M_j \, \sigma_{ij}, \qquad\qquad\qquad i,j = K,L,E,M$$

where the M_i are the cost shares of the ith input in the total cost C of producing gross output, obtained by logarithmically differentiating (32) and using Shephard's Lemma:

$$(36) \qquad M_i \equiv \frac{\partial \ln C}{\partial \ln P_i} = \frac{P_i X_i}{C} = \beta_i + \sum_j \beta_{ij} \ln P_j, \quad i,j = K,L,E,M$$

The price elasticities in (35) are of course net price elasticities. It is of interest to rewrite the net price elasticities (35) in terms of gross price elasticities $\bar{\varepsilon}^*_{ij}$ and scale elasticities. Using (33) and the fact that

$$(37) \qquad M_K = \beta_K^* N_K, \quad M_E = \beta_K^* N_E, \quad M_L = \beta_L^* N_L, \quad M_M = \beta_L^* N_M$$

where from (28) and (29)

$$(38) \qquad N_K = \frac{\partial \ln (C_K^*/K^*)}{\partial \ln P_K} = \frac{P_K K}{C_K^*} = \alpha_K + \gamma_{KK} \ln P_K + \gamma_{KE} \ln P_E$$

$$N_E = \frac{\partial \ln (C_K^*/K^*)}{\partial \ln P_E} = \frac{P_E E}{C_K^*} = \alpha_E + \gamma_{KE} \, n \, P_K + \gamma_{EE} \ln P_E$$

$$N_L = \frac{\partial \ln (C_L^*/L^*)}{\partial \ln P_L} = \frac{P_L L}{C_L^*} = \alpha_L + \gamma_{LL} \ln P_L + \gamma_{LM} \ln P_M$$

$$N_M = \frac{\partial \ln (C_L^*/L^*)}{\partial \ln P_M} = \frac{P_M M}{C_L^*} = \alpha_M + \gamma_{LM} \ln P_L + \gamma_{MM} \ln P_M$$

we can rewrite the net price elasticity ε_{ij} in terms of the gross price elasticity ε^*_{ij} and the scale elasticity. In the context of the E-K net price elasticities, we have

$$(39) \qquad \varepsilon_{KE} = \varepsilon^*_{KE} - N_E \beta_L^*, \qquad \varepsilon_{EK} = \varepsilon^*_{EK} - N_K \beta_L^*$$

$$\varepsilon_{KK} = \varepsilon^*_{KK} - N_K \beta_L^*, \qquad \varepsilon_{EE} = \varepsilon^*_{EE} - N_E \beta_L^*$$

Equation (39) indicates that, for example, the net price elasticity ε_{KE} along a gross output isoquant is equal to the gross price elasticity $^*_{KE}$ along a utilised capital isoquant plus the scale elasticity term $(-N_E \beta_L^*)$ which reflects the cost share of energy in the K^* subfunction times the price elasticity of demand for K^* which in this particular model is $\beta_K{}^* - 1 = -\beta_L{}^*$. Since $-N_E\beta_L{}^*$ is negative, $\varepsilon_{KE} < \varepsilon_{KE}^*$. Even though the gross substitution term ε_{KE}^* is positive, the sign of the net elasticity ε_{KE} depends on whether the absolute value of the scale elasticity is larger or smaller than the positive gross price elasticity ε_{KE}^*.

Empirical research is therefore necessary in order to determine the relative magnitudes of the gross substitution and scale elasticities. An interesting feature of (33) is that it constitutes a set of parametric restrictions on the more general four input KLEM translog unit cost function with Hicks-neutral constant exponential technical change (32). E.R. Berndt and David O. Wood (1975) have called the set of restrictions (33) on (32) linear separability restrictions for $/(K,E),(L,M)/$ separability. They report these restrictions could not be rejected with their data - annual Unites States manufacturing, 1947-71.[1] Furthermore, Berndt-Wood tested for many different types of separability among the K, L,E and M inputs; all forms except that represented in (33) were rejected. These results therefore provide some empirical support for our simple model specification.[2]

In Table 1 we present maximum likelihood estimates of the parameters in the share equations (36) with the utilised capital separability restrictions (33) imposed based on the Berndt-Wood data. Each of the coefficients is significantly different from zero, and the fit, as measured by the generalised R^2 measure, is good.[3]

Based on these parameter estimates, we compute maximum likelihood estimates of selected net, scale, and gross elasticities for the last

---

1) In Berndt-Wood (1975), we presented estimated elasticities based on iterative three stage least squares estimation. In the present paper, we assume input prices and gross output quantity are exogenous, and estimate the parameters of the share equations (36) using maximum likelihood procedures; the 13SLS and maximum likelihood estimates are virtually identical. The likelihood ratio test statistic of the four independent restrictions in (33) using maximum likelihood estimation is 10.326, while the .01 chi-square critical value is 13.277; under 13SLS estimation, the Wald test statistic is 9.038 and the .01 chi-square critical value remains 13.277.

2) We have also tested for the validity of a related utilised capital separability specification using the three input K,L,E data of Griffen-Gregory, which they kindly provided us. Based on their data, the chi-square test statistic for the two restrictions is 2.2505, while the .01 chi-square critical value is 9.210. Hence, using the Griffen-Gregory KLE data, we cannot reject the null hypothesis of non-linear $/(K,E),L/$ separability.

3) For further discussion, see Berndt and Wood (1977) and E.R. Berndt (1977).

year of our sample (1971).(1) These are presented in the top panel of Table 2. There it is seen that although E and K are gross substitutes in United States manufacturing, they also are net complements. The gross substitution effect (.133) is dominated by the scale effect (-.462), resulting in a value for the net elasticity ε_{EK} of -.329. These elasticities are significantly different from zero.

The above results were based on United States manufacturing time series data. To investigate the robustness of our net complementarity findings within a utilised capital framework, we now estimate a slightly

Table 1

MAXIMUM LIKELIHOOD PARAMETER ESTIMATES WITH THE SEPARABLE UTILISED CAPITAL SPECIFICATION $\underline{/}(K,E),(L,M)\underline{7}$ LINEAR SEPARABILITY RESTRICTIONS IMPOSED, U.S. MANUFACTURING, 1947-1971

Parameter	Parameter Estimate	Ratio of Parameter Estimate to Asymptotic Standard Error
β_{K*}	.0983	89.78
β_{L*}	.9017	823.46
α_K	.5702	231.46
α_E	.4298	174.48
α_L	.2800	118.16
α_M	.7200	303.84
$\gamma_{KK} = -\gamma_{KE} = \gamma_{EE}$	.1851	15.16
$\gamma_{LL} = -\gamma_{LM} = \gamma_{MM}$	.0868	12.87
Implied Estimates		
β_K	.0561	75.79
β_L	.2525	124.42
β_E	.0423	93.13
β_M	.6492	352.84
$\beta_{KK} = -\beta_{KE} = \beta_{EE}$	.0182	15.38
$\beta_{LL} = -\beta_{LM} = \beta_{MM}$	.0782	12.88
Fit Statistic		
$\tilde{R}^2 = .9844$		

1) Since all our elasticity estimates are very stable over the 1947-71 time period the year 1971 can be interpreted as representative. It is also worth noting that all our fitted shares were positive and that the strict quasi-concavity curvature conditions were satisfied for all years in our sample.

Table 2

NET, SCALE, AND GROSS SUBSTITUTION ELASTICITIES IN UTILISED CAPITAL MODEL
/⁻(K,E),(L,M)⁻7 SEPARABILITY RESTRICTIONS IMPOSED
U.S. AND CANADIAN MANUFACTURING, 1971
(Estimated Asymptotic Standard Errors in Parentheses)

Net Elasticity	Gross Substitution Elasticity	Scale Elasticity	Value of Net Elasticity
U.S. Manufacturing, 1971			
ε_{KK}	-.126 (.024)	-.462 (.003)	-.588 (.026)
ε_{EE}	-.133 (.026)	-.440 (.003)	-.573 (.024)
ε_{KE}	.126 (.024)	-.440 (.003)	-.314 (.027)
ε_{EK}	.133 (.026)	-.462 (.003)	-.329 (.026)
Canadian Manufacturing – Ontario, 1971			
ε_{KK}	-.039 (.009)	-.765 (.238)	-.744 (.238)
ε_{EE}	-.505 (.115)	-.054 (.018)	-.559 (.117)
ε_{KE}	.039 (.009)	-.054 (.018)	-.015 (.020)
ε_{EK}	.501 (.115)	-.705 (.238)	-.200 (.264)
Canadian Manufacturing – British Columbia, 1971			
ε_{KK}	-.121 (.011)	-.664 (.206)	-.705 (.206)
ε_{EE}	-.650 (.052)	-.123 (.038)	-.773 (.066)
ε_{KE}	.121 (.011)	-.123 (.038)	-.002 (.040)
ε_{EK}	.650 (.052)	-.664 (.206)	-.014 (.213)

generalised model using pooled cross-section time series data for Canadian manufacturing, by region, 1961-1971. Recently Melvyn A. Fuss (1977) published estimates of substitution elasticities for Canadian manufacturing based on a non-homothetic KLEM translog cost function.(1) Because Fuss

1) Fuss' principal findings for Canadian manufacturing were similar to the Berndt-Wood results for United States manufacturing. In particular, Fuss' price elasticity estimates (calculated at the mean values for Ontario) are ε_{EE} = -.486, ε_{KK} = .762, ε_{EK} = ٠050, ε_{KE} = .004, ε_{EL} = .554, and ε_{LE} = .043. Hence the Berndt-Wood (1975), Fuss finds E-K complementarity and E-L substitutability.

specified a non-homothetic translog cost function and estimated using an error components estimation procedure (the "covariance method"), the conditions (33) for separability are not directly applicable or testable.(1)

To preserve the distinguishing features of Fuss' paper - a non-homothetic translog specification and the covariance estimation method - we proceed with separate estimation of the gross, net, and scale elasticities as follows. First the energy-capital (K^*) and labour-materials (L^*) subfunctions are again specified as (28) and (29). Using Canadian manufacturing data, 1961-71, by region,(2) we estimate the N_K and N_L equations in (38) using the covariance method. We then insert the resulting parameter estimates into (28) and (29) and form fitted data series for P_K^* and P_L^*; these values are unique up to a multiplicative scaling, since the intercept terms in (28) and (29) cannot be identified.(3) Secondly, for our master function we follow Fuss and specify a non-homothetic translog cost function and then estimate the resulting "master" cost share equations using the covariance method.(4) Finally, we compute the associated gross, scale, and net substitution elasticities; 1971 estimates for two provinces - Ontario and British Columbia - are presented in the bottom two panels of Table 2.

As shown in Table 2, for Canadian manufacturing, E and K are gross substitutes but net complements. The net substitution effect for Ontario in 1971 (-.200) is negative, while for British Columbia the gross substitution and scale effects almost offset each other, resulting in only a very slight net complementarity value of -.014.(5) Unlike the United States manufacturing case, however, for Canadian manufacturing the negative net elasticity estimates are statistically insignificantly different from zero.

---

1) We note that in the context of a translog gross output function, it is not possible to test for non-homothetic separability; see Blackorby, Primont, and Russell (1977) and Denny-Fuss (1977).

2) The Canadian manufacturing data was kindly provided us by Melvyn Fuss. Using this data, we successfully replicated the KLEM results reported by Fuss (1977), Table 4, p. 109), except for a typographical error on his reported estimate of γ_{MM}. The correct estimate of γ_{MM} is .0618, and the correct standard error estimate for γ_{MM} is .0140.

3) This approach is equivalent to that used by Fuss in forming an aggregate P_E series from data on constituent fuel prices.

4) Further details are presented in Berndt and Wood (1977).

5) It should be noted that with this specification and Fuss' data, the estimated parameters and fitted cost shares satisfy the strict quasi-concavity conditions in all regions for all years, 1961-71.

V. TOWARDS A RECONCILIATION OF SEEMINGLY DISPARATE ECONOMETRIC FINDINGS

In the preceding paragraphs we have emphasized that results of engineering process analysis are consistent with our E-K complementarity econometric findings. In effect, we have shown that the Griffen-Gregory intuitive argument for E-K substitutability can be misleading: the engineering notion of E-K substitutability does not necessarily imply net E-K substitutability in the sense of Hicks-Allen. Griffen-Gregory (hereafter, GG) have, however, published econometric findings which appear to report Hicks-Allen E-K substitutability. Thus it remains to reconcile our econometric findings with those of GG.

GG have estimated a three input (K,L, and E - but not M) translog cost function based on data for the manufacturing sector of nine industrialised OECD countries in four benchmark years - 1955, 1960, 1965, and 1969.(1) Parameters are estimated using a maximum likelihood procedure. For reasons unspecified, however, GG do not use maximum likelihood methods in estimating elasticities. Recall that Allen partial elasticities of substitution σ_{ij} (34) and price elasticities ε_{ij} (35) are functions of the translog parameters and the cost shares (36). Maximum likelihood (ML) estimates of the price elasticities are of course obtained by inserting the ML parameter estimates into (36) and then using these "fitted" or "predicted" shares along with the ML parameter estimates in (34) and (35) to compute estimates of ε_{ii} and ε_{ij}. Instead of using predicted shares, GG insert actual data shares into (34) and (35). The difference between their procedure and the ML approach would be negligible if the estimated model fitted their data closely, but unfortunately this is not the case with the GG model - especially for the United States. The $\widetilde{R}^2$ for the GG preferred Model I is .41, and the difference between fitted and actual cost shares for the United States is considerable. For example, in 1965 - the year for which GG report elasticity estimates - the GG predicted capital cost shares in the United States is .2205, while the actual share is a 35 per cent lower .1436; the corresponding predicted (actual) cost shares for labour and energy are .6622 (.7311) and .1174 (.1253). It should be noted, however, that the difference between the GG and ML price elasticity estimates is quite small for all countries except the United States.(2) Since GG compared their results with our United States manufacturing findings, it is useful to examine their 1965 United States estimates more closely. The GG (ML) estimates for ε_{KK} in the United States

---

1) The nine countries are Belgium, Denmark, France, West Germany, Italy, Netherlands, Norway, United Kingdom, and the United States.

2) For further details, see Berndt and Wood (1977).

are -.18 (-.34), for ε_{KL}, .05 (.22) and for ε_{KK}, .15 (.23). These differences in the ML and GG estimates reflect the rather poor fit of the GG model to their United States data, and ought to make one cautious in comparing their United States results with those of Berndt-Wood.(1) With respect to the ε_{EK} elasticity, it might also be noted that the crucial β_{KE} parameter reported in GG has a very large standard error estimate. This leads to two standard error confidence intervals for ε_{EK} which include E-K complementarity; for the United States the confidence intervals are $-.05 \leq \varepsilon_{EK} \leq .51$ (ML) and $-.13 \leq \varepsilon_{EK} \leq .43$ (GG). The above comments suggest then that differences between the econometric results of Griffen-Gregory and Berndt-Wood may well be statistically insignificant.

An even stronger analytical argument can be made that all σ_{KE} estimates based on three input KLE models are upward biased. GG noted that comparison of their elasticity estimates with those of Berndt-Wood might be questioned since, unlike Berndt-Wood, they omit M and justify this omission by assuming weak separability of the form $\underline{/}(K,L,E),\underline{M}\underline{/}$. GG correctly note that "....this omission may bias our findings if our weak separability assumption...is invalid" $\underline{/}\overline{G}G$, (1976), p. 852$\underline{7}$. We now show that even if this weak separability assumption were valid, all the GG elasticity estimates reflect gross substitution elasticities and therefore all are upward biased.

A number of authors - among them Griffen-Gregory, Robert S. Pindyck (1977) and Jan R. Magnus (1977) - have been unable to obtain sufficiently reliable data on M, and for this reason they have estimated substitution elasticities among K,L, and E assuming that these three inputs are weakly separable from M, i.e. that $Y = F(K,L,E,M) = h\underline{/}\overline{h}^{*}(K,L,E),\underline{M}\underline{/} = h^{**}(V,M)$ where V is the output of the $h^{*}(K,L,E)$ production subfunction. Even if this untested restrictive KLE specification were valid, however, the resulting elasticity estimates are not in general directly comparable to those based on four input KLEM models.

The economic intuition on this issue is similar to that utilised in Section IV. Suppose that the price of energy increases, other input prices remaining fixed. If within the KLE subfunction K and E are gross substitutes, then the increased energy prices will induce substitution toward capital, holding fixed the output of the KLE subfunction. But the increased energy prices will raise the cost of producing this output V, and this will induce a substitution away from $V = h^{*}(K,L,E)$ and toward M,

---

1) This computational nuance and not the speculative "rather intriguing explanation" by GG $\underline{/}\overline{s}$ee GG (1976), p. 853$\underline{7}$ may explain their "unexpected" low σ_{KL} estimate for the United States.

holding the $Y = f(K,L,E,M)$ output fixed. This latter scale effect reduces the derived demand for all three inputs in the $V = h^*(K,L,E)$ subfunction. The net effect of the energy price increase on the demand for capital is therefore indeterminate; the sign of ε_{KE} depends on whether the positive gross substitution effect or the negative scale effect is dominant.

Hence by analogy with the gross substitution and scale effect argument developed in Section IV, we can relate the three input K-E gross substitution elasticities ε^*_{ij} to the four input K-E net substitution elasticities ε_{ij} as

$$\begin{aligned}
\varepsilon_{EK} &= \varepsilon^*_{EK} + N_K \Sigma \varepsilon_{VV} \\
(41) \\
\varepsilon_{KE} &= \varepsilon^*_{KE} + N_E \Sigma \varepsilon_{VV}
\end{aligned}$$

where now N_K and N_E are the cost shares of K and E, respectively, in the total cost of producing the $N = h^*(K,L,E)$ output and ε_{VV} is the price elasticity of demand for the output V along a four input $Y = f(K,L,E,M) = h^{**}(V,M)$ isoquant. Since N_K and N_E are positive cost shares and ε_{VV} is non-positive, $\varepsilon^*_{EK} \geq \varepsilon_{EK}$ and $\varepsilon^*_{KE} \geq \varepsilon_{KE}$, implying that unless the output price elasticity $\varepsilon_{VV} = 0$, the K-E substitution elasticity estimates based on a three input KLE specification are upwards biased. Furthermore ceteris paribus, this upward bias will be larger the more capital and energy intensive is the industry.

It would of course be interesting to obtain some idea of the potential quantitative magnitude of this upward bias. The three input KLE studies by Griffen-Gregory and Pindyck report estimates for the United States of about 1.1 and .8, respectively, while Magnus' KLE study for the Netherlands finds a σ^*_{KE} estimate of about -4.4. Hence the three-input study reporting the greatest amount of E-K substitutability is Griffen-Gregory. Let us insert into (41) reasonable values of N_K, N_E, ε_{VV} and the GG "high" estimates of ε^*_{EK} and ε^*_{KE}. The 1965 values of N_K and N_E in GG's United States data are about .14 and .13 respectively, while their reported ε^*_{EK} and ε^*_{KE} estimates are .15 and .13. Reasonable estimates of ε_{VV} are more difficult to obtain. We can proceed by letting ε_{VV} take on three alternative values: -0.5, -1.0, and -1.5. Inserting these alternative estimates of ε_{VV} into (41), and using the GG values for N_K, N_E, ε^*_{EK} and $-\varepsilon^*_{KE}$, we obtain three alternative net elasticity estimates: .08, .01, and -.06 for ε_{EK}, and .065, .0, and -.065 for ε_{KE}. Notice that even with the high ε^*_{EK} and ε^*_{KE} estimates of GG, we obtain values for the net elasticities ε_{KE} and ε_{EK} in the United States that include negative (complementary) estimates. Thus the poritive GG estimates for ε^*_{EK} and ε^*_{KE} are not necessarily inconsistent with the E-K complementarity estimates obtained by Berndt-Wood, Berndt-Jorgenson, and Fuss. Moreover, since GG's positive estimates of ε_{KE}^{*} are the largest of those reported in the various three-input KLE studies, the K-E elasticity estimates of the various KLE and KLEM studies can be reconciled.

We conclude that the seemingly inconsistent Berndt-Wood energy-capital complementarity and Griffen-Gregory energy-capital substitutability econometric results may simply be due to the fact that different elasticities are being compared; when the distinction between net and gross elasticities is acknowledged and the same output is held constant, the various net elasticity estimates are reasonably consistent with one another. Any remaining discrepancies are likely to be statistically insignificant, especially since standard errors for the GG energy-capital elasticity estimates are large.

VI. A RECONCILIATION WITHOUT THE SEPARABILITY ASSUMPTION

In Section IV we showed that if one assumes $\underline{/}(L,E),(L,M)\underline{7}$ separability, the resulting "utilised capital" specification enables us analytically and empirically to reconcile the engineering notion of E-K substitutability with the Hicks-Allen concept of E-K complementarity. Then in Section V we demonstrated that the Griffen-Gregory $\underline{/}(K,L,E),\underline{M7}$ separability assumption and their use of only KLE data implies that the GG elasticity estimates are not directly comparable with those of Berndt-Wood, since different outputs are being held constant; when the elasticities are properly compared, the seemingly disparate empirical findings can be reconciled. Separability has played a prominant role in both of these discussions. This raises the issue of whether our reconciliation of engineering E-K substitutability with Hicks-Allen E-K complementarity is dependent on the separability assumption.

We shall now show that the separability assumption, although useful for purposes of exposition and pedagogy, is not necessary for the reconciliation of engineering E-K substitutability with Hicks-Allen E-K complementarity. Thus our analytical findings are considerably generalised and strengthened.

It is well known that separability places restrictions on the Hicks-Allen partial elasticities of substitution σ_{ij} and price elasticities ε_{ij}.[1] In the present context, $\underline{/}(K,E),(L,M)\underline{7}$ separability implies

---

1) For further discussion of this point, see E.R. Berndt and
   L.R. Christensen (1973).

$\sigma_{KL} = \sigma_{EL} = \sigma_{KM} = \sigma_{EM}$.(1) Moreover, if we hold the $K^* = f(K,E)$ output constant and use only the K-E data, this separability specification implies that the Allen gross substitution elasticity σ^*_{KE} is independent of L and M.(2) Hence, under $\underline{/}(K,E),(L,M)\underline{7}$ separability, σ^*_{KE} can be computed without any consideration of inputs L and M.

In Section II we defined the direct elasticity of substitution d_{ij} (14) and the shadow elasticity of substitution S_{ij} (15). There we also noted that in general with more than two inputs, σ_{ij}, d_{ij} and S_{ij} differ from one another. In the present $Y = F(K,L,E,M)$ context, d_{KE} and S_{KE} are

$$(42) \quad d_{KE} = \frac{\partial \ln (K/E)}{\partial \ln (P_E/P_K)}, \qquad Y,\ L,\ \text{and M fixed,}$$

$$(43) \quad S_{KE} = \frac{\partial \ln (P_K/P_E)}{\partial \ln (E/K)}, \qquad \widetilde{C},\ \widetilde{P}_L,\ \text{and } \widetilde{P}_M \text{ fixed,}$$

where $\widetilde{C}$, $\widetilde{P}_L$, and $\widetilde{P}_M$ are the shadow total cost and shadow prices of L and M, respectively. Note that the direct elasticity d_{KE} is computed <u>conditional</u> on the values of Y, L, and M, in contrast to the gross substitution elasticity σ^*_{KE} which is computed completely independent of L and M. Whether the recent economic engineering literature on energy-capital trade-offs totally ignores independent inputs such as L or M or merely assumes they are fixed is of course a matter of interpretation. The important role of separability is that if $\underline{/}(K,E),(L,M)\underline{7}$ separability holds, then $\sigma^*_{KE} = d_{KE} = S_{KE}$. Hence, under this type of separability, the "conditional" and "independent" elasticity measures coincide.

When the production function $Y = f(K,L,E,M)$ is not separable, the various elasticty measures will generally differ from one another, i.e. conditional and independent elasticity measures will usually not coincide. In particular the d_{KE} elasticity which measures E-K substitutability conditional on L and M will of course be positive, even though K and E may be complementary inputs in the sense of Hicks-Allen. If one interprets engineering-economic studies as measuring E-K substitutability conditional on L and M, then the resulting positive d_{KE} estimates can be completely consistent with negative, complementary values for the "unconditional" σ_{KE} estimates. Hence in this sense engineering E-K substitutability and Hicks-Allen complementarity can be reconciled without the assumption of separability.

---

1) Our use of the translog function in the empirical application also constrained these σ_{ij} to equal unity.

2) The gross elasticity σ^*_{KE} would of course be computed applying (13) to the utilised cost subfunction (22).

We now illustrate these points empirically. When $\underline{/}(K,E),(L,M)\underline{/}$ separability is imposed on the Berndt-Wood data for total United States manufacturing, 1947-71, the estimates of σ^*_{KE}, d_{KE}, and S_{KE} coincide; their common value in 1965 is .243. However, without imposing these $\underline{/}(K,E),(L,M)\underline{/}$ separability restrictions, the conditional elasticities d_{KE} and S_{KE} and the unconditional Hicks-Allen elasticities of substitution σ_{KE} differ; with the Berndt-Wood data, the 1965 values are .308, .320, and -3.193, respectively, while with the Fuss data the 1965 values for Ontario are .465, .501, and -.207. Thus the positive conditional d_{KE} and S_{KE} estimates are completely consistent with negative σ_{KE} estimates. We conclude that a reconciliation of engineering E-K substitutability with Hicks-Allen E-K complementarity does not depend on the $Y = F\underline{/}(K,E),(L,M)\underline{/}$ separability assumption.

VII. CONCLUDING REMARKS

The purpose of this paper has been to provide a rather general analytical and empirical foundation for E-K complementarity consistent with the process engineering-technological view of E-K substitutability. Along the way we have developed the notion of utilised capital and have reconciled some of the seemingly disparate econometric findings. Above all this research has emphasized to us that care must be taken in interpreting and properly comparing alternative elasticity estimates. We note that our preferred net elasticity estimates hold gross output fixed and that yet another set of elasticity estimates could be derived that allow this output to be price responsive.

A particularly interesting implication of our analytical framework is that if one is willing to assume $Y = F\underline{/}(K,E), L,\underline{M}\underline{/}$ separability, then one can use either econometric or engineering estimates of gross E-K substitution elasticities in modelling industrial demand for energy. The engineering estimates might be preferable for longer term forecasts involving new technologies, or for regions or countries which utilise known technologies but lack reliable economic data.

At this point in time there appears to be a substantial and growing body of econometric evidence supporting the notion of Hicks-Allen E-K complementarity.(1) If our E-K complementarity finding is true and

---

1) A recent study by William Hogan (1977) argues for the plausibility of a different type of E-K complementarity; using a three input KLE model of the aggregate United States economy, Hogan argues that if L and P_K are fixed, then the recent increases in P_E will result in reduced rates of capital formation.

sufficiently general, then the implications could be significant. For example, in the economic growth literature Robert M. Solow (1974) and Robert M. Solow and Frederic Y. Wan (1976) note that if the capital-natural resource (in our context, capital-energy) Allen partial elasticity of substitution is less than unity, then unless technical change is of the resource (energy) saving form, a positive level of consumption is no longer sustainable for ever. On the other hand, if this elasticity of substitution is greater than unity, then resources are not even essential to production. In order to reflect the fact that resources may be indispensable to production but that economic growth can continue indefinitely, Solow and Solow-Wan employ the Cobb-Douglas production form in which the capital-resource elasticity of substitution equals one. This unitary elasticity of substitution assumption is clearly inconsistent with our finding of E-K complementarity. In our view, however, the empirical issue of E-K complementarity is still far from settled. A number of data and basic economic model specification problems are worthy of particular attention in future research.

First, all the econometric evidence available to date is based on data that does not include the post-1973 dramatic energy price increases. It would be useful to examine the robustness of the E-K complementarity findings with the more recent data. Nonetheless, it is worth noting that E-K complementarity is perfectly consistent with the following analysis of a bank economist regarding current economic conditions:

> "Finally, on the list of deterrents to capital spending, there is the significantly increased cost of building and operating new plants and equipment because of the higher price of energy of all types. The average economist may have forgotten his micro-economics, but the average businessman has not; he pays close attention to the relative cost of factors of production. And over the past three years it has become more expensive to increase capacity by adding machinery and equipment than it has by adding workers."(1)

Second, there remains the concern, expressed vigorously by GG, that E-K complementarity estimates based on annual time series data actually reflect short-run variations in capacity utilisation, and that the "true" long-run relationship is one of E-K substitutability. For this reason GG and Pindyck (1977) prefer pooled international cross-section time series elasticity estimates to estimates based solely on time series data. We have shown, however, that the GG and Pindyck pooled cross-section time

---

1) Irwin Kellner (1977), p. 3.

series elasticity estimates must be interpreted carefully, and that in particular they are not inconsistent with E-K complementarity. Moreover, Fuss' pooled cross-section time series data yields E-K complementarity. Hence the short-run, long-run E-K substitutability-complementarity issue does not seem to be simply one of pooled versus time series data.(1) In our view, this empirical issue cannot be resolved at this time, for even if extremely reliable data were available, we still would need an economic model of the disequilibrium or adjustment process. At the present time, the econometric literature on dynamic adjustment processes relies largely on ad hoc, constant coefficient adjustment specifications, rather than on explicit dynamic optimisation.(2)

Third, a number of data problems arise. In their international pooled cross-section time series studies, both GG and Pindyck were unable to take into account variations in effective corporate and property tax rates among OECD countries and over time. Also, both studies computed the value of capital services as value added minus the wage bill. This procedure has been criticised by, among others, E.R. Berndt (1976), for the resulting residual captures not only the return to capital equipment and structures, but also the returns to land, inventories, economic rent, working capital, and any errors in the measurement of value added or wage bill. Berndt (1976) finds that elasticity estimates are very sensitive to such data errors and to the choice of rate of return. Interestingly enough, in a recent unpublished KLE study, Barry Field and Charles Grebenstein (1977) use total United States manufacturing cross-section data for states in 1971 and obtain E-K substitutability when the return to capital is computed as value added minus wage bill, but find E-K complementarity when the capital rental price measure refers only to plant-equipment. Clearly, research on this topic merits additional attention.

Finally, we wish to emphasize that we are not rigidly beholden to the view that in all industries over any time period, E and K are comple- ments. We expect variations in σ_{KE} estimates across industries and over

---

1) In background work for his (1976b) paper, Edwin Kuh /(1976a), Table 6c7 finds that when the six "recession" or "excess capacity" years of 1949, 1954, 1958, 1961, 1970 and 1971 are dropped from the Berndt-Wood 1947-71 data set, E-K complementarity still results, albeit in a smaller absolute magnitude. Also, E.R. Berndt and M.S. Khaled (1977) use the Berndt-Wood data and find that E-K complementarity is robust even when the assumptions of constant returns to scale and Hicks neu- tral technical change are relaxed, and when flexible functional forms other than the translog cost function are employed.

2) Research on this topic is presently underway; see E.R. Berndt, M.A. Fuss, and L. Waverman (1977).

time. The post-1973 tripling in energy prices may in fact induce a change in the relative magnitudes of the E-K gross substitution and scale elasticities. Such changes are of course compatible with our analytical framework that permits variable elasticities. Although ambiguities still remain in model specification and data, we believe this paper has provided a useful framework for analysing energy conservation and capital formation. Engineering and econometric approaches are mutually consistent, and seemingly disparate econometric estimates can be reconciled.

REFERENCES

Allen, R.G.D. (1938), Mathematical Analysis for Economists, London: MacMillan, pp. 503-509.

Berndt, E.R. (1976), "Reconciling Alternative Estimates of the Elasticity of Substitution", Review of Economics and Statistics, Vol. 58, February, pp. 59-68.

Berndt, E.R. (1977), "Notes on Generalized R^2", unpublished manuscript, Vancouver: University of British Columbia, Department of Economics, July.

Berndt, E.R. and L.R. Christensen (1973), "The Internal Structure of Functional Relationships: Separability, Substitution, and Aggregation", Review of Economic Studies, Vol. 40, July, pp. 403-410.

Berndt, E.R., M.A. Fuss, and L. Waverman (1977), "Factor Markets in General Disequilibrium: Dynamic Models of the Industrial Demand for Energy", paper presented to the European Meetings of the Econometric Society, Vienna, Austria, September.

Berndt, E.R. and D.W. Jorgenson (1973), "Production Structure", Chapter 3 in D.W. Jorgenson, E.R. Berndt, L.R. Christensen, and E.A. Hudson, U.S. Energy Resources and Economic Growth, Final Report to the Ford Foundation Energy Policy Project, Washington, D.C., October.

Berndt, E.R. and M.S. Khaled (1977), "Energy Prices, Economies of Scale and Biased Productivity Gains in U.S. Manufacturing, 1947-71", Vancouver, University of British Columbia, Department of Economics, Discussion Paper 77-23, August.

Berndt, E.R. and D.O. Wood (1977), "Engineering and Econometric Approaches to Industrial Energy Conservation and Capital Formation: A Reconciliation", MIT Energy Laboratory Working Paper No. MIT-EL 77-040WP, November; also issued as University of British Columbia, Programme in Natural Resource Economics, Working Paper No. 14, November.

Berndt, E.R. and D.O. Wood (1975), "Technology, Prices, and the Derived Demand for Energy", Review of Economics and Statistics, August, Vol. 56, pp. 259-268.

Blackorby, Charles, Daniel Primont, and R. Robert Russell (1975), "Budgeting, Decentralization, and Aggregation", Annals of Economic and Social Measurement, Vol. 40, No. 1, Winter, pp. 23-44.

Cowing, T.G. (1974), "Technical Change and Scale Economies in an Engineering Production Function: The Case of Steam Electric Power", Journal of Industrial Economics, Vol. 23, No. 2, December, pp. 135-152.

Darmstadter, Joel, J. Dunkerly, and J. Alterman (1977), How Industrial
    Societies Use Energy, Baltimore: Johns Hopkins University Press,
    1977.

Denny, M.G.S. and C. Pinto (1976), "The Energy Crisis and the Demand for
    Labour", Discussion Paper 36, Research Projects Group, Strategic
    Planning and Research Division, Department of Manpower and Immigra-
    tion, Ottawa, August.

Denny, M.G.S. and M. Fuss (1977), "The Use of Approximation Analysis to
    Test for Separability and the Existence of Consistent Aggregates",
    American Economic Review, Vol. 67, No. 3, June, pp. 404-418.

Diewert, W.E. (1974), "Applications of Duality Theory", in M.D. Intriligator
    and D.A. Kendrick, eds., Frontiers of Quantitative Economics, Vol. II,
    Amsterdam: North-Holland Publishing Co.

Field, Barry C. and C. Grebenstein (1977), "Substituting for Energy in
    U.S. Manufacturing", mimeo, Amherst, Massachusetts: The University
    of Massachusetts, Department of Food and Resource Economics, May.

Ford Foundation Energy Policy Project (1974), A Time to Choose: America's
    Energy Future, Cambridge, MA: .Ballinger Publishing Co.

Fuss, M.A. (1977), "The Demand for Energy in Canadian Manufacturing: An
    Example of the Estimation of Production Structures with Many Inputs",
    Journal of Econometrica, January, Vol. 5, pp. 89-116.

Griffen, J.M. and P.R. Gregory (1976), "An Intercountry Translog Model of
    Energy Substitution Responses", American Economic Review, December,
    Vol. 66, pp. 845-857.

Gyftopoulos, E.P., L.J. Lazaridis, and T.F. Widmer (1974), "Potential for
    Effective Use of Fuel in Industry", Report to the Energy Policy
    Project, Cambridge, Mass: Ballinger Press.

Gyftopoulos, E.P. and Thomas F. Widmer (1977), "Energy Conservation and a
    Healthy Economy", Technology Review, Vol. 79, No. 7, June, pp. 31-40.

Hausman, J.A. (1978), "Energy Conservation and Consumer Demand", paper to
    be presented at Urbino, Italy, January.

Hicks, J.R. (1933), "Notes on Elasticity of Substitution", Review of
    Economic Studies, October.

Hirst, Eric (1976), "Residential Energy Use Alternatives: 1976 to 1000",
    Science, Vol. 194, 17th December, pp. 1247-1252.

Hirst, Eric, and Jane Carney (1977), "Residential Energy Use to the Year
    2000: Conservation and Economics", Oak Ridge Tennessee: Oak Ridge
    National Laboratory, ORNL/CON-13, September.

Hirst, Eric, Jane Cope, Steve Cohn, William Lin, and Robert Hoskins (1977),
    "An Improved Engineering-Economic Model of Residential Energy Use",
    Oak Ridge, Tennessee: Oak Ridge National Laboratory, ORNL/CON-8,
    April.

Hogan, W.W. (1977), "Capital Energy Complementarity in Aggregate Energy-
    Economic Analysis", Stanford: Stanford University, Institute for
    Energy Studies, Energy Modeling Forum, August.

Kellner, Irwin (1977), "Quarterly Business Conditions Analysis", New York: Manufactures Hanover Trust, Office of Vice President and Economics, March.

Kuh, E. (1976a), "Background Material for Research on Translog Cost or Production Functions", unpublished xerolith, 4th February.

Kuh, E. (1976b), "Preliminary Observations on the Stability of the Translog Production Function", in J. Daniel Khazzoom, ed., Proceedings of the Workshop Modeling the Interrelationships Between the Energy Sector and the General Economy, Palo Alto: Electric Power Research Institute, Special Report 45, July 1976.

Magnus, Jan R. (1977), "Substitution Between Energy and Non-Energy Inputs in the Netherlands, 1950-1974", Amsterdam: Universiteit van Amsterdam, Instituut voor Actuariaat en Econometrie, Report AE3/76, Revised, August 1977.

McFadden, D. (1963), "Constant Elasticity of Substitution Production Functions", Review of Economic Studies, June, pp. 73-83.

Moody, C.E., Jr. (1974), "The Measurement of Capital Services by Electrical Energy", Oxford Bulletin of Economics and Statistics, Vol. 36, No. 1, February, pp. 45-52.

National Petroleum Council (1971), U.S. Energy Outlook: An Initial Appraisal, 1971-1985, Vol. 1-3, Washington, D.C.

Ohta, M. (1975), "Production Technologies of the U.S. Boiler and Turbo-generator Industries and Hedonic Price Indexes for their Products: A Cost Function Approach", Journal of Political Economy, Vol. 83, No. 1, February, pp. 1-26.

Pindyck, R.S. (1977), "Interfuel Substitution and the Industrial Demand for Energy: An International Comparison", MIT Energy Lab. Working Paper MIT-EL 77-026WP, August.

Rosenfeld, A.H. (1977), "Some Potentials for Energy and Peak Power Conservation in California", mimeo, Lawrence Berkeley Laboratory, Berkeley, 26th March.

Samuelson, P.A. (1974), "Complementarity - An Essay on the 40th Anniversary of the Hicks-Allen Revolution in Demand Theory", Journal of Economic Literature, Vol. 21, December, pp. 1255-1289.

Sato, Ryuzo and Tetsunori Koizumi (1973), "On the Elasticities of Substitution and Complementarity", Oxford Economic Papers, Vol. 25, No. 1, March, pp. 44-56.

Schipper, L. and J. Darmstadter (1977), "What is Energy Conservation?", manuscript prepared for the National Academy of Sciences - National Research Council, CONAES Demand and Conservation Panel, Berkeley, Lawrence Berkeley Laboratory Report LBL-5919, 7th April.

Socolow, Robert H. (1977), "The Coming Age of Conservation", Annual Review of Energy, Vol. 2, pp. 239-289.

Solow, Robert M. (1974), "Intergenerational Equity and Exhaustible Resources", Review of Economic Studies, Symposium Issue, pp. 29-45.

Solow, R.M. and F.Y. Wan (1976), "Extraction Costs in the Theory of Exhaustible Resources", Bell Journal of Economics, Vol. 7, No. 2, Autumn, pp. 359-370.

Swain, P. and G. Friede (1976), "Die Entwicklung des Energieverbrauchs der Bundesrepublik Deutschland und der Vereinigten Staaten von Amerika in Abhangigkeit von Preisen und Technologie", Karlsruhe: Institut fur Angewandte Systemanalyse, June.

United States Environmental Protection Agency (1973), "Energy Conservation Strategies", Office of Research and Monitoring, Washington, D.C. ERA-R5-73-021, July.

United States Federal Energy Administration (1974), Project Independence Report, Washington, D.C.: U.S. Government Printing Office, November.

United States National Academy of Sciences – National Research Council (1977), "Report of the Demand/Conservation Panel", Committee on Nuclear and Alternative Energy Systems, Washington, D.C. forthcoming.

United States Office of Emergency Preparedness (1972), The Potential for Energy Conservation: A Staff Study, Washington, D.C., Executive Office of the President, October.

Uzawa, H. (1962), "Production Functions with Constant Elasticities of Substitution", Review of Economic Studies, Vol. 29, October, pp. 291-299.

Zarembka, Paul (1974), "On Determinants of Investment in a Short-Run Macroeconomic Model", Buffalo: State University of New York, Department of Economics, Disc. Paper No. 300, February; to appear in The Stability of Contemporary Economic Systems, Proceedings of the Third Reisenberg Conference, July 1975.

METHODOLOGY FOR STUDYING WORLD ENERGY DEMAND

by

R. Eden(*)

*) This report is an extract from the chapter on world energy demand to
2020 prepared by Dr. Eden and co-authors for the Conservation Commission
of the World Energy Conference and published with chapters on world
energy resources in World Energy Resources 1985-2020 by the WEC and the
IPC Science and Technology Press (in 1978).

THE PROBLEM

The main question at issue in this report is the possible range of
response in energy demand to an impending scarcity of oil relative to other
energy carriers, and the possibility that the average real price of energy
to the consumer may rise substantially relative to other goods and services. The method adopted aims to describe the essential features of the
problem through broad, simple assumptions, which are consistent with each
other, and then with the help of computer models to identify those assumptions that are critical to the main question and to analyse the consequences of varying these critical assumptions.

The "projections" are only conditional - conditional on the assumptions intended to contribute to the discussion of a limited problem. The
projections of total energy demand and the distribution of demand between
different fuels or energy carriers are uncertain, but they are believed
to provide typical "scenarios" for high or low economic growth that illustrate the magnitude of the problems of energy supply and energy conservation that must be faced during the coming decades. The projections are
based on the world regions listed in Exhibit 5, and the methodology is
structured in Exhibit 6.

Table 1

WEC WORLD REGIONS

Group	Region
OECD(1, 2, 3) CP(4, 5) (Centrally planned) Developing(6 to 11) OPEC developing(6) Non-OPEC developing(7 to 11)	1. North America 2. West Europe 3. Japan, Australia, New Zealand 4. USSR/East Europe 5. China and centrally planned Asia 6. OPEC 7. Latin America 8. Middle East and North Africa 9. Africa South of the Sahara 10. East Asia 11. South Asia

CONSTANT ASSUMPTIONS COMMON TO ALL CASES

1 Basic data on energy demand in each world region by fuel (including wood fuel) and sector in 1972, which is taken as the base year for the projections. This is illustrated and summarised in Exhibit 7.

2 Regional variations in GNP (gross national product) corresponding to low (3.0 per cent per annum) and high (4.1 per cent) world economic growth for the period 1975-2020. These are summarised in Exhibit 8 and their effect on the distribution of world total GNP is shown in Exhibit 9. In the low growth case the world average GNP per capita in 2020 is nearly twice today's level, and in the high growth case it is nearly three times (we assume the UN median projection for population to 2000 and extend it to 2020 at the average growth rate for 1990-2000).

3 No political, economic or physical disasters. Any of these might alter the smooth extrapolation to the year 2020, though it may be noted that the average economic growth for the past 50-year period containing the world depression of 1929-39 is higher than that assumed in the low growth projections.

4 A preference for the use of indigenous energy supplies (if they are available) in the world regions shown in Exhibit 5, representing the idea of a cost penalty for transporting fuels.

Figure 1
ENERGY DEMAND METHODOLOGY

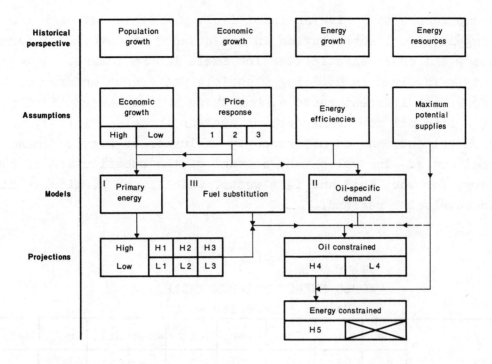

CONSTANT ASSUMPTIONS RELATING TO FUEL SUBSTITUTION

5 A preference against the manufacture of synthetic fuels such as oil and gas until the total world demand for those fuels enters the range of values where a potential scarcity of supplies is expected.

6 In regions 1 to 3 nuclear power is assumed to grow rapidly from 1972, in region 4 from 1977, and in regions 5 to 11 from 1985.

7 Rules for substitution between fuels as follows:

a) A preference order within each economic sector for particular fuels (see Exhibit 7). This represents the differences in convenience and cost to the user between one fuel (or electricity) and another, and takes account of the world supply potential for different fuels in future years.

b) A concept of inertia, restricting the rate at which one fuel may be substituted for another within a sector, indicating that plant for using a new fuel will penetrate a market only slowly as methods change and old plant is retired. The modelling assumptions restrict both the rate of growth and the rate of decline of each fuel for the final demand sectors. A faster rate of growth is

317

permitted for fuels in the energy supply sectors, but here also the rate is related to the potential development of new plant.

Assumptions 5 to 7 give an intuitively plausible set of rules for projecting interfuel substitution in the absence of detailed information about costs and cross elasticities for fuels in the future. A degree of judgement is involved in deciding plausible preference orders for individual fuels in different sectors, but these are either explicitly discussed (as with oil for transport, or nuclear for electricity), or their effects cancel out, and do not strongly influence aggregate demand for individual fuels. The rules take account of the substitution of commercial energy for wood fuel and farm waste, which is of particular significance in developing regions.

Table 2

WORLD ENERGY BALANCE TABLE 1972

Units: ExaJoules = 10^{18} J

	Coal	Oil	Gas	Elec.	Heat[*]	Wood	Sol.	Nucl.	Hydr.	Total
Transport	1.9	41	0	0.5	0	–	–	–	–	43.4
Industry	22.3	21.7	18.9	9	3.5	5.2	–	–	–	80.6
Domestic	8.1	17.9	12.7	7.6	1.3	20.9	–	–	–	68.9
Feedstocks	0.1	9.7	0.7	–	–	–	–	–	–	10.5
Total final consumption	32.4	90.3	32.3	17.1	4.8	26.1	–	–	–	203
Electricity generation	27.4	15.1	9.1	-20.3	-5.8	–	–	1.6	13.8	40.9
Synthetic gas	2.7	1.9	-3.8	–	–	–	–	–	–	0.8
Synthetic oil	0	-0	–	–	–	–	–	–	–	0
Energy sector own use	3.1	6.8	8.4	3.2	1	–	–	–	–	22.6
Primary energy input	65.6	114.1	46	–	–	26.1	–	1.6	13.8	267.2
Indigenous supply	65.8	115.2	46	–	–	26.1	–	1.6	13.8	268.5
Net imports(**)	-0.2	-1.1	0	–	–	–	–	–	–	-1.3

*) "Heat cogenerated with electricity.

**) And items unaccounted.

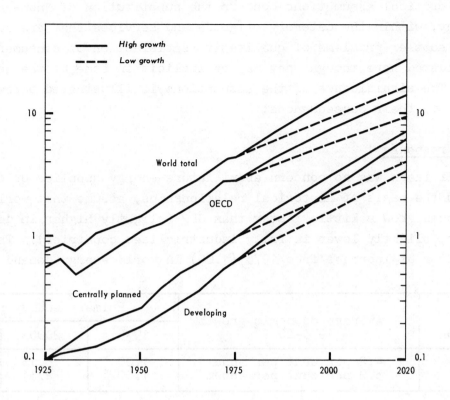

Figure 2
WORLD AND WORLD GROUP TOTAL GNP/$ 10^{12} US (1972)

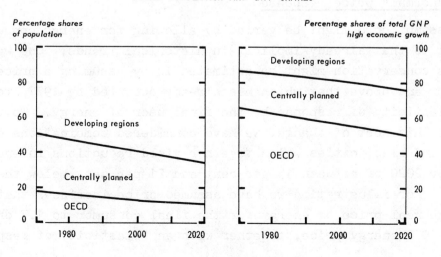

Figure 3
WORLD POPULATION AND GNP SHARES

CRITICAL ASSUMPTIONS

The critical assumptions concern the substitution of "non-energy" for energy, within the category of goods and services that are counted in GDP (the subtler problems of qualitative satisfaction of consumers are not considered here though they may be implicit in some of the assumptions). The significance of the assumptions is illustrated below by brief reference to their consequences:

No price response

8 If there was no concern about future energy supplies or costs we could, on the basis of historical relationships, assume that world energy demand would grow a little faster than GDP (slightly higher in developing countries, slightly lower in fully industrialised countries). The consequent factor of increase from 1975 (1.00) in world energy demand would be:

Projection	Average economic growth 1975-2020	Primary energy ratio		
		1975	2000	2020
H1	4.2 per cent per annum	1.00	3.18	7.00
L1	3.0 per cent per annum	1.00	2.31	4.12

Under this assumption, only an average rate of economic growth to 2020 of less than 3 per cent would fall within the range of expected supplies. Such a growth would be consistent with a growth of 5 per cent for 10 to 15 years followed by a prolonged depression in which economic growth was no faster than population growth. This would be the reverse in time of the growth pattern over the past 50 years.

Constant "normal" price response

9 Assumption 8 might be varied by allowing for energy conservation additional to that already implicit in historical trends. One way in which this conservation might be estimated is by assuming a price increase over and above that which has already occurred by 1977, together with an elasticity of response by the final user of energy. To test the effects of this sort of change, we have considered combinations of price increases and elasticities which together yield reductions in energy use by the year 2020 of between 15 per cent and 21 per cent below the cases L1 and H1. For illustration we have assumed price increases that raise the average real price of energy for the final consumer to 1.7 or 2.2 times the 1972 energy price, together with an "elasticity of response"

of minus (0.3). These price increases give projections L2, H2, and L3, H3, which (by 2020) are respectively 15 per cent and 21 per cent below the corresponding energy demand in L1 and H1. Under these assumptions, world energy demand associated with "low" (3.0 per cent) economic growth, would rise from 268 EJ in 1972 to a value in 2020 in the range 830 to 970 EJ, which is within the estimated range of supply. No significantly higher economic growth would be feasible unless greater supplies were available at these prices, or unless the change in energy demand exceeded these "normal" responses.

Premium uses for oil

10 Assumptions 8 and 9 follow from relatively simple ideas about energy as a whole. However, we cannot sensibly avoid considering individual fuels, particularly oil. If the demand for oil increased at the same rate as energy, even in the low growth case L3 oil demand would increase over three times by 2020 and would be likely to exceed the maximum potential supply before the year 2000. We therefore assume that scarcity of oil will increase its price relative to other fuels. The cost of using any fuel other than oil in many forms of transport is likely to exceed the cost of similar substitution for oil in domestic heating or industrial heating and steam raising, so we assume that oil would be gradually priced out of the heat market (except for specialist applications), in line with our "substitution" assumptions (7 above). The use of oil as a chemical feedstock is assumed also to maintain its premium value, as with transport.

11 We assume that the demand for energy in transport in the long run is modified in response to the relative price increase for oil by two developments:

a) The "saturation" of the demand for private motoring in the more wealthy countries will have a lower saturation level than might otherwise have been expected. At the margin, motorists will substitute other forms of transport and other forms of consumption for private motoring.

b) Technical improvements, which are already targeted (for example, in the United States of America), and changes in load factors, will in the long run significantly increase the efficiency with which oil inputs deliver ton miles or passenger miles of transport.

The effect of these assumptions has been studied in sub-models of the transport sector of the main regions. While much of the detail must be speculative, the overall effects seem clear. Compared to the total

energy for transport, the share required for road transport will fall
(Exhibit 15), though there is likely to be a substantial increase in the
transport energy requirements of the developing regions. The total demand for oil will start to become "constrained" (Exhibits 16 and 17),
within the period 1985 to 1995.

General long-run responses

The special case of oil as a relatively scarce fuel has led us naturally to expect its allocation to premium uses such as transport and
chemical feedstock. In the transport sector we have identified a variety
of long-run adjustments to the increasing relative cost of oil that could
enhance already approaching saturation of demand and accelerate the development of better technical efficiencies, better operational efficiencies
(such as load factors) and some modal changes. All of these contribute
to changed relationships between oil demand, transport activity, and economic growth.

We cannot be so specific about the potential for similar long-run
adjustments in the domestic and industrial sectors, but we can reasonably
assume that such adjustments could take place in the face of an increasing scarcity of energy itself and an increase in the price of energy compared to all other goods and services.

With all other assumptions unchanged, and keeping within the estimated range of supply, an energy demand of the order of 990 EJ by 2020
(compared with 276 EJ in 1975) could be associated with high economic
growth averaging 4.2 per cent per annum (instead of low at 3.0 per cent)
if we assume:

12 An improvement in the efficiency of industrial and domestic
energy use giving a reduction by 2020 of the order of 30 per cent on the
oil constrained, high growth projection H4, which is equivalent to an
annual reduction of 0.8 per cent over a 40-year period compared with the
H4 trend. The resulting projection, denoted H5, will be described as the
"energy constrained" high growth projection. These efficiency improvements are assumed to be in addition to those arising from the "normal"
price response which is already included in H4. Whether the degree of
adjustment should be considered very large (in comparison to long-run
productivity trends), or very small (because it involves only a small
annual percentage change), depends on information which either does not
exist, or is sparse, or is dispersed. This exercise shows the importance
of that information. Some of the potential for adjustments towards
"energy constraint" can be assessed from savings through measures for
energy conservation that have already been identified, but caution is

322

required since many measures which are clearly cost effective will already be contained implicitly in the historical trends and in the price responses already considered.

ASSUMPTIONS CONCERNING POTENTIAL ENERGY SUPPLY

The magnitude and the necessity of the long-run improvements in efficiency stated in assumption 12 depend on an assessment of the future potential for energy supply. This will be discussed in reports by the WEC supply and resource groups. However, guidelines were required for the demand projections.

13 The guidelines for the potential supply of energy are summarised in Exhibit 10. The ranges shown are intended to reflect not only the constraints for maximum potential supply indicated by the WEC resource studies but also the expectation that the potential supply for each fuel will depend on the pressure from demand. Thus potential supply will be higher for a high growth scenario than for a low growth scenario. The ranges shown are used as guidelines to influence the pattern of energy demand and fuel substitution by means of the fuel preferences in the demand model that are adjusted in an attempt to keep world demand for each fuel within the guidelines for supply. Due to the constraints noted in assumption 7, the maximum potential supply in particular fuels is not always taken up by the projections of energy demand.

14 "Fast Development" projections (FD) for the Developing Regions have been examined in combination with alternative assumptions for OECD and with the assumption that the group of centrally planned economies remains in net energy balance with the rest of the world. Two fast development projections are considered, FD4 and FD5, in which the relations between energy and economic growth are analogous to those for H4 (medium conservation) and H5 (high conservation).

15 The "transition" scenario T4 represents an attempt to evaluate possible economic assumptions that would be consistent with:

    i) an initial period of high growth (to 2000),
   ii) medium levels of energy conservation similar to those for H4
        and L4,
  iii) the guidelines assumed for potential energy supply to 2020.

Thus T4 follows the path of H4 to 2000 and thereafter economic growth assumptions are adjusted so that energy demand remains within the supply guidelines to 2020.

Table 3

GUIDELINES ASSUMED FOR POTENTIAL ENERGY SUPPLY

Units: ExaJoules

	Coal	Oil	Gas	Nuclear	Renewable	Total
OECD						
1972	28	28	32	2	11	101
2000	50-70	32-44	20-35	50-80	20-30	190-240
2020	70-110	26-44	10-20	100-160	35-70	280-360
Centrally Planned						
1972	33	19	12	0	13	77
2000	70-100	20-40	20-40	20-40	20-35	160-230
2020	110-150	20-40	25-45	50-100	30-60	280-360
Developing						
1972	5	67	9	0	16	97
2000	15-30	100-140	25-45	7-15	20-35	180-240
2020	20-50	70-130	30-70	30-60	30-60	220-330
World						
1972	66	114	53	2	40	275
2000	145-190	165-210	70-110	85-125	65-90	580-680
2020	220-290	135-200	70-120	200-300	105-175	820-1010

For illustrative purposes it may be assumed that the upper and lower ends of the ranges in Exhibit 10 correspond approximately to the upper and low quartiles in a probability distribution.

STRUCTURE AND TECHNOLOGY AS DETERMINANTS OF
ENERGY DEMAND IN POST—WORLD WAR II ITALY

by

O. Bernardini

INTRODUCTION(1)

Energy demand growth in Italy in the quarter century or so following
the Second World Conflict has been the highest in the industrial world
with the exception of Japan. Leaving out the five years immediately following the war which were reconstruction years, characterised by abnormally
high rates of growth as capacity and distribution systems were brought back
on line, Italy averaged an annual growth rate in energy consumption exceeding 8.5 per cent up to the first years of the 1970s.

This greatly sustained growth was not however only a simple linear
consequence of the rapid economic development experienced by Italy during
this period, (exceeding 5 per cent average per year) but also of the remarkable increase which occurred in the energy intensity of gross national
product. This is show in differential form in Figure 1 for both overall
final energy demand and for electricity demand. Far from retaining roughly
constant or even decreasing, as one might expect by analogy with other industrial countries, this figure reveals that the elasticity of energy demand growth relative to GNP growth has been driven instead by a rapidly
increasing long-term component.

The increase is shown in integral form in Figure 2 where the energy
intensity of gross national product is compared for Italy, and the major
world regions North America, Western Europe, Japan and finally the Rest of

----

1) The work presented is a continuation of research begun under the aus-
    pices of the Workshop on Alternative Energy Strategies (WAES).

the World outside Comecon and China, comprising mostly - though not entirely - countries that are frequently identified as less developed or
developing.(1)

Figure 1
INCOME ELASTICITY OF ENERGY DEMAND IN FINAL USES
(10 year moving average centred on mid point)

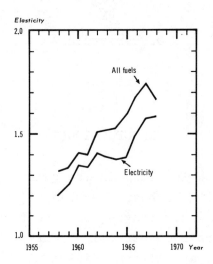

The striking differences in time behaviour between North America and
Western Europe on the one hand and Italy, Japan and the Rest of the World
on the other, naturally provide food for thought. It is a fact however
that very little effort is ever dedicated to their explanation whereas considerably more research is made in trying to understand the differences in
energy intensity between countries in any one year. Yet the peculiarities
of time dependence may hold the key to secrets for a better understanding
of energy demand growth in the World at large and their study open up new
frontiers in the problem of energy demand management.

This paper is in line with this spirit and provides an attempt to
understand the underlying causes of increase in the energy intensity of
Italian gross national product in the period 1953 to 1972.

---

2) Gross national product used in the calculation of energy/GNP ratios in
Figure 2 is obtained directly from World Bank tabulations without cor-
recting for purchasing power differences between countries. Whereas
the absolute magnitude of the curves may thus need to be corrected somewhat, nevertheless the trend in time dependence appearing in this
figure can be considered as well established.

Figure 2

NET ENERGY REQUIRED TO PRODUCE
A UNIT OF GROSS NATIONAL PRODUCT
(extension to the year 2000 based on WAES projections)

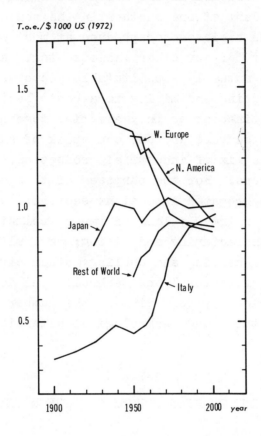

THE MODEL(1)

The period 1953 to 1972 was chosen because a consistent and stable
energy data base is available only beginning with 1953 and because 1972
is probably the last year which can be considered more or less unaffected
by oil price increases and rapid inflation.

The model used to cover this historic period is superficially an
accounting model in the sense that relations between relevant physical,
economic and technological parameters are of a very simple nature consisting only of additive and multiplicative operations. However, essentially all of the information used was obtained through application of
mathematical-statistical models to existing data so that in effect the

1) A partial description of the model is available in U. Colombo,
    O. Bernadini, R. Galli and W. Mebane, Il rapporto WAES-Italia: le
    alternative strategiche per una politica energetica, Franco Angeli,
    1978.

327

underlying model is of a very involved econometric form carrying with it
a high degree of sectoral disaggregation and within each sector, an especially high degree of structural and technological detail.

Basic sectors considered in the model are 25 in number. These are
defined to be independent of one another in the sense that ideally changes
in activity of any one basic sector produce at most only second order
changes in the activity of any other basic sector. Moreover they are defined to contain only mutually substitutable sectors. Thus to take just
one example, manufacturing and mining is a basic sector whereas the various
component industries, insofar as they are tied together by an input-output
matrix, are not basic sectors and one can speak of substitution between
them as changes in the mix of industrial production. In all, the model
considers 257 subsectors. For the purposes of this discussion the 25
basic sectors are reaggregated into eight sectors: urban passenger travel;
long-distance passenger travel; long-distance freight transport; local
freight transport; manufacturing and mining; agriculture, livestock, forestry and fishing; space-heating and cooling; other civilian uses. An indication of the degree of sectoral and technological detail involved in the
model is given in Table 1 by referring to the number of subsectors and to
other essential structural and technological dimensions.

Table 1

INFORMATION CONTENT OF THE MODEL

Major sectors	Number of subsectors	Average number of parameters per subsector	Total information available on a yearly basis
Urban passenger travel	18	10.5	189
Long-distance passenger travel	17	10.5	179
Long-distance freight transport	16	9.1	145
Local freight transport	6	9.1	55
Manufacturing & mining	70	9.2	642
Agriculture livestock, forestry and fishing	4	4.0	16
Space heating and cooling	98	7.9	774
Other civilian uses	28	5.6	157
Total	257	8.4	2157

Thanks to the highly detailed accounting nature of the model, it is possible to pick up significant effects on energy demand issuing from structural and technological changes within the system as it develops over time. The approach, as will be shown, allows a fairly accurate attribution of cause and effect and provides a method of tracking changes in energy demand with reference to the various contributing factors of change.

Writing the general demand equation for basic sector j as:

$$E_j = \sum_i E_{ij} = \sum_i A_{ij} \cdot T_{ij}$$

where E_{ij} is the energy demand in the ith subsector and A_{ij} and T_{ij} are the corresponding activities and energy intensities, then the overall energy intensity T_j of sector j is given by:

$$T_j = \frac{E_j}{A_j} = \frac{\sum_i A_{ij} \cdot T_{ij}}{\sum_i A_{ij}}$$

What is of interest in the present context is determining the changes in energy intensity T_j resulting from variations occurring in the factors of change X_k in a given interval of time, t. These changes can be calculated as the partial derivatives:

$$\left(\frac{\partial T_j}{\partial t} \right)_k = \frac{\partial T_j}{\partial X_k} \cdot \frac{\partial X_k}{\partial t}$$

Since the model is available on a discrete time basis the changes in T_j are calculated from the corresponding finite difference expression:

$$\frac{\Delta T_{jk}}{\Delta t} = \frac{T_j(X_k)_t - T_j(X_k)_{t-1}}{\Delta t}$$

The terms $\Delta T_{jk}/\Delta t$ are in some cases so large, however, that the finite difference approximation is generally only acceptable for a one-year discretisation interval which has therefore been used throughout the study.

The factors X_k are various in number stretching from such concepts as fuel substitution and changes in mix, to scale factor and capacity utilisation. In all cases the change in energy intensity is calculated as indicated in the above expression, by introducing the value of the factor X_k at the beginning and end of the period. The approach neglects interaction between different factors, which are second order differences of the type

$\Delta^2 T_j / \Delta X_k$, $\Delta X_k''$, and generally quite negligible over a one-year interval when compared with first order terms. It follows that the over-all change in T_j due to all factors is thus simply:

$$\frac{\Delta T_j}{\Delta t} = \sum_k \frac{\Delta T_{jk}}{\Delta t}$$

In this work the results are stated in integral rather than differential form. Energy intensity trajectories $T_{jk_1 k_2}^t$ ($t = 1, 2, \ldots 19$) are defined as:

$$T_{jk_1}^t = T_j^{1953} + \int_{1953}^{1953 + t} \frac{\partial T}{\partial \tau} jk_1 \; d\tau$$

$$T_{jk_1 k_2}^t = T^{1953} + \int_{1953}^{1953 + t} \left(\frac{\partial T}{\partial \tau} jk_1 + \frac{\partial T}{\partial \tau} jk_2 \right) d\tau$$

. . . .
. . . .

$$T_{jk_1 k_2 \cdots k_i}^t = T_j^{1953} + \sum_{k=k_1}^{k_i} \int_{1953}^{1953 + t} \frac{\partial T}{\partial \tau} jk \; d\tau$$

which represent the time evolution of the energy intensity of sector j as factors $X_{k_1}, X_{k_2}, \ldots X_{k_i}$ are included in succession.

DEFINITION OF ENERGY DEMAND

Energy demand is accounted in terms of final use. That is primary energy conversion and distribution losses are not considered in this part of the model. This choice may be questionable in some respects and more particularly in a short-term view, but it is methodologically correct. In fact, insofar as really effective energy demand management policies are concerned, the time horizon is generally sufficiently long that the question of fuel accounting in final uses can be considered essentially de-coupled from primary energy conversion. Electricity for example, which is the major cause in point, and which is today predominantly of fossil fuel origin, could within 20-30 years, be predominantly of nuclear origin. To include conversion losses in electricity demand accounting can thus be

fundamentally misleading and incorrect if the very nature of conversion
changes over time.

The energy content of electricity in this work is therefore calculated as 860 kcal/kWh consumed. Clearly in this approach care is needed in
making value judgements over the short and medium term for which conversion systems remain essentially unchanged. Thus to say that an electric
tram is more energy saving than a diesel bus is not correct in a short-
term view but it can be in a long-term view. The energy intensity of these
two travel modes using the adopted method of accounting is presently about
65 and 148 kcal/pass. km respectively. Thus in the short or medium term
during which electricity will continue to be produced predominantly from
fossil fuel, the diesel bus is undoubtedly more efficient in primary energy
terms. On the other hand, the electric tram or trolley is probably more
efficient in the long term over which fossil fuel can be and will be essentially eliminated as an input to electricity production.(1)

THE DATA BASE(2)

Evidently the work presented would not have been possible without reference to a very ample data base. The data used in the model is not always directly available from official compilations and it has therefore
largely been reconstructed from significant information available from all
source.

Standard statistical techniques were used to obtain the best fit between the model parameters and structural and technological information.
Since the model parameters are generally directly measurable and good estimates are usually available from either official sources or from private
studies, the statistical techniques usually amounted to obtaining the best
first order corrections to available data to give the best overall fit to
existing national energy accounts, the Bilanci Energetici which collect
information on final energy use on the basis of 22 sectors and 23 fuels.

---

1) The question is however complicated by the definition used for primary
   energy.
2) For further information see reference 1, page . The complete data
   base has not yet been published and indeed is presently in the process
   of expansion both in depth and in time.

THE RESULTS

Urban passenger travel

This sector is defined to include commuter as well as metropolitan travel and covers all such activities as travel to and from work, education, hospital and other social services, shopping and entertainment.

Modes delivering urban passenger travel vary from the very low energy intensity commuter tram (41 kcal/pass.-km in 1972) to larger capacity class automobiles which averaged 1660 kcal/pass.-km in urban traffic in 1972 (Table 2). With the exception of subway travel which did not exist in Italy prior to 1955, modes prevalent in 1972 were all present at the beginning of the period considered although their average characteristics had changed substantially.

Table 2

MODAL DISTRIBUTION OF URBAN PASSENGER TRAVEL (%)

	1953	1963	1972	Specific energy consumption in 1972 (kcal/pass.-km)
Motorbicycle	4.7	6.0	7.4	134
Lightweight motorcycle	10.7	8.8	1.8	165
Motorcycle	2.4	4.3	1.4	377
Automobile 800 cc	5.4	19.4	18.9	535
Automobile 800/1000 cc	0.9	2.3	10.9	609
Automobile 1000/1500 cc	4.5	14.9	20.2	835
Automobile 1500/2000 cc	0.9	1.2	5.2	1053
Automobile 2000 cc	0.5	0.9	0.8	1658
Taxi	1.4	0.7	0.4	870
Metropolitan bus	7.5	9.8	10.0	148
Metropolitan trolley	6.1	3.0	0.9	62
Metropolitan tram	12.3	3.5	0.8	65
Subway	–	0.1	0.4	122
Commuter bus	19.2	14.6	10.7	136
Private and rental bus	0.7	1.3	4.4	91
Commuter trolley	0.7	0.8	0.5	53
Commuter tram	4.2	0.6	0.1	41
Rail	17.9	7.8	5.2	112
Total	100	100	100	468

Between 1953 and 1972 specific energy consumption for the sector as a whole increased more than two-fold from 197 to 468 kcal/pass.-km

(Figure 3). The initial period is characterised by a small increase due largely, but not entirely, to a decrease in the capacity utilisation of steam trains, which at that time delivered around 10 per cent of the traffic and consumed up to as much as 35 per cent of the sectoral energy use. The specific energy consumption came back down to the starting value in 1958 as a consequence of the displacement of steam trams by electricity and diesel which occurred very rapidly beginning with the mid-fifties.

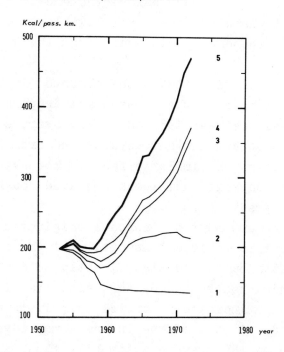

Figure 3

ENERGY INTENSITY OF URBAN PASSENGER TRAVEL

Trajectories due to successive inclusion of :

1. Fuel substitution
2. Shift between public and private travel
3. Changes in capacity utilisation
4. Changes in specific energy consumption of individual modes
   (at specified full capacity)
5. Modal shift within public and private travel

Essentially all the increase in sectoral energy intensity thus occurred after 1958 and can be attributed almost entirely to three effects evidence in Figure 3: a massive shift from public to private forms of travel, a decrease in capacity utilisation and a shift to more intensive modes within both private and public travel.

Between 1953 and 1964 urban travel delivered by private means doubled from 31 per cent to 63 per cent of the total (Figure 4). Most of this impressive increase occurred after 1957 in just seven years during which the total sector activity also doubled, from 56 to 111 x 10^9 pass.-km, so that private travel actually went through a four-fold multiplier in terms of activity. The shift to private means slowed down markedly after 1965 and actually inverted after 1970 probably as a result of the improving quality of public travel as opposed to the deteriorating quality of private travel, related to increasing traffic congestion and scarcity of parking space.

Whereas in the early stages private travel was dominated by two-wheel modes and particularly by scooters, these were very rapidly displaced by the more energy intensive automobiles which came to cover well over 80 per cent of urban travel delivered by private means after the mid-sixties. As the displacement continued, the modal mix within private travel moved to more powerful models (Figure 5) contributing another important component of increase to the average sectoral energy intensity.

A similar trend to greater energy intensity is evident in the modal shifts observed for public travel. The very low intensity trams and trolleys reached a maximum development before the end of the fifties and thereafter were rapidly displaced by the more flexible bus, while a minor component of increase in the same period came from the development of subway travel.

The major factor contributing to the observed increase in energy intensity after 1960 was however the substantial drop in capacity utilisation of private means. Between 1953 and 1972, average load factors dropped from 1.3 to 1.1 for scooters and motorcycles and from 2.1 to 1.2 for automobiles in urban traffic. A similar effect is not evident for public travel for which average capacity utilisation stayed roughly constant throughout the period considered.

Finally, a minor component of increase originated in the changes occurring in the specific vehicle consumption of the individual modes. It is difficult to attribute this significantly to different causes. Most of the effect was undoubtedly due to increasing traffic congestion accompanying the exponential increase in travel by private means observed throughout the period. A part of the increase may have come from the increasing average age of the private automobile fleet after 1960 and to slight increases in power. As regards public modes, scale factor effects coming from the use of larger vehicles actually determined a component of decrease in average specific energy consumption which is however completely lost in the average over the whole sector.

Long-distance passenger travel

This sector is defined to include all passenger travel outside city limits, excluding commuter travel, and within Italian territory and territorial waters. Modes delivering this type of activity partly overlap those

Figure 4

DISTRIBUTION OF URBAN PASSENGER TRAVEL
BETWEEN PUBLIC AND PRIVATE MEANS

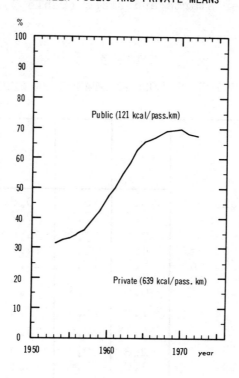

Figure 5

MODAL SHIFT WITHIN AUTOMOBILE TRAVEL

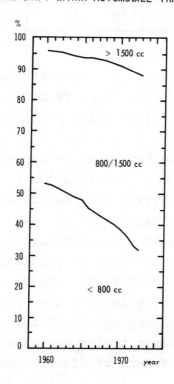

involved in urban transport but generally with considerably lower energy intensity on account of higher capacity utilisation and higher engine cycle rating due to fewer starts and stops (Table 3).

Table 3

MODAL DISTRIBUTION OF LONG-DISTANCE PASSENGER TRAVEL

	1953	1963	1972	Specific energy consumption in 1972 (kcal/pass.-km)
Lightweight motorcycle	10.0	6.5	0.8	167
Motorcycle	2.5	3.8	0.7	373
Automobile 800 cc	7.9	25.0	24.6	191
Automobile 800/1000 cc	1.2	2.9	14.3	217
Automobile 1000/1500 cc	6.5	19.2	26.4	299
Automobile 1500/2000 cc	1.2	1.6	6.9	377
Automobile 2000 cc	0.6	1.1	1.1	592
Rental automobile	7.8	5.4	3.5	346
Rail	56.1	28.7	15.4	72
Public bus	2.8	2.1	0.9	136
Private and rental bus	1.0	1.6	3.3	72
Sea	2.3	1.7	1.2	1310
Air	0.1	0.4	1.0	1530
Total	100	100	100	254

Specific energy consumption for the sector as a whole increased by over 30 per cent between 1953 and 1966, from 208 to 273 kcal/pass. km, and thereafter declined again to a value of 254 kcal/pass. km in 1972 (Figure 6).

Many of the remarks made for urban passenger travel also apply to long distances. Here again fuel substitution is responsible for a size-able component of decrease particularly before 1960 in relation to the rapid displacement of steam trains. In the early fifties rail travel as a whole was much more important over long distances than it was for urban travel, covering for example in 1953, respectively 56 per cent and 18 per cent of the total activity. On the other hand, steam trains were much more important in urban travel, where in 1953 they covered 65 per cent of the activity delivered by rail as opposed to 21 per cent over long distances. Overall the role played by steam trains was about the same in the two sectors and moreover the displacement by diesel and electric trains went ahead at much the same rates, facts which are reflected in the very similar decreasing component due to fuel substitution witnessed in Figures 3 and 6.

Figure 6

ENERGY INTENSITY OF LONG-DISTANCE PASSENGER TRAVEL

Trajectories due to successive inclusion of :

1. Fuel substitution
2. Shift between public and private travel
3. Changes in capacity utilisation
4. Changes in specific energy consumption of individual modes
   (at specified full capacity)
5. Modal shift within public and private travel

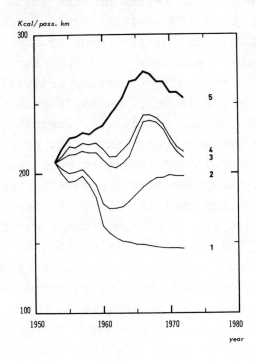

Kcal / pass. km

Evidently in the case of rail travel fuel substitution is not just the substitution on one fuel for another within a given technology. It is in fact intimately related to substitution between entirely different technologies.

Electric trains were effectively displacing steam trains in Italy well before World War II, the railway system could already in the early fifties be considered as basically planned for electric trains. Rapidly ageing steam locomotives were usually relegated to secondary routes or functions where in the absence of electrification they were anyway being rapidly displaced by diesel locomotives. They were rarely used at full capacity, which generally speaking was considerably lower than that of the more modern trains.

This fact is of importance in understanding the markedly different efficiencies of rail travel by steam and by diesel and electricity

337

(Figure 7) which is quite out of proportion with the effective thermo-dynamic efficiencies and which explain the strong effect of "fuel substi-tution" as steam trains were phased out of service in the late fifties.

In this respect the very different time dependence of energy effi-ciency of railway travel by the three technologies apparent in Figure 7 is rather remarkable. Whereas the longer term trend for diesel and elec-tric trains is towards a decrease, due largely to a scale factor effect, and better management, in the case of steam trains there is a very evident tendency to increase. The maxima in steam train efficiency interestingly enough, occur during periods of recession when quite naturally it is ex-pected that the slower growth in activity would tend to be taken out of the older technology undergoing displacement by such circumstances as greater engineering neglect and even lower capacity utilisation as steam trains were allocated to even more secondary activities.

As in the case of urban travel, the shift from public to private means occurring throughout the period considered, contributed substantial in-creases to the average overall energy intensity of the sector. During the latter part it represented indeed the major component of increase. The percentage of long-distance passenger travel effected by private means in-creased steadily from 38 per cent in 1953 to 77 per cent in 1968. There-after the relative growth slowed down but there was no inversion as for urban passenger travel and private means continued to capture increasing portions of long-distance passenger travel at least until 1972. It might be pointed out that although the doubling in percentage took considerably longer than for urban travel (15 as opposed to seven years), in actual fact the overall sectoral activity grew much faster over long distances so that in absolute terms the growth rates of private travel are more com-parable: 15 per cent for long distances between 1953 and 1968 and 19 per cent for urban travel between 1957 and 1964. Indeed during the fastest period of growth, between 1957 and 1965 private travel over long distances grew at an average annual rate of 19 per cent as in urban travel.

In this sector automobiles represented the dominant mode of private travel right from the beginning of the period considered and thus any dis-placement of two-wheel travel which occurred led to only minor increases in energy intensity. Beginning with the late fifties however, changes in modal mix of automobiles in favour of higher engine capacity led, as for urban travel, to consistent increases in the average energy intensity of the sector.

Within public travel, rail continued to represent the principal mode throughout the period considered although bus travel, particularly private and rental bus, capture rapidly increasing share during the second half of

Figure 7

ENERGY INTENSITY OF RAIL TRANSPORT
(statistical average)

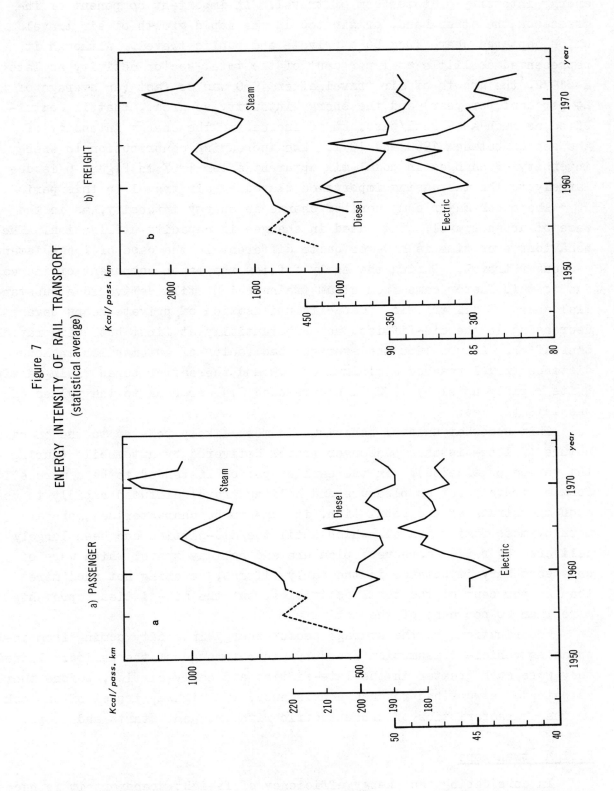

a) PASSENGER

b) FREIGHT

the sixties, however without leading to significant changes in final use
energy intensity of this form of travel. An important component of increase on the other hand, originated in the rapid growth of air travel
which captured share from both private and public travel. Although it
represented as little as 1 per cent of the total sector activity as late
as 1972, the growth of air travel after 1960 was so fast (an average of
24 per cent per year) and the energy intensity so great, that it contributed as much as 9 kcal/pass. km to increasing the energy intensity of
the sector between 1960 and 1972. The increasing contribution to energy
intensity of changes in modal mix apparent after 1967 in Figure 6 is due
largely to the increasing importance played by air travel in this period.

The other major component of change in energy intensity, as in the
case of urban travel, originated in changes in capacity utilisation. The
behaviour over time is however quite different in the case of long-distance
passenger travel. During the early fifties the major component of increase
due to this factor came from steam trains which still delivered a substantial share of all traffic. Capacity utilisation of private means began
decreasing in the mid-fifties, not substantially at first but quite rapidly
thereafter. Around 1966 the average load factor of automobiles in long-
distance travel reached a minimum of 1.9 and thereafter began rising again
quite rapidly until by 1972 it had reached 2.4, or back to the values of
the late fifties.

This somewhat unusual behaviour is most likely related to the changing
nature of long-distance passenger travel delivered by automobile (during
the period considered). In the earlier period it tended to be a more elite
form of travel, but as pleasure and holiday travel diffused rapidly to most
economic strata of the population, it came to be characterised more and
more by mass family travel, which until the mid-sixties had been largely
delivered by rail. Moreover, pleasure and holiday travel which were of
only secondary importance in the early-fifties, covering not much more
than 25 per cent of the total activities, but the mid-sixties represented
more than 60 per cent of the activity.

Contributions to the average sector energy intensity coming from increasing vehicle consumption are very minor throughout the period. Indeed
they were much greater in the late-fifties and early-sixties, before the
autostrada network had reached consistency, and highway travel often took
on the characteristics of urban traffic with frequent starts and stops.

Freight transport

In considering the energy efficiency of freight transport it is necessary to distinguish between its long-distance and local aspects. For

journeys over land and for distances up to about 100 km road carriers are generally the most effective means of delivery, although in special circumstances rail transport can be an important competitor (Table 4). For journeys over distances greater than 100 km, other modes such as sea and rail become progressively more competitive (Table 5). Long-distance and local freight transport are therefore essentially different systems absolving generally different functions.

Table 4

MODAL DISTRIBUTION OF LOCAL FREIGHT TRANSPORT

	1953	1963	1972	Specific energy consumption in 1972 (kcal/ton-km)
Rail	20.2	7.3	6.1	649
Three-wheeled van	4.7	8.5	6.1	1100
Special purpose truck	1.2	2.5	5.6	446
Truck 2.5 t	30.2	35.6	32.8	2435
Truck 2.5/5.0 t	29.9	26.9	28.4	1145
Truck 5.0 t	13.8	19.2	19.3	975
Total	100	100	100	1462

Table 5

MODAL DISTRIBUTION OF LONG-DISTANCE FREIGHT TRANSPORT

	1953	1963	1972	Specific energy consumption in 1972 (kcal/ton-km)
Air	--	-	0	8480
Sea	7.9	11.8	16.2	220
Rail	30.3	21.5	14.7	111
Truck 2.5 t	1.6	1.3	1.0	2150
Truck 2.5/5.0 t	6.4	4.0	3.4	1010
Truck 5.0 t	12.0	11.4	9.4	860
Medium trailer truck	5.9	1.7	1.1	304
Heavy trailer truck	33.2	40.7	32.9	292
Tractor semi-trailer	-	2.9	5.6	398
Special purpose truck	1.5	2.7	4.7	343
Pipeline	0.2	1.2	10.2	45
Inland waterway	1.0	0.9	0.8	78
Total	100	100	100	350

The energy intensity of these two sectors show a remarkably different behaviour with time over the period from 1953 to 1972 (Figures 8 and 9). Whereas in the case of local freight transport the energy intensity almost doubled during this period, from 835 to 1462 kcal ton-km., in the case of long-distance freight transport it dropped by about 20 per cent, from 439 to 350 kcal/ton-km.

By far the greatest effect in reducing the energy intensity of long-distance freight travel can be attributed to the substitution of electric for steam trains. In 1953 rail transport accounted for 30 per cent of all long-distance freight deliveries and 43 per cent of this activity was driven by steam locomotives with an average specific energy consumption of 1350 kcal/ton-km. Thus while steam trains were responsible for barely 13 per cent of the overall sectoral activity, they actually accounted for more than 40 per cent of the energy demand.

Steam trains were phased out of freight transport more slowly than from passenger travel and this basically explains why a strong decreasing component due to fuel substitution in Figure 8 lasts all the way into the seventies. In the case of local freight travel, the use of steam trains stabilized at between 0.3 and 0.4 x 10^9 ton-km., after 1957, but rail travel is sufficiently unimportant in this sector (less than 6 per cent in 1972) than even if they had been rapidly displaced this would not have contributed a very significant reduction in energy intensity of the sector.

Another important factor acting to decrease the energy intensity of both long-distance and local freight transport was the modal shift to more efficient carriers. In local freight transport this occurred basically through an increase in the size of carriers determining a decrease in energy intensity at full capacity. Unfortunately, as will be discussed below, this decreasing component was in practice totally outweighed by decreasing capacity utilisation. A substantial shift to larger and more efficient road carriers also occurred in long-distance freight transport but in this sector the major decreasing component due to modal shift came from the very rapid growth of sea travel and pipeline characterised by very low energy intensities (respectively 220 and 45 kcal/ton.-km. in 1972). These two modes increased their share in total long-distance activity from less than 8 per cent in 1953 to over 26 per cent in 1972.

The major increasing component of enery intensity in both freight transport sectors originates in the loss in capacity utilisation, particularly of road transport. There exist as yet no complete explanation of why capacity utilisation of road transport should have dropped between 1953 and 1972 by 35 per cent in local transport and by 21 per cent over

Figure 8

ENERGY INTENSITY OF LOCAL FREIGHT TRANSPORT

Trajectories due to successive inclusion of :

1. Fuel substitution
2. Modal shift
3. Changes in capacity utilisation
4. Changes in specific energy consumption of individual modes
   (at specified full capacity)

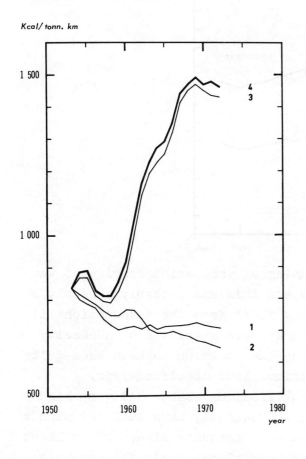

Figure 9

ENERGY INTENSITY OF LONG-DISTANCE FREIGHT TRANSPORT

Trajectories due to successive inclusion of :

1. Fuel substitution
2. Modal shift
3. Changes in capacity utilisation
4. Changes in specific energy consumption of individual modes
   (at specified full capacity)

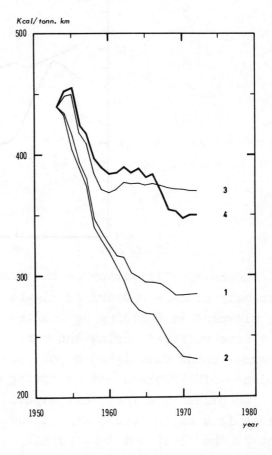

long distances (Figure 10). It should be borne in mind that the effect on energy intensity was actually much greater than that immediately apparent from this Figure on account of the fact that the greatest loss in capacity utilisation occurred in the smaller sized, more energy intensive carriers.

Figure 10
CAPACITY UTILISATION IN FREIGHT TRAVEL BY ROAD
(Statistical average over all carriers excluding
special purpose trucks)

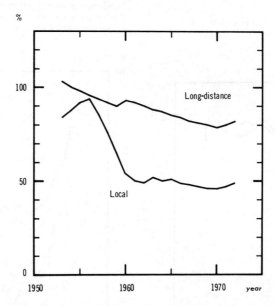

Capacity utilisation at the beginning of the period considered was extremely high as appears in Figure 10 and this was probably a result of the slowness in building up the trucking fleet from the devastations it must have suffered during the war. Trucks were overladen by necessity because demand outweighed supply and the law is known to have been quite lenient with truckers not observing maximum load specifications.

Doubtless the fact that capacity utilisation started at very high values is a major reason why it should subsequently have suffered such a drop as the fleet built up rapidly to a more adequate size, but at least three other factors mentioned here must have played a significant role. Foremost among these is the very high dispersion of the trucking sector. In 1966 for example, 74 per cent of all companies owned only one truck, 18 per cent owned only two trucks and fewer than 1.5 per cent owned more than five trucks, with an overall average of 1.5 trucks per company. Legislation enacted in the early-sixties essentially prohibiting issue of new truck licenses to third customer trucking companies, tended to favour this dispersed structure or at least to slow down the inevitable trend to larger economies of scale. Moreover, there seems to have been a mismatch between the optimal size vehicles and those actually available or demanded on the market on the basis of excessively optimistic forecasts of freight transport growth.

Towards the end of the period considered there seems to be some saturation in the magnitude of the effect and this may be due in part to the modal shift to larger capacity road carriers, to the phasing out of over-sized trucks as they ran out their useful lifetime, and to the slow but inevitable absorption of small trucking companies into larger organisations disposing of more efficient techniques for time and space allocation of their resources.

Changes in the specific vehicle consumption of individual modes have consistently tended to provide for an increase in energy intensity of local freight transport, probably due to the aggravating conditions of traffic congestion. Over long distances a similar change occurred but only up to the mid-sixties after which changes in specific vehicle consumption led to a decrease in energy intensity of the sector as a whole probably in relation to the rapidly increasing autostrada network.

Manufacturing and mining

The specific energy consumption of manufacturing and mining increased by 30 per cent between 1953 and 1972, from 20.5 to 26.5 kcal/Lira of value-added measured at factor cost and in 1972 prices (Figure 11). The increase would however have been much greater had it not been for the role played by two important decreasing components of change and these will be considered first.

Fuel substitution contributed a minor but not insignificant component of decrease, particularly during the first seven years of the period considered, as coal was rapidly displaced by oil and natural gas from most activities other than iron and steel production. Whereas coal substitution continued unabated throughout the sixties and into the seventies, penetration by natural gas which had been extremely dynamic in the early-fifties, came to a half around 1958 due to the much more rapid growth in consumption of petroleum products. And indeed, as a consequence of the considerably lower thermodynamic efficiency of liquid petroleum products as opposed to natural gas, the component of change in energy intensity coming from fuel substitution became slightly positive, beginning around 1960. The share of natural gas in the industrial system declined quite steadily until it began rising again in the late-sixties contributing a new negative component to the overall energy intensity of the sector. Electricity lost share slightly throughout the period considered (Figure 12).

Figure 11

ENERGY INTENSITY OF MANUFACTURING AND MINING

Trajectories due to successive inclusion of :

1. Changes in specific energy consumption of individual sectors
2. Fuel substitution
3. Changes in product mix
4. Economies of scale and changes related to the management of production

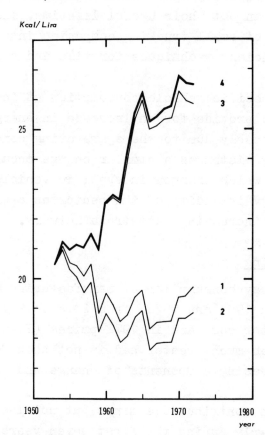

Kcal/ Lira

The major decreasing component of energy intensity originated in technological and basically engineering changes in the unit processes of production occurring within most individual sectors. Unfortunately the model is not as yet sufficiently refined to allow separation between this important effect and the superposed contributions coming from fluctuations in capacity utilisation of industrial plant related to economic cycles and to the general process of growth. The component in question is shown isolated in Figure 13, from which it is possible to discern the impact of poor capacity utilisation on increasing the overall energy intensity of manufacturing and mining.

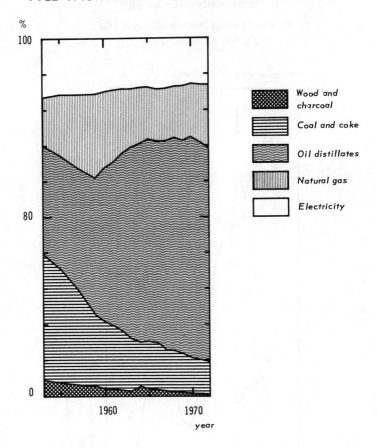

Figure 12
FUEL SUBSTITUTION IN MANUFACTURING AND MINING

Wood and charcoal

Coal and coke

Oil distillates

Natural gas

Electricity

Examples of the technological changes involved are numerous. Perhaps
the best known and certainly among the most important, are related to the
iron and steel industry and to cement production. In the former case much
of the decrease in specific energy consumption per unit of production
(Table 6) came from such technological changes as the displacement of
Bessemer and Thomas converters by the L/D process in steel production, improvements in refractory coating and better process control of the blast
furnace, and the introduction and rapid spread of continuous casting. In
the case of cement production the decrease in energy intensity came about
largely through the almost complete displacement of the wet process and
the traditional vertical kilns (respectively 1485 and 1150 kcal/kg of clinker) prevalent in the early-fifties by the much less intensive dry process
(990 kcal/kg of clinker). In all industries heat recovery and other energy
savings measures were increasingly practiced throughout the period
considered.

Figure 13

ADVANCES IN TECHNOLOGY AS DETERMINANTS OF THE
ENERGY INTENSITY OF MANUFACTURING AND MINING
(arrows indicate recession years)

a) decrease in energy intensity relative to 1953
b) rate of growth of industrial production

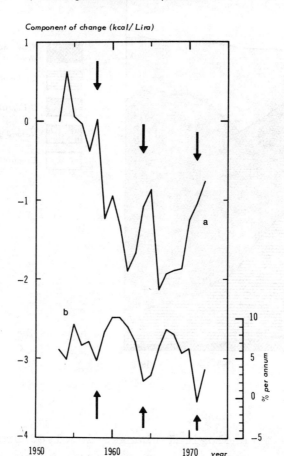

A final factor of importance contributing to the decreasing specific energy consumption of individual sectors is related to the renewal of usually inefficient pre-war industrial plant still existing in the early-fifties. In the chemical industry for example it is estimated that this effect contributed as much as a 10 per cent reduction in energy intensity between 1953 and 1960.

Improvements in machine and manpower management between 1953 and 1972, in a word, the sleeker and more efficient organisation of production accompanying both economies of scale, increased mechanisation, automation and more recently the advent of the computer, have obviously had an impact on energy intensity. Whereas all these effects in general led to an increase in the energy consumed per physical unit produced, the question of whether

they led to an increase or decrease in the energy consumed per unit of
value added is highly complex and as yet incompletely resolved. There is
evidence that many if not most of these effects can be captured by considering as indicator the long-term changes in the size of productive units
in terms of number of employees. The effect shown in Figure 11 determined
by reference to the energy impact of changes in the distribution by size
of productive units, indicates that improvements in the organisation of
production has led to an overall increase in intensity although in many
subsectors of manufacturing and mining the inverse appears to be true.

Table 6

SPECIFIC ENERGY CONSUMPTION IN THE IRON AND
STEEL INDUSTRY (10^3 kcal/ton)

	1955	1966	1972
Pig iron	4.530	3.574	3.430
Steel	1.100	723	585
Finished products	1.630	1.276	860

By far the strongest component of change in the energy intensity of
manufacturing and mining in the period considered, has been the very marked
variation in product mix (Table 7). Although a number of low intensity
sectors, among which the most important was transport equipment (6.7 kcal/
Lira in 1972), grew considerably faster than manufacturing and mining as
a whole, two very energy intensive sectors, iron and steel and chemicals
and allied, also grew equally rapidly and clearly dominated the energy
picture. The change in mix in the gross sectors of manufacturing and
mining (Table 8) was of course accompanied by significant changes in the
finer product structure which here again led to an overall increase in
energy intensity. Thus, while in some sectors like textiles and clothing,
changes in mix led to a definite decrease in intensity, by far the most
important components of change originated in energy intensive sectors like
chemicals and allied where the product structure moved to even greater
average intensity in relation to the very rapid growth of petrochemicals
particularly during the late fifties and sixties.

Table 7

DISTRIBUTION OF VALUE-ADDED IN MANUFACTURING
AND MINING AMONG ENERGY INTENSITY CLASSES (%)

kcal/Lira	1953	1960	1967	1972
0-3	31.2	28.4	26.4	26.8
3-9	35.9	35.7	34.1	35.0
9-27	16.0	15.3	13.9	13.2
27-81	15.7	18.9	21.4	20.4
81	1.2	1.7	4.2	4.6
Total	100	100	100	100

In calculating the distributions, the energy
intensities of individual sectors were assumed
constant at 1972 values.

Table 8

COMPOSITION OF VALUE-ADDED IN MANUFACTURING AND MINING (%)

Sector	1953	1963	1972	Specific energy consumption in 1972 kcal/Lira
Mining	3.28	2.57	2.08	6.1
Iron and steel	2.20	3.98	4.56	82.1
Nonferrous metals	1.65	1.39	1.26	43.4
Paper and allied	6.53	5.65	5.31	18.3
Chemicals and allied	5.10	9.06	10.55	103.6
Mineral products	4.55	5.63	5.29	84.5
Food and related	11.88	11.75	11.12	13.9
Transport equipment	4.29	6.78	7.54	6.7
Mechanical	20.34	20.41	19.37	7.3
Textiles and clothing	17.72	14.58	13.09	10.1
Miscellaneous Manufacturing	22.46	18.20	19.83	4.4
Total	100	100	100	26.5

Value-added expressed in 1972 Lire at factor cost.

Agriculture, livestock, forestry and fishing

The specific energy consumption of this sector presented in Figure 14, does not need much comment. Almost the entire increase between 1953 and 1972, a factor of 2.5, comes from capital/labour substitution which can be largely identified with the diffusion of agricultural machinery, such as tractors, harvesters, etc. The very strong positive component finds no analogue in individual sectors where capital/labour substitution of comparable magnitude also occurred. This is presumably due to the extensive use of animals, mostly oxen, whose "labour" and equivalent energy use is generally not accounted for in the statistical estimates of value-added. Substantial changes in the mix of activity in favour of the much less energy intensive livestock sector occurring particularly during the last five years of the period, induced a minor decreasing component, amounting to about - 0.2 kcal/Lira of value-added.

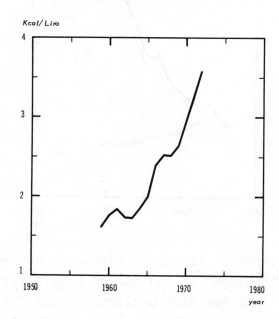

Figure 14

ENERGY INTENSITY OF AGRICULTURE, LIVESTOCK, FORESTRY AND FISHING

Space heating and cooling

The energy intensity of this sector has undergone more than a four-fold increase in the period from 1953 to 1972, from 9.6 to 40.1 x 10^3 kcal/m^3 (Figure 15). Only a very small portion of the energy consumed goes to the air cooling of buildings (about 0.5 per cent in 1972), thus the following discussion will be dedicated essentially to space heating uses in residential, commercial and public buildings.

Figure 15

ENERGY INTENSITY OF SPACE-HEATING AND COOLING

Trajectories due to successive inclusion of :

1. Fuel substitution
2. Change in distribution by housing type
3. Change in the mix of heating systems
4. Change in housing materials
5. Migration from the South
6. Change in specific energy consumption of individual heating systems

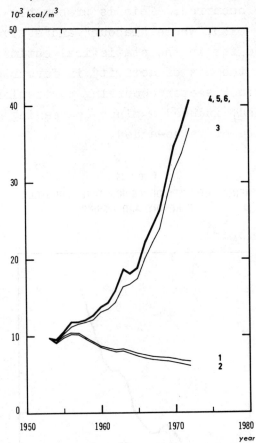

The impressive increase in energy intensity can be traced almost entirely to fundamental changes in the mix of heating systems occurring throughout the period and reaching a climax during the second half of the sixties. In considering these changes it is useful to refer to five basic types of heating system. In order of increasing "comfort" and energy consumption these are: rudimentary or no heating systems, fireplaces and other traditional systems, stoves, single-household radiators and centralised apartment heating. The inside temperature characterising each of these systems of course varies substantially according to circumstances, but as a general indication it can be taken that - during the heating period and averaged over the entire internal volume - it varies from as

352

high as 23°C or more in the case of central apartment heating, to around
19°C for single household radiators, 14°C for stoves, to less than 12°C
for traditional heating systems and to close to outside temperatures for
rudimentary or no heating system.

The essential factors leading to these differences are the increasing
volume over which the heating is carried out and the decreasing degree of
discrimination between heated and unheated volumes in passing from rudimentary to central heating systems. These factors are often assimilated
to an abstract quality referred to as "comfort level" which really amounts
in final instance, to having a relatively high inside temperature throughout the building at all times. Whereas in the case of rudimentary and
traditional heating systems and generally also of stoves, the heating is
generally located in living quarters, often only the kitchen where fortui-
tous heat sources, particularly from cooking, play an important role, in
the case of single household radiators and central apartment heating the
entire volume benefits from the heating system. The difference between
these two latter units lies essentially in the degree of control remaining
to the individual user. Central apartment heating is relatively indif-
ferent to the living habits in terms of heating period, and desired temperature, whereas single household radiators can be turned on or off to
suit individual tastes and needs. Incidentally, the choice of unit for
energy intensity, $kcal/m^3$ rather than $kcal/m^3$ degree.day reflects the fact
that the inside temperature and therefore the degree.days, depend so much
on the heating system.

A complete treatment of the evolutionary dynamics of the space heating sector is complicated by such factors as migration between climatic
zones and from rural to urban areas, which result in changes in degree.day,
in heating systems and in distribution by type of building. As such it is
beyond the scope of this work and is referred to amply elsewhere.

The change in distribution by heating system evidence in Figure 16
and aggregated over Italy as a whole gives a sufficient indication of the
predominant cause of increase. Taking the average inside temperature indicators referred to above for the individual systems and an average outside temperature of 7°C which is a close proxy to the average winter temperature, the changes in distribution by heating system indicated lead in
themselves to an increase in average temperature difference from 5°C in
1953 to 10°C in 1972, a factor of 2 increase. The full effect however,
can only be investigated at a much greater level of disaggregation because
of the great difference in degree.days between the North and the South
(almost a factor of 3) and because of the different dynamics in these two
regions.

Figure 16

DISTRIBUTION OF HOUSEHOLD AMONG HEATING SYSTEMS

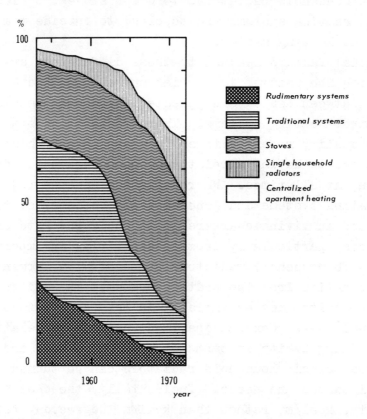

Other effects contributing to changes in energy intensity of the sector are in comparison relatively minor. Fuel substitution occurred generally in relation to substitution between heating systems. Wood and other
vegetable fuels were all-important in the fifties as also other solid fuels
such as coal and coke (Figure 17). These remained predominant into the
early sixties when they were displaced by fuel oil as central heating systems diffused rapidly. Liquid petroleum gas and kerosene also grew rapidly
in relation to the growth of single household radiators and to stoves which
were the only traditional systems to retain importance into the seventies.
Natural gas began to diffuse quite rapidly in relation to central heating
and single-household radiators in the late-sixties but it was not until
after 1973, beyond the time period of this study, that this fuel reached
its maximum rate of substitution. Electricity as a whole played a very
minor role in the sectors throughout the period considered (< 2.5 per cent
of total consumption).

Migrations from Southern to North Italy occurring throughout the
period, as is to be expected on account of the much greater degree.days in
the North, led to a component of increase in overall energy intensity but
this was mitigated by the fact that most of the migration occurred from

single family rural households or from building with few apartments to large apartment buildings in the metropoles with relatively lower heating demands. Indeed the increase in energy intensity due to these migrations cannot have exceeded 0.5 kcal/m^3 during the entire period considered.

Figure 17
FUEL SUBSTITUTION IN SPACE HEATING AND COOLING

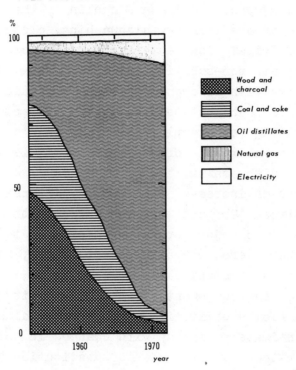

In the model five classes of household were taken into consideration defined according to the number of independent apartments in each building: 1, 2, 3-15, 16-30 and > 30. The change in distribution of families between household classes occurring between 1953 and 1972 was largely related to migrations between rural and urban areas and led to a 10 per cent increase, from an average of 4.1 apartments/building in 1953 to about 4.5 in 1972. This implies of course a decreasing component of energy demand on account of the increasing number of shared surfaces, between households which is however not very large as indicated in Figure 15.

Another minor component of increase comes from changes in building materials which almost passed through a discontinuity during the Second World War, an increase of about 30 per cent in average thermal conductivity of new construction. This effect is however minor in the period considered (less than 0.5 kcal/m^3) on account of the slow turnover of the housing stock.

A sizeable component of increase appearing particularly in the latter part of the period and amounting to about 2.5 kcal/m^3 originates in the general degeneration of the efficiency of central and single radiator systems as average age rapidly increased during the sixties.

Other civilian uses

This is a residual sector comprising all residential, commercial and public uses other than space heating and cooling. The major subsectors included in this sector are cooking, dishwashing, clotheswashing, personal hygiene, lighting, refrigeration, entertainment and other minor uses. By definition these are all non-substituting uses. When speaking of changes in mix it is thus only with reference to changes within each of the principal sectors indicated, and not between them. Faster growth in dishwashing relative to lighting for example, is not in this context considered a change in mix whereas an example of a change in mix is the shift from water heating to automatic dishwashing.

Other civilian uses increased their energy intensity from 390 to 1000 kcal/person between 1953 and 1972, an increase of 2.5 times (Figure 18). Whereas the major forces of change led to an increase in energy intensity, there were two basic effects which, had they acted alone, would have determined a substantial decrease.

The major decreasing component of energy intensity came as in most other sectors, from fuel substitution. However, as also in most other sectors, this was intimately tied up with changes in technology. Its nature is best understood by reference to substitution between different appliances, for example, between the traditional stoves, running on solid or liquid fossil fuel to heat water, and the natural gas or electric water heater. The distribution by fuel which was heavily weighted on the solids in 1953 had moved to a basically liquid and natural gas structure in 1960 and then developed rapidly into an electric and natural gas dominated structure through the sixties and early-seventies (Figure 19).

The other basic decreasing component of energy intensity originated in the overall shift from the residential to the less intensive commercial and public category. Whereas in some instances, for example, cooking and dishwashing, the energy intensity in the residential category is only some 20 per cent greater than in the commercial and public category, there are some uses where the difference is very much greater. This is particularly true of entertainment where the residential category, typified by television, was in 1972 some 20 times more energy intensive than the commercial and public categories. In this particular sector however as also for refrigeration, the shift has been markedly away from commercial and public

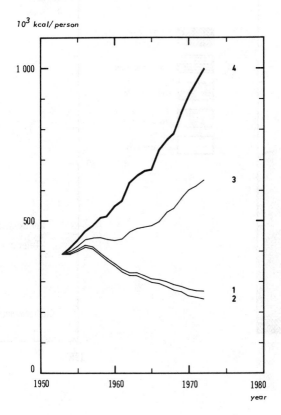

Figure 18

ENERGY INTENSITY OF OTHER CIVILIAN USES

Trajectories due to successive inclusion of :

1. Fuel substitution
2. Shift between categories
3. Change in mix of appliances
4. Change in specific energy consumption of individual appliance uses

categories and towards residential, contributing to mitigate the otherwise quite strong decrease coming from the basic intercategory shift. The overall effect is indeed quite minor reaching barely -25 kcal/person over the period to 1972.

Changes in the mix of appliances concern practically all subsectors of other civilian uses. Dishwashing, which in the early fifties used to be largely effected with cold water, or water heated on the kitchen fire or stove, in the early-seventies was increasingly carried out by electric or gas water heaters and by highly energy intensive automatic appliances. The same is of course true of clotheswashing. An extreme case is refrigeration which diffused into the residential sector with competition coming only a stage removed from the commercial sector which had previously almost entirely fulfilled this function. The diffusion of highly energy intensive appliances was most rapid during the sixties (Figure 20) and this is clearly reflected in the curves of Figure 18.

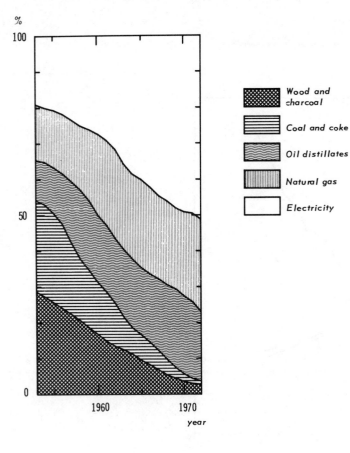

Figure 19

FUEL SUBSTITUTION IN OTHER CIVILIAN USES

%
100
50
0
1960
1970
year
Wood and charcoal
Coal and coke
Oil distillates
Natural gas
Electricity

Figure 20

DIFFUSION OF MAJOR APPLIANCES IN THE RESIDENTIAL SECTOR

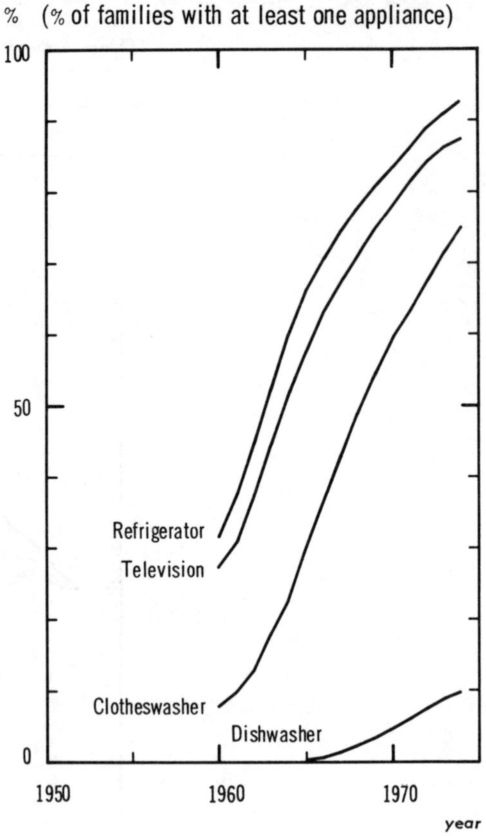

% (% of families with at least one appliance)
100
50
0
Refrigerator
Television
Clotheswasher
Dishwasher
1950
1960
1970
year

An equally important effect contributing almost the same component
of increase – 360 kcal/person – as the change in mix between 1953 and 1972,
came from changes in the specific energy consumption of individual appliances. This component originated in three basic effects, the increasing
use of appliances (number of showers/day), the increasing energy intensity
of appliances (larger refrigerators, and television sets) and the diffusion of multiple appliances (two or more televisions per household). No
attempt has been made to separate out these three components in Figure 18,
however it can be said that the first played the most important role during the fifties and early-sixties, both the first and second were about
equally important during most of the sixties and all three effects were of
roughly equal importance during the late sixties and early-seventies.

CONCLUSIONS

Energy intensity and activities of the basic sectors defined in the
study are given in Table 9 for 1953 and 1972. Table 9 also compares actual
final energy demand in 1972 with what it would have been in the hypothetical case that the energy intensities had remained constant at the 1953
levels. The significance of this test is to estimate the contribution of
each of the sectors considered, to the energy demand increase of the economy
as a whole. Referring to the last column of the Table it is possible to
conclude that in this hypothesis final energy demand in 1972 would have
been lower by 374.5×10^{12} kcal, that is 578.8 instead of 953.3×10^{12} kcal.
Reaggregating the various basic sectors it is possible to conclude that
the most important contributions came from space heating followed by industry and agriculture and only at a distance, by transport and other
civilian uses.

Space heating and cooling	173.9×10^{12} kcal
Industry and agriculture	119.6
Transport	47.6
Other civilian uses	33.4
Total	374.5

These results are significantly different from the picture which is
usually advanced that the major contributions to increasing energy demand
in Italy have come from industry.

Table 8

FINAL ENERGY DEMAND BETWEEN 1953 AND 1972

	Final energy demand		Activity		Energy intensity		Increment in energy demand due to in- crease in intensity
	1953	1972	1953	1972	1953	1972	
Urban passenger travel	8.3	68.9	42.3	146.9	197	469	40.0
Long-distance passenger travel	6.8	49.6	32.7	195.1	208	254	9.0
Long-distance freight travel	16.0	39.0	36.4	111.3	439	350	-9.9
Local freight travel	5.6	19.7	6.7	13.5	835	1462	8.5
Manufacturing & mining	96.9	471.6	4.73	17.8	20.5	26.5	106.8
Agriculture, livestock forestry & fishing	5.7	21.2	4.05	5.91	1.42	3.58	12.8
Space heating & cooling	39.0	228.6	4.06	5.70	9.6	40.1	173.9
Other civilian uses	18.8	54.7	48.1	54.7	390	1000	33.4
Total economy	197.1	953.3	26.8	78.7	7.35	12.1	374.5

1) Energy in 10^{12} kcal; transport activity in 10^9 pass.-km. (10^9/ton-km); industrial and agricultural activity and gross national product in 10^{12} Lire (at 1972 prices); space heating and cooling activity in 10^9 m^3; other civilian uses in 10^6 inhabitants.

2) Increment in energy demand ΔE due to increase in intensity I, between 1953 and 1972 calculated from:

$$\Delta E = A^{1972} (I^{1972} - I^{1953}).$$

Two different cross-sections are given in Figures 21 and 22 with reference to the energy intensity of the economy as a whole. In producing these figures the activities used in the calculations were those actually verified in the economy year by year. Since the different sectors grow at substantially different rates it was necessary to provide a base line (the dashed line in both Figures) which represents the energy intensity of gross national product assuming the actual measured activities but the 1953 sectoral intensities. As the base line indicates, different growth rates of the various basic sectors have had only a small effect on the overall energy intensity of the economy.

Figure 21 provides a pictorial, time dependent representation of what has already been said in regard to the contribution of the major sectors to overall increase in energy demand. Contributions from freight transport are not included. Indeed the decrease in energy intensity of long-distance freight transport discussed previously, and the increase in local freight transport practically balance one another in the overall energy

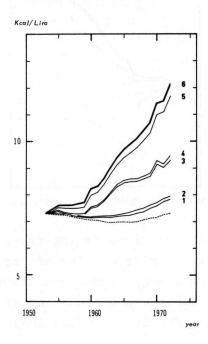

Figure 21

ENERGY INTENSITY OF GROSS NATIONAL PRODUCT

Trajectories due to successive inclusion of changes
in the energy intensity of :

1. Urban passenger travel
2. Long-distance passenger travel
3. Manufacturing and mining
4. Agriculture, livestock, forestry and fishing
5. Space-heating and cooling
6. Other civilian uses

Dashed line represents base line obtained assuming
sectoral intensities constant at 1953 values.

picture with combined effects on the energy/GNP ratio never greater than 0.03 kcal/Lira in absolute magnitude.

Figure 22 assesses the contributions arising from the major factors of change aggregated over all sectors. Not surprisingly from what has already been said, by far the greatest contribution comes from changes in the mix of activities within the basic sectors. This is some four times greater than that coming from the overall decrease in capacity utilisation occurring mainly in the transport sectors. Contribution from other changes studied are not insignificant but tend to balance one another as indicated by the size of the increment between trajectories 3 and 4 of the Figure.

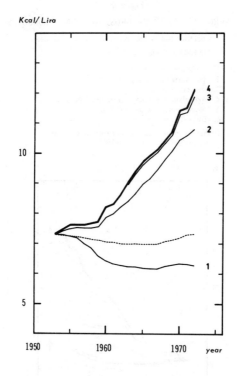

Figure 22

ENERGY INTENSITY OF GROSS NATIONAL PRODUCT

Trajectories due to successive inclusion of :

1. Fuel substitution
2. Changes in mix
3. Changes in capacity utilisation
4. Other changes

Dashed line represents base line obtained assuming sectoral intensities constant at 1953 values.

In conclusion, the picture given in the previous pages is not intended to be exhaustive but simply to give an idea of the complexity in reality underlying the evolution of energy demand over time. It is pointed out that insofar as a fairly complete understanding of the system has been obtained, this has required a fairly extensive in-depth analysis of structure

and technology which goes quite beyond conventional econometric techniques
focussing on income and price changes and based on gross aggregates of demand. Quite evidently these or equivalent techniques need to be applied
when it comes to projecting energy demand into the future. But it seems
equally evident from the present analysis, that they need to be applied at
a much greater level of disaggregation and detail than is usually the
case.(1)

1) No attempt has been made in the present context to relate changes in
    energy intensity through the factors of change, to income and price
    variations at an appropriate level of disaggregation. This is of
    course possible and indeed forms the basis of the WAES-Italy projec-
    tions reported in reference 3 and in the WAES technical reports I and
    III, Energy Demand Studies: Major Consuming Countries and Energy
    Supply - Demand Integrations to the Year 2000, MIT Press, 1977 and
    respectively 1978.

ENERGY BALANCES AND ELECTRICITY IN TERMS
OF FUEL EQUIVALENTS WITH REFERENCE TO OECD BALANCES

by

J.R. Frisch   J. Lacoste

It is well known that accountants have no easy job. Nevertheless, they enjoy at least the precious advantage to have always at their disposal one single and undisputed standard of measurement to describe the economic flow of the activity they try to analyse, that is the franc, the dollar or any other monetary unit.

Enery economists and statisticians have not this advantage. In addition to the accountant's problems, they have to face the difficulties resulting from the diversity of the primary energy sources mobilised within the framework of the complex conversion machinery which down the line makes it possible for the end user (industry, households, transport etc.) to operate an enormous variety of machines, to have heating or to see TV at home.

At least in the industrial countries, the collection of the basic statistical data concerning the primary input of the energy machinery (tonnes of coal or oil, cubic meters of gas, hydraulic or electronuclear kwh) works rather well. But problems arise as soon as these primary sources are converted into secondary energies of a sometimes different nature (coal into gas, fuel oil into electricity, etc.). It is then necessary to resort to coefficients which make it possible to re-establish the balance between consumptions and resources.

By their very nature, these systems of coefficients are conventional. Human ingenious inventiveness has been given free rein. It might be said that there are about as many systems as there are countries or organisations. The situation has even reached a point where it should be asked if the game is worth the candle. In fact, the only real problem is electricity in terms of fuel equivalents.

For example, France and the European Economic Community apply a standard of 1,000 kWh = 0.222 t.o.e., while the United Nations and OECD — on the level of end-uses — apply a standard of 1,000 kWh = 0.086 t.o.e. This gives a factor of 2.6 between the two equivalents.

This divergence is considerable. Moreover, the choice of one or the other method changes not only the global volume of energy consumption but also electricity's real role in energy supply.

The critical examination of electricity's fuel equivalents is, therefore, not an academic exercise. In fact, this only apparently technical question covers a number of highly important implications in the field of energy strategy.

ENERGY CHAINS AND COEFFICIENTS OF EQUIVALENCE

In the field of energy, no useful discussion of any coefficient of equivalence is possible without reference to the exact structure of the energy chains which are the subject of the comparison.

An energy chain is composed of a series of transformation and transfer units which make it possible to adapt a given resource progressively to the end use required by the consumer and to transfer this resource down the whole line to the consuming unit.

The nature of each energy chain (cf. Figure 1) is defined by its interconnected links which constitute the different conversion or transfer stages of energy.

An energy system is composed of a series of independent or interconnected chains which may be competing or complementary. A simplified representation of an energy system will be found in Appendix 1.

Figure 2 represents the two chains which are supposed to be competing with each other, the oil chain and the oil-electricity chain.

The choice of energy coefficients depends upon their final purpose.

- To measure the primary energy consumption of a country, e.g., to determine its supply policy, interest will be concentrated on the beginning of the chain, that is level "L1".

To estimate the energy quantities necessary to really meet the specific needs of end users, energy accounts will be established in optimal terms for the end of the chain, that is level "L5".

- The two purposes are equally useful but entirely different. The methods applied will also be different.

There are no particular problems as far as the first purpose is concerned. In fact, it is possible to compute the global consumption of primary energy of a country without running the risk of major mistakes of appreciation. The different fuels will be integrated in accordance with a more or less differentiated range of coefficients. Primary electricity (hydraulic and nuclear) will be included analogically by computing the

Figure 1
REPRESENTATION OF AN ENERGY CHAIN

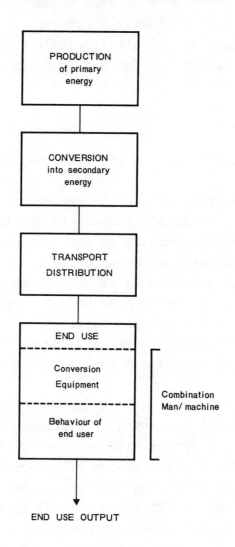

quantity of fuel necessary for the production of one kWh in a thermal
power station. This is a generally admitted practice. For electricity
generation, an average coefficient of 1 kWh = 2.22 Mcal(1) will be applied.
According to the progress of technology, this coefficient may be variable.

On the other hand, the problems of the second case are almost unsol-
vable. In fact, it would be necessary to measure case by case, at every
moment, and for every end-use unit, the quantity of effective (or "useful")
output energy. It is obvious that this quantity does not only depend on
the equipment employed but also on the way it is operated. This would soon
result in an overwhelming mass of figures and calculations and, needless
to say, any such measuring operation would be impossible in practice. It

1) 1 Mcal = 10^6cal = 3968 BTU.

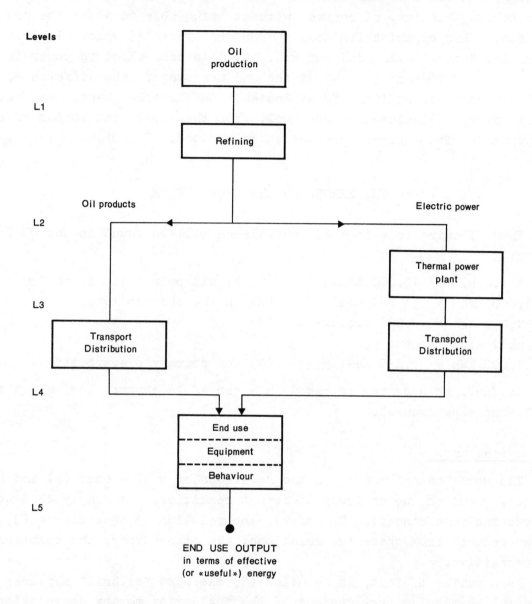

Figure 2

REPRESENTATION OF THE OIL CHAIN AND THE OIL-ELECTRICITY CHAIN

is, therefore, inevitable to forget about accuracy and to resort to simplification. For this reason, numerous Anglo-Saxon organisations – and also and in particular OECD – apply uniformly a coefficient of 1 kWh = 0.86 Mcal for electricity. This coefficient is based on the theoretical physical equivalence of the Joule effect. By the way, it should be noted that this thermal equivalent is uniformly applied to all uses of electric power (motive power, electrolysis, lighting, information transmission). As an example: these consumptions represent today about 75 per cent of French electricity demand.

As a matter of fact, France and the EEC have always adopted the first solution, that is a coefficient of 1 kWh = 2.22 Mcal. OECD balances try to go beyond this but, of course, without being able to adopt the second solution. This organisation uses a compound system in which electricity production "costs" 2.22 Mcal per kWh, but this same electric power is "worth" only 0.86 Mcal per kWh at the end use stage. The difference, that is 60 per cent, is written off as losses. On the other hand, one fuel Mcal is always calculated as one Mcal, from the top to the bottom of the energy chain, from primary production to end use.

THE ERRORS OF THE OECD SYSTEM

The following relations of equivalence will be found in the OECD system:

1 t.o.e.	=	10,000 Mcal	(1) at all points of the chains
1,000 kWh	=	860 Mcal	(2) on the end use level
1,000 kWh	=	0.086 t.o.e.	(3)

and, since recently,

1,000 kWh	=	2,220 Mcal	(4) for primary electricity

In fact, this system is based on a threefold error: logical, theoretical, and experimental.

a) Logical error

The word "calorie" used in the relations of equivalence (1) and (2) is, in fact, based on two entirely different realities. It simply does not express the same concept. Therefore, the relation of equivalence (3) cannot be deduced from these two relations. In other words, the relation is not transitive.

As a matter of fact, in relation (1) the word "calorie" measures the potential chemical energy content of the fuel prior to any degradation, loss of efficiency, or real conversion into any "effective" form of energy.

In relation (2), the same word "calorie" is applied to a "quasi-effective" energy, almost without any additional conversion at the end-user stage.

Figure 3 makes the difference between the two notions very clear.

In the case of electricity, the essential operations of the conversion of chemical energy into effective energy take place very far up the line of the chains and on centralised levels (L2 – L3). In the case of

368

Figure 3

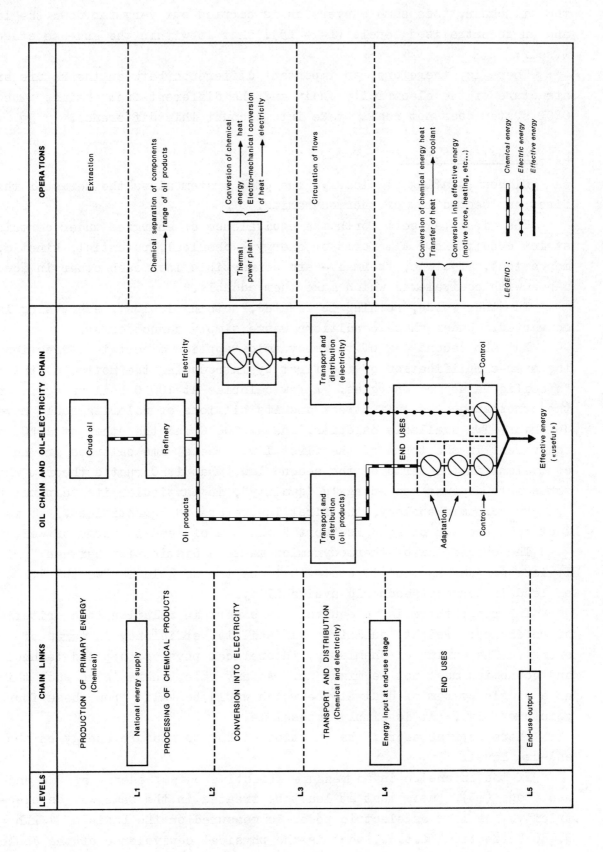

the oil chain, this same conversion is carried out very far down the line and on decentralised levels (L4 - L5), that is within the end-use stage itself.

There is, therefore, an important difference between the nature and structure of the electricity chain and the different fuel chains. The OECD system does not really take into account this difference.

b) Theoretical error

Without stating it clearly, the OECD system mixes the terms of the first and second law of thermodynamics.

The first law sets forth the equivalence of energies under certain strict conditions. All forms of energy - chemical, potential, kinetic, mechanical, electric, thermal - are convertible into each other in accordance with equivalents which make them addable.

In other words, nothing is created, nothing is lost, everything is converted. Under these conditions waste simply cannot exist.

But the second law of thermodynamics applies a certain correction to the over-simplification of the first by introducing the notion of the "specific value" of energies. A few calories at 100°C boil an egg. On the other hand, the Gulf Stream and its billions of calories could never do this. Two available calories, one at 100°C and the other at 20°C, are equivalent in the terms of the first law. These same calories are no longer equivalent in the terms of the second law, that is Carnot's theorem which takes into account the energy's "quality", in particular its temperature.

By economic analogy, the first law refers to "quantities". It is a fact that one ton of gold is "worth" one ton of lead in terms of weight.

The second law of thermodynamics makes a distinction between "prices". It is thus just as true to state that one ton of gold is "worth" 1,000 tons of lead in terms of economic availability.

In brief, there is no equivalence except in relation to a criterion of reference: weight, quality, availability, efficiency in terms of energy. The worlds of chemistry, theoretical physics, applied technology, and economics must not be mixed up. At any rate, there is no such thing as a single system of equivalence which would be valid from these four points of view. It is simply impossible.

There is, therefore, the question of how to measure energy on the end-use level.

If the intention is to measure effective output energy at the end-use stage (L5), fuels must at least be treated in the same way as electricity. As long as electric power is measured on the basis of 1 kWh = 0.086 "effective" t.o.e., that is the physical equivalence of the Joule effect, the evaluation of fuels has to be corrected proportionally in order to convert potential chemical energies into effective energies.

For example, the average efficiency of a car engine on the road amounts to 15 per cent. By way of simplification, it could thus be said that one tonne of gasoline delivers 0.15 t.o.e. of effective energy, instead of one t.o.e. as it is stated in the balances.

On the other hand, for lack of a valid reference, it is impossible to assess the relation kWh/effective kOE for motive force.

It is also possible to measure the input energy consumed at the end-use stage (L4). This method offers the advantage that energy flows at this level are the subject of commercial operations, that is statistical data. But there remains the problem of comparing energy chains which are not really comparable. In fact, fuels at this stage are still in the form of potential chemical energy, while electricity, an already processed energy, is by its very nature a quasi-effective energy, which can be adapted to the specific requirements of end uses by means of very few transformation operations. As soon as the two energies are added, the result is a mixture that does not make any sense. In fact, it is like running with the hare and hunting with the hounds. Under the pretext of pertinence, OECD tries to get around the problem by fixing for electricity a theoretical coefficient of thermal efficiency of 1 kWh = 0.86 Mcal for all its end uses. At the same time, OECD continues to assimilate potential and effective energy for fuels. The result of this method is an extremely serious distortion of the true facts and figures in the field of energy. In fact, it consolidates the initial theoretical error and widens the gulf between OECD energy statistics and the true situation as proved by practical experience.

c) Experimental error

It is possible to try to approach the real facts and figures in the field of the equivalence of different energies by the examination of a certain number of clearly defined cases. It should, however, be understood that the purpose of this operation is to obtain a practical idea of the relation of equivalence between electricity and fuels. There cannot be any exact comparison because the notion of "equal end-use output" is bound to be always more or less approximate. For this reason the question is not to measure theoretical but technological equivalences. On the one hand, the kWh consumed on level "L4" will be counted and, on the other hand, the number of fuel calories mobilised for the same result in a corresponding process.

First of all, a few examples for our guidance.

A 1975 sample survey in the United States covers 10 million all-electric housing units with a total power consumption for heating of 100 TWh, that is 10,000 kWh per unit. On the other hand, the average consumption for heating purposes of other housing in the United States is about 4 t.o.e. per annum and unit. This means that the practical equivalence in the United States is at present 1 kWh = 0.4 kOE for residential heating.

In industry, according to the branches, practical equivalences varying very often between 2 to 5 or 6 Mcal (Appendix 2) have been observed, with peaks of 10 or 40 Mcal.

In view of the very particular qualities of electricity, these results should not be surprising. In fact, these qualities compensate – and more than compensate – for numerous end-uses the incidence of Carnot's efficiency handicap of electricity generation in power plants.

In fact, the substitution of fuels by electricity does not mean that electric power will be used like a fuel.

Electric power is an energy which is perfectly controlled and organised. It is infinitely divisible. It can be used exactly when and in the quantity it is needed. For example, the physical mass itself of things can be heated, the faithful can be heated instead of the cathedral. In addition, it is possible to control electric power with utmost precision in time and space. It is easy to programme and to cut off. It offers greater facilities for integration in energy recovery and conservation technology than other energies. Electricity is, therefore, able to compete very efficiently with fuels within the framework of thermal processes.

In addition, the production of heat is neither an ideal in itself nor a compulsory or inevitable performance.

Electricity frequently makes it possible to use neater and, above all, more efficient solutions than heating. The utilisation of its other possibilities – mechanical, magnetic, radiation, bombardment of the electrons – may cut out the thermal process, the only one available with fuels.

The different forms of radiation make it possible to boil or to de-freeze food, to polymerise latex, to dry testile fibres or inks.

Mechanical membrane technology is used in the food and chemical industries for effluent treatment, the fractionation, purification and de-mineralisation of the liquid components.

The use of heat pumps means considerable primary energy saving for all drying processes.

There is nothing in all this to indicate any thermodynamic disability of electric power in comparison with its competitors. On the contrary!

Electricity is, therefore, far from the theoretical equivalence of 1 kWh = 0.86 Mcal. In fact, its equivalence is much nearer 1 kWh = 2 or 3 Mcal. To simplify matters and to facilitate the establishment of consistent energy balances, an identical coefficient of 1 kWh = 2.22 Mcal from production (L1) to end use (L4) would logically appear to be the obvious solution for electricity. This solution is convenient as well as realistic. It is, of course, understood that this coefficient will vary in accordance with technologies employed, investments made and, last but not least, with the progress of know-how in the management of energy systems.

A NEW ACCOUNTING METHOD FOR ENERGY

Let us take the OECD energy balances for 1975 (excluding non-energy consumptions) for the whole of the Organisation and computed according to the two methods.

OECD Method	Fuels	Electricity	Total	Electricity in %
Standard	Potential chemical energy on level L4	Theoretical thermal value for all end uses (L5?)		
Equivalence	1 kg = 1 kOE	1 kWh = 0.086 kOE		
Final consumption 1975 (Mt.o.e.)	2 080	323	2 403	13.4%

"French" method	Fuels	Electricity	Total	Electricity in %
Standard	Potential chemical energy on level L4	Fuel equivalent on level L3 (generation in power plants)		
Equivalence	1 kg = 1 kOE	1 kWh = 0.22 kOE		
Final consumption 1975 (Mt.o.e.)	2 080	833	2 913	28.6%

This comparison shows:

1. Total final consumption (measured on L4) is more than 20 per cent higher according to the "French" method. Knowing in addition that total primary energy demand (measured on L1) is estimated at 3,315 Mt.o.e. (excluding non-energy consumptions), the difference between the two levels (L1 - L4), which represents the losses and the consumptions of the energy conversion stages, is reduced proportionally from 912 Mt.o.e. to 402 Mt.o.e. In fact, the flow of energy "losses" down to end-use input is only 50 per cent of what the OECD computation makes believe. These "losses" represent only 12 per cent of initial primary energy input instead of 28 per cent.

2. As far as end uses are concerned, electricity's share is, in fact, much higher than the OECD computation makes believe (28.6 per cent versus 13.4 per cent).

But let us go further on. Despite the difficulties of the operation, let us try to estimate very roughly the flows of effective energy on level L5. To do this, we maintain for electricity the theoretical thermal equivalence of 1 kWh = 0.086 t.o.e. and, in the interest of simplification, we proceed like OECD, that is we assume that this equivalence is applicable to all end uses. This assumption is, of course, wrong and unfavourable because for motive power the coefficient is certainly much higher, without mentioning the other captive consumptions (lighting, electrolysis, electronics).

Electricity input for end uses will be attributed an average control efficiency of, for example, 95 per cent. Under these conditions, the flow of effective energy, measured on level L5 and coming from the power line, would be 323 x 0.95 = 307 "effective" Mt.o.e.

Now fuels will have to be treated according to the same principles to calculate their "efficiency" in effective energy on the basis of the potential chemical input energy on level L4. Within the end-use stage, fuels will have a flow of energy losses resulting from the equipment, the successive exchanges, and the control of the installations.

Transport represents about one-third of final fuel consumption (2,080 Mt.o.e.). Applying uniformly to these consumptions the average efficiency of 15 per cent shown by car engines, the equivalent of 665 x 0.15 = 100 Mt.o.e. of effective energy would be left.

For the other uses (motive power, but above all thermal uses), that is two-thirds of consumption, let us estimate the average efficiency at 50 per cent. This would leave on level L5 the equivalent of

1,415 x 0.50 = 707 Mt.o.e. of effective energy. On these assumptions, the mean efficiency of fuels in terms of effective energy and for all consumptions would, therefore, be about 38 per cent.

Under these conditions, the total consumption of effective energy on level L5 would be the following:

Estimation of effective energy	Fuels	Electricity	Total	Electricity in %
Transport Other consumptions	100 707	{ 307		
Final consumption 1975 (effective Mt.o.e.)	807	307	1 114	27.6%

This very rough computation does not claim to be exact, and yet it leads to an interesting result. The global share of electricity is not far from the figure computed by means of the "French" system of standards. This is very significant. Leaving aside the simplifications applied, this share is, at any rate, far from the figure computed by OECD.

To summarize, it seems that the "French" system is at the same time less complicated, more realistic, and fairer than the OECD system. At any rate, it is much more representative as far as the true facts and figures in the field of equivalences are concerned. This statement may be made without any excessive risk of error.

IMPLICATIONS IN THE FIELD OF ENERGY STRATEGY

Beyond the technical aspects, energy strategy is at stake.

In fact, by measuring electricity on the basis of a theoretical thermal equivalence of 1 kWh = 0.86 Mcal, its market share is artificially and unrealistically reduced. This might easily lead to underestimate considerably the importance and the efficiency of this vital vector of energy that is electric power.

Therefore, in prospective it does not seem likely that electricity will be able to replace massively fuels in the market of final consumptions, even if its progress will be fast and consistent.

Let us have a look at the year 2010. According to the OECD system, electricity will by then represent for example 25 per cent of final energy consumption. The obvious conclusion is that fuels will cover by far the major share of energy demand, that is 75 per cent. In particular, the direct use of hydrocarbons will still be inevitable. On the other hand, the "French" method leads to a very different outlook as far as constraints in the field of energy are concerned. For the same quantity of kWh, electricity would represent, in fact, 55 per cent of consumption and the share of fuels would be reduced to 45 per cent. This means that electric power would cover then already the major part of demand. This shift is due to the proper estimation of electricity's substitution capacity and efficiency.

But up the line of the different energy chains, consequences are just as serious. For a long time, nuclear energy will essentially be available only in the form of electric power. But according to OECD, nuclear energy appears to be very inefficient because it is transmitted to final consumption by means of an energy vector which itself is very inefficient. This concept is at the origin of the idea to look at any price for other secondary vehicles to "optimise" the economic production of nuclear energy: hot water, steam, hydrogen.

In fact, as soon as you take the trouble to work with more realistic coefficients of equivalence - like those proposed by the French method - it becomes quite obvious that electricity is the optimal available means to mobilise the so-called new energies (nuclear as well as solar and geothermal energy). The same applies to such conventional resources as coal which, without electricity, would only have a limited market. It will also become clear that electricity is the optimal available means to take over in time a large share of the market of final consumptions from fossil fuels.

Apart from bad habits, there is no reason to maintain the famous coefficient of 1 kWh = 0.86 Mcal for end uses. It was chosen as the result of a confusion of theoretical and technological equivalence.

On the other hand, the proposed coefficient of 1 kWh = 2.22 Mcal offers the advantages of simplicity and consistency. At the same time, it is much more representative as far as the equivalences at the production end-use stages are concerned.

It has been shown that the consequences and implications of a system of equivalence are very real and may have a considerable incidence on alternative energy policies and long-term strategies.

Any mistaken appreciation of the real role of electricity in the field of fuel substitution and as a vehicle for the mobilisation of new and conventional energy resources would mean to favour and to perpetuate our dependence on hydrocarbons. And yet we all know that hydrocarbon reserves will progressively be depleted and that the price of hydrocarbons can only increase.

This outlook leads us inevitably to dramatise our perspective of the future, as it seems that there is no real alternative solution capable of lessening the tensions ahead.

This is when the temptation of resignation becomes greater and greater, and resignation is the worse of all policies.

Appendix 1

SIMPLIFIED SCHEME OF THE ENERGY SYSTEM FROM PRIMARY RESOURCES TO END USES

Appendix 2

INDUSTRIAL EQUIVALENCES

	Total consumptions		Efficiency coefficients
	(Mcal)*	(kWh)	
	fuel process (1)	electric process (2)	(1)/(2)
Steel industry			
Heating of steel billets prior to hot working (by ton of steel)			
induction heating		400	1 kWh = 3.3 Mcal
Conduction heating		300	1 kWh = 4.3 Mcal
fuel heating	1,300		
Cast iron industry			
Cast iron melting (by ton)			
coreless induction furnace		650	1 kWh = 2.5 Mcal
arc furnace		550	1 kWh = 3.0 Mcal
cupola	1,655		
Aluminium melting (by ton)			
electric furnace		600	1 kWh = 3.9 Mcal
fuel furnace	2,350		
Timber			
Timber drying (by m^3)			
resistance electric dryer		240	1 kWh = 2.1 Mcal
dehumidification dryer		100	1 kWh = 4.9 Mcal
fuel dryer	492		
Chemistry			
Sea water distillation			
fuel process	60		1 kWh = 6.0 Mcal
turbocompressor		10	
reverse osmosis		5.5	1 kWh = 11 Mcal
electrodyalisis of brine water		1.5	1 kWh = 40 Mcal

*) 1 Mcal = 10^6 cal = 3968 BTU.

EMPIRICAL APPROACHES TO ENERGY DEMAND

RESULTS OF POLICY SIMULATIONS TO 2000

by

J.M. Griffin(*)

INTRODUCTION

Recently, attention in the energy arena has focussed on energy supply and demand relationships in the year 2000. The Ford Foundation's 1972 prognostications for the United States(1) have been followed by studies by the OECD(2) and most recently by the Report of the Workshop on Alternative Energy Strategies (WAES Report) entitled Energy: Prospects 1985 - 2000. The WAES Report is particularly important in that it shows a shortage of crude oil over the report period under a variety of scenarios. The shortages range from one-quarter to one-third of desired oil imports.

From a policy-making perspective, the WAES Report is particularly pessimistic, suggesting that conventional price responses are inadequate to achieve the requisite energy production or energy conservation. For example, the WAES Report concludes:

"Many people believe that the price mechanism could play a major role in producing and allocating energy efficiently and effectively. Yet it is widely recognised that price is insufficient for a variety of reasons."(3)

Although the WAES Report does not describe the "variety of reasons", it must assume that the necessary price increases to achieve the desired

*) The author gratefully acknowledges the support of the National Science Foundation for its support of this research. The views expressed herein are solely those of the author and not those of the foundation.

1) Energy Policy Project, A Time to Choose (Cambridge, MA., Balinger Books, 1972).

2) OECD, Energy Prospects to 1985: An Assessment of Long-Term Energy Developments and Related Policies (OECD, Paris, 1974.

3) Workshop on Alternative Energy Strategies, Energy: Global Prospects 1985-2000 (New York: McGraw-Hill, 1977), p. 40.

level of conservation and/or increased production would have to be so large as to have severe impacts on inflation rates, international trade, and macro-economic activity.

To exacerbate the air of pessimism surrounding the WAES Report, even the scenario based on slower economic growth leads to only a slightly smaller shortage. While we agree with the Report's principal thesis that there exists a serious problem with long-run oil supply, does it follow that conventional market responses and policy tools using market forces[1] are inadequate? Is the problem so severe and/or conventional forces so ineffectual that an energy bureaucracy must oversee the production and consumption of energy?

My forthcoming book describes an econometric model of energy and fuel demand for 18 OECD countries.[2] The OECD Energy Demand Model provides a vehicle by which the effects of economic growth and fuel prices on energy consumption can be assessed. Through simulation exercises, the effect of government fuel taxes on fuel conservation can be analysed. Rather than dismissing prices responses via OPEC action and government fuel tax policy as inadequate for a "variety of reasons", we view it as an empirical question to be examined within the context of the model. Slower economic growth rates resulting from slower population and labour productivity growth may also tend to attenuate the gap between desired and actual oil consumption. Does it follow that the problem will persist effectively undiminished in severity? The economic growth rates in the model are varied to assess their effects on energy consumption.

The limitations of the model must also be emphasized. The OECD Energy Demand Model is purely a demand-side model describing the market response of energy users to exogenously determined fuel prices. It does not characterise the market responses of energy producers to prices. The OECD Energy Demand Model will show for a given set of fuel prices, etc., what quantities would be demanded. It does not assure they will be produced! Consequently, without a detailed supply side in which prices are determined endogenously, the model is quite limited as a forecasting device. Generally, the set of exogenously assumed fuel prices need not correspond to the prices necessary to bring forth production equal to the quantity demanded. In sum, the OECD Energy Demand Model is not intended as a forecasting model, but rather as a policy simulation vehicle to show the impact of price and economic growth on desired energy consumption.

---

1) For example, fuel taxes.
2) J.M. Griffin, Energy Conservation in the OECD: 1980 to 2000, Ballinger: Cambridge, Mass., 1979.

This paper outlines five scenarios in Section B, designed to show the effects of certain key variables on energy patterns. Section C presents the energy consumption paths to 2000 resulting from these assumptions. These scenario results are of direct importance to the issues raised by the WAES Report. Section D provides sectoral comparisons to examine whether certain fuel using sectors are more fertile for conservation than others. Section E returns to the policy implications of these results and present six major policy conclusions.

A DESCRIPTION OF THE FIVE SCENARIOS

This section describes five scenarios chosen to elicit the response of the model over the period 1976 to 2000. No one scenario is chosen as "most likely", as the results are not intended as forecasts. Rather the scenarios are selected to show the response of energy consumption to changes in key fuel prices and economic growth variables. The magnitude of the changes to these key exogenous variables is selected on grounds of plausible ranges of variation.

Scenario 1: Baseline

As a baseline against which to measure changes in key exogenous variables, a simple set of assumptions are selected. Stated concisely, the baseline simulation assumes historical growth rates in productivity, historical inflation rates, and constant real prices of energy. Growth rates, in real GDP, are based on summing the country's historical labour productivity growth rate plus the labour force growth rate projections of the International Labor Office.(1) The same growth rate for real GDP is also applied to real value added in manufacturing and real agricultural output. For purposes of comparison, the historical average growth rate in GDP over the period 1955 to 1970 is given in column 4. It should be noted that the assumed GDP growth rates are ½ per cent to 1 per cent lower than in the period 1955 to 1970 (column 3 vs 4) because of the projected decline in the labour force growth rates. Shown in column 5 are the assumed population growth rates which also reflect a decline from historical trends. As in the case of labour force growth projections, the International Labor Office is the source of these data.

---

1) International Labor Office, <u>Population and Labor Force Projections, 1975-2000</u>. (International Labor Office, Geneva, 1975.)

Perhaps the most difficult variable to forecast, other than the price
of oil products, is the general inflation rate measured by the GDP deflator.   In this mode, it is changes in relative prices which induce substitution responses; since fuel prices are tied to the inflation rate the
actual rate of inflation is not critical.   For simplicity, the inflation
rates prevailing over the more recent period, 1965 to 1973, were averaged
to obtain those values in column 6 of Table 1.   By selecting the more recent period, the inflation rates are considerably higher than a period en-
compassing the postwar period.   The higher rates would appear more appropriate, but they too may well turn out to be underestimates.   Nevertheless,
as noted earlier, the assumption is that all fuels escalate in price at
the same rate as inflation.   Consequently, the rate of inflation is
irrelevant.(1)

Table 1

BASELINE SIMULATION ASSUMPTIONS BY COUNTRY: 1973 - 2000

	1. Labour Produc- tivity Growth Rate (%)	2. Growth Rate in Labour Force (%)	3. Assumed Growth Rate (%)	4. Historical GDP Growth Rate (%)	5. Population Growth Rate (%)	6. Infla- tion Rate (%)
Canada	2.28	1.49	3.77	4.70	1.31	4.39
U.S.	1.82	1.13	2.95	3.39	.85	4.06
Japan	7.23	.87	8.10	9.06	.22	5.68
Austria	5.17	.63	5.80	4.71	.30	4.60
Belgium	3.67	.50	4.17	3.87	.36	4.61
Denmark	3.54	.45	3.99	4.30	.26	6.62
France	5.23	.90	6.13	5.67	.64	4.96
Germany	5.46	.39	5.85	5.93	.29	4.48
Greece	6.14	.22	6.36	6.40	.30	4.45
Ireland	3.62	1.20	4.82	3.12	.98	7.90
Italy	5.61	.43	6.04	5.47	.40	3.63
Netherlands	3.59	.80	4.39	4.83	.65	5.94
Norway	3.48	.68	4.16	3.84	.45	5.63
Spain	5.43	.92	6.35	5.80	.95	6.39
Sweden	3.49	.64	4.13	4.35	.50	5.40
Switzerland	2.16	.59	2.75	4.12	.48	5.91
Turkey	4.80	2.31	7.11	6.27	2.40	8.54
U.K.	1.95	.56	2.51	2.50	.43	5.66

1)   A macro-economic model with energy detail is needed to examine the in-
     flation effects on real GDP growth rates.

It should also be noted that the exchange rates were projected to reflect inflation rate differences between countries. For example, with an inflation rate of 5.66 per cent in the United Kingdom compared to 4.06 per cent in the United States, the exchange rate of pounds in terms of dollars will increase at 1.6 per cent annually. Also, price increases in manufactured goods were assumed to be similar to the GDP inflation rate.

The assumption that the price of all forms of energy escalate with the general inflation rate is indeed at variance with experience over the 1950s and 1960s, when real energy prices declined by 1 per cent to 3 per cent per year. The question arises as to what conditions might cause energy prices to remain constant in real terms. First, and most obvious, is the existence of OPEC acting as a monopolistic price setter for crude oil. Price increases over the period 1975 to 1977 have indicated a tendency for the cartel to maintain a roughly constant real price of oil. Justification given for price increases is typically couched in terms of adjustments to reflect the declining purchasing power of oil revenues. With the raw material prices escalating at a rate either consistent with or slightly in excess of inflation rates, one must ask if other cost components of the final price will also match inflation. Processing and distribution costs tend to reflect some inflationary effects but typically escalate at slower rates than inflation due to productivity gains. On the other hand, government tariffs and taxes on petroleum products could well offset these factors and lead to overall increases in petroleum product prices similar to inflation rates.

Natural gas prices are likewise assumed to escalate with inflation rates. In this case, increasing scarcity rather than a cartel or government taxes could provide the stimulus for increased prices.

The assumption that coal and electricity prices escalate with inflation rates is subject to considerable speculation. Historically, coal has enjoyed considerable labour productivity growth while electricity generation has benefited both from productivity gains and scale economies. As a result, both coal and electricity prices have fallen in real terms. However, for several reasons, it is plausible that these fuels may escalate in price with oil and gas products. It would appear that most new electricity generation units take full advantage of scale economies. In addition, air pollution, safety, and increased nuclear generation all imply sharply rising electricity prices. Over the last few years, labour productivity growth in coal mining has declined. If slow productivity growth continues and is coupled with the pressure placed on coal prices by other fuels, the price of coal could also escalate with the general inflation rate. It should be noted that the fuel price escalation factors begin in 1976. Prices prior to 1976 are actuals, and therefore include the substantial increases following the Arab Oil Embargo of 1973.

In addition to these assumption for the exogenous variables, there are several remaining variables about which assumptions must be made. The rate of hydro-electric and nuclear generation growth is an important assumption. Suffice it to say at this point that the WAES Report's "minimum likely" case was assumed. Later in Scenario 5, these assumptions are discussed in detail. A time trend in the thermal efficiency of electricity generation required modification. To reflect the fact that heat rates are approaching their technical minimum, the time trend was reduced so as to increase by only .5 per year instead of 1. In the energy sector, it is necessary to postulate crude oil input to refineries, and coal and gas consumption by the energy sector. All of these variables are arbitrarily assumed to increase by 2 per cent annually.

Scenario 2: Slow Growth

The economic recovery following the Arab Oil Embargo in 1973 has been less than complete in most of the OECD countries. Since then, economic growth rates have fallen well below historical trends and the growth rate in labour productivity has fallen appreciably. This has led some forecasters to assume that the growth in labour productivity may be considerably slower than past trends. Thus, the slow growth scenario assumes 1 per cent slower growth in labour productivity and, consequently, 1 per cent slower GDP growth rates than shown in column 3 of Table 1. Also, real GDP, manufacturing and agricultural output are assumed to expand at a 1 per cent slower growth rate than in the Baseline Scenario. All other assumptions remain unchanged from the Baseline Scenario. For example, constant real fuel prices, similar inflation rates, etc., are assumed.

Scenario 3: Cheap Energy

Our third scenario is designed to capture the effect of returning to a world of declining real energy prices, such as occurred during the 1950s and 1960s. Even though the OPEC price increases are maintained, beginning in 1975, all fuels increase in price at a rate 2 per cent less than the inflation rate.

While this scenario may not rank high in plausibility, it is, nevertheless, extremely valuable as it isolates the substitution response between energy and non-energy inputs. Since all fuel prices are increasing at similar rates, relative fuel prices are constant and interfuel substitution effects are made inoperative. With the general inflation rate at 2 per cent greater than energy prices, the price of non-energy goods became increasingly expensive over time. The extent of substitution between energy and non-energy goods can be analysed. Simulations of this type are particularly germaine to energy conservation policies and the effectiveness of Btu taxes.

In the Cheap Energy Scenario, all other exogenous assumptions remain unchanged relative to the Baseline.

Scenario 4: Rising Relative Oil Prices

The fourth scenario is designed to capture interfuel substitution. It hypothesises that oil products increase at the general inflation rate while all other fuels increase at a rate 2 per cent less than the inflation rate. This implies that oil products increase in price over time relative to coal, gas, and electricity.

Again the motivation for this assumption is illustrative – to test for significant interfuel substitution against oil products. The scenario may contain a good deal of realism in that oil products might increase in price relative to other fuels. There are large quantities of coal, with the principal determinant of price being labour costs. A resumption of productivity gains or rapid escalation of strip mining could restrain coal price increases. For electricity, once the costs of air pollution control equipment and nuclear equipment are reflected in electricity prices, subsequent price increases could be less than the general inflation rate. The price of natural gas at the wellhead is likely to increase faster than the general price level; however, the consumer pays a delivered price which includes a large, relatively fixed distribution cost. With some improvement in the supply situation, gas prices could escalate at a 2 per cent slower rate than oil products and the general inflation rate.

Scenario 5: High Nuclear

The WAES Report was careful to distinguish a "minimum likely" and "maximum likely" nuclear generation capacity for the years 1985 and 2000.(1) The former would require deliberate policy stimulus to achieve, but nevertheless, is attainable. In view of the discussion in the United States and elsewhere on the relative merits of nuclear power, it is instructive to ask if nuclear power makes much difference between now and 2000. While this study does not attempt to analyse the safety and security

---

1) Workshop on Alternative Energy Strategies, op. cit., page 203.

consequences of nuclear power, the model will show the effects of nuclear power on the use of alternative fossil fuels.

GENERAL RESULTS FOR 5 SCENARIOS

Baseline Scenario

Figure 1 summarizes the aggregate results for the five scenarios simulated over the period 1973 to 2000. The figures show total primary energy consumption, and consumption of primary coal, primary gas, primary oil, and electricity. The Baseline Scenario indicates that aggregate energy growth should fall well below the growth rates of the past quarter century. Several factors explain the 3.2 per cent growth rate. First, the Arab Oil Embargo of 1973 resulted in sharply higher energy prices in 1974 and 1975. These price increases resulted in very slow growth in energy consumption in the late 1970s and early 1980s as a consequence. After the lagged price effects have impacted fully, growth rates accelerate in the mid 1980s, but even then are slower than historical patterns for two reasons. First, the slower growth in the labour force implies slower economic growth rates by .5 per cent to 1 per cent in most OECD countries. By comparing the Slow Growth Scenario growth rate of 2.3 per cent versus 3.2 per cent in the Baseline Scenario, we see that slower economic growth could account for a reduction in aggregate energy growth by .5 per cent to 1 per cent. Second, unlike the historical pattern in which energy prices declined relative to other goods, the Baseline posits a constant real price of energy. The Cheap Energy Scenario indicates that this could account for a reduction in the growth rate by .9 per cent. In sum, future energy growth rates may fall well below historical growth rates due to a combination of demographic, economic growth, and price phenomena. Combining the effects of these phenomena results in a decline in the growth rate by over 2 per cent per year - from in the range of 5 per cent to 6 per cent per year to 3.2 per cent per year.

One obvious question is how do these energy growth rates compare to the projections of WAES and the Energy Policy Project. Depending on the scenario, WAES projects energy demand growing from 2.4 per cent to 3.4 per cent per year. Our Baseline Scenario with a growth rate of 3.2 per cent falls in the upper end of the WAES range and our Slow Growth Scenario growth rate of 2.3 per cent corresponds closely to the WAES Scenarios, emphasizing slower growth and higher energy prices. Thus the aggregate energy consumption of the WAES Report agree closely with the findings here.

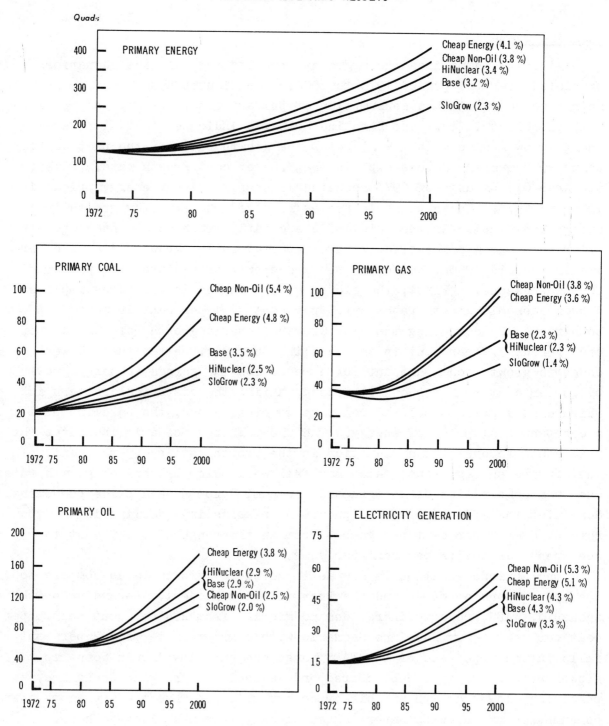

Figure 1
SUMMARY SCENARIO RESULTS

The United States results for the Baseline Scenario agree closely with the Technical Fix Scenario in the Energy Policy Project's study. The Baseline Scenario project energy growth at 1.7 per cent per year while the Technical Fix Scenario posits 1.9 per cent per year growth. However, the latter postulates GDP growth of 3.3 per cent as compared to 2.95 per cent in the Baseline Scenario.(1) If our Baseline Scenario were adjusted to the higher 3.3 per cent GNP growth, then the Baseline Scenario would project energy growth at 2 per cent - slightly higher than the Technical Fix Scenario. It should also be noted that the Energy Policy Project's Zero Energy Growth Scenario is clearly incompatible with the results here.

The growth paths for individual fuels differ appreciably, even though in relative terms they are constant in real prices. In Figure 1 coal grows at 3.5 per cent per year and electricity grows at 4.3 per cent per year - both rates being well above the 3.2 per cent for all energy. In contrast, the growth rates for gas and oil are 2.3 per cent and 2.9 per cent, respectively. The reasons for the different paths can be seen by studying growth over the period 1972 to 1985 for the four fuels. Coal and electricity grow at an accelerated growth rate while gas and oil even decline slightly over the period 1972 to 1980.

The greater growth of coal and electricity over this period relative to gas and oil can be attributed to relative fuel price changes from 1973 through 1975. Over the period, all fuel prices rose sharply, but oil and gas experienced greater relative increases than did coal and electricity. Substantial interfuel substitution then occurred from oil and gas to coal and electricity. By 1985, the model indicates these substitutions were complete, and subsequently fuel consumption grew at similar rates. It should be remarked that the decline in oil and gas consumption is due to the tendency of the model to exhibit too much price responsiveness in the early years following a price increase (1973 - 1976) and too little responsiveness over the intermediate term (1977 - 1980). While the net effects of these distortions on long-run growth are offsetting, it does suggest that absolute declines from one year to the next in fuel consumption are implausible over the period 1972 to 1980.

Even over the period 1985 to 2000, the consumption of all fuels need not expand at similar rates. From 1985 to 2000, aggregate energy consumption grows at 4.3 per cent annually. Oil and electricity consumption grow relatively faster than coal or gas use (4.6 per cent and 4.8 per cent vs 4.0 per cent and 3.6 per cent respectively). As discussed later in the sectoral analysis, different fuels serve as principal fuels in the various

---

1) Energy Policy Project, op. cit., page 502

sectors. With energy consumption in the various sectors proceeding at different rates, aggregate fuels will grow at different rates even though no appreciable interfuel substitution is occurring within any given sector.

The WAES Report offers an interesting basis on which to compare the growth of oil products. Depending on the WAES scenario, the growth rate in oil demand varies from 1.8 per cent to 2.6 per cent.[1] The Baseline Scenario projecting oil growth at 2.9 per cent per year indicates somewhat higher oil demand. The WAES growth rate of 1.8 per cent is predicated on slow economic growth and constant (real) energy price. The Slow Growth Scenario appears roughly comparable as it projects oil growth at 2 per cent per year.

Slow Growth Scenario:

Comparison of the Slow Growth Scenario with the Baseline Scenario reveals several interesting conclusions about the relationship between economic growth and energy consumption. The Slow Growth Scenario, positing 1 per cent slower economic growth than in the Baseline Scenario, indicates a growth rate in aggregate energy consumption of 2.3 per cent as compared to the 3.2 per cent annual rate in the Base Scenario. It is apparent, ceteris paribus, that energy and economic growth are closely tied. It is not simply a random occurrence that energy consumption parallels economic growth.

The Slow Growth Scenario indicates significant, but non-proportionate, declines in growth rates for individual fuels. Compared to the Baseline Scenario, the coal consumption growth rate declines from 3.5 per cent to 2.3 per cent. Electricity declines by 1 per cent – from 4.3 per cent to 3.3 per cent. Both gas and oil experience declines in their growth rates of .9 per cent. Paradoxically, coal has the highest income elasticity! Casual empiricism might conclude that coal is an inferior good due to cleanliness and convenience problems. The explanation is found in the manner in which the fuels are combined. Coal's principal market is in electricity generation, which in turn has a greater income elasticity than oil or gas. Coal serves as a residual supplier to electricity generation, being the principal fuel to meet requirements not met by hydro or nuclear generation. The decline in electricity generation from 4.3 per cent to 3.4 per cent leads to an even greater percentage for coal inputs into electricity generation because hydro and nuclear generation are unaffected. The sectoral results presented subsequently elaborate on this phenomena.

---

1) Workshop on Alternative Energy Strategies, op. cit., page 25.

Even though the decreases in the growth rates differ according to
fuels, the strong ceteris paribus relationship that exists between aggregate energy and economic growth also holds vis-à-vis individual fuels and
economic growth. Thus, the primary force driving fuel consumption growth
is economic growth.

Cheap Energy Scenario:

While the Slow Growth Scenario points to a strong casual relationship
extending from economic growth to energy demand, other factors can affect
the relationship as well. The Cheap Energy Scenario shows that how we
choose to price energy can have a significant effect on future consumption
rates. The Cheap Energy Scenario is predicated on a return in 1976 of the
historical relationship between energy prices and other goods. Measured
in relative terms, the price of energy increases at a 2 per cent/year
slower rate than the prices of all other goods. The effect on aggregate
energy consumption of cheap energy is marked; the growth rate of energy
consumption rises from 3.2 per cent to 4.1 per cent. As shown in subsequent sectoral comparisons, in the transportation, industrial, and residential sectors, significant substitution exists between energy and other
goods. In the transportation sector, higher gasoline prices led to adjustments in the number and design of the auto fleet. In the industrial sector, cheap energy leads to the adoption of more energy intensive production using relatively less capital and labour. In the residential sector,
energy expenditure is a part of the consumer budget and consumers adapt
their size and type homes in response to energy prices.

Just as the Slow Growth Scenario underscored the link between energy
consumption and economic growth, the Cheap Energy Scenario provides support to those favouring energy conservation. In the long run, higher
energy prices will reduce energy consumption relative to what it would have
been otherwise. Consequently, it is technically possible to decouple
within limits the relationship between energy consumption and economic
growth. The real questions are: (1) What are the costs of this decoup-
ling? (2) Given these costs, what are the limits within which the relationship can be altered? Chapter I of my forthcoming book showed that
energy conservation, rather than being something to be viewed as good in
itself, is in fact a perpetrator of welfare losses. Whether accomplished
through higher prices or quantity rationing, energy conservation will lead
to real changes in lifestyle. The chapter showed that the welfare losses
increased with the square of the percentage of desired energy conservation,
thus the desirability of conservation decreases sharply with the ambitious-
ness of the goal.

Based on detailed assumptions,(1) the present value of welfare losses over the 25-year period 1976 to 2000 is approximately $220 billion, or about $9 billion annually, on average. The reader should be reminded that welfare losses of this magnitude occur for a rather modest conservation policy, aimed at reducing aggregate energy growth from 4.1 per cent to 3.2 per cent. For example, more ambitious programmes to achieve zero energy growth would result in far greater welfare losses. As noted above the welfare losses rise with the square of the degree of conservation required.

In sum, the welfare loss calculations would appear to suggest that within bounds, energy conservation is an attractive policy instrument. A continuous increase of 2 per cent in real energy prices reduces aggregate growth rates by .9 per cent. The welfare loss on the demand side of the market of $220 billion is large, but not overwhelming. More vigorous conservation policies to achieve zero energy growth may be technically feasible, but would result in progressively larger welfare losses, income redistribution, and general dislocations in the economy.

Cheap Non-Oil Scenario:

Some policy analysts would question whether the benefits to energy conservation exceed its costs. If the supply of energy is infinite at somewhat higher prices, there may be little economic justification for conservation by this generation for future generations, which will face an abundant supply at higher energy prices in the future. These same policy analysts might argue, however, that we should conserve oil because the OECD is highly dependent on high-risk imported oil. In effect, imported oil bears a risk premium due in part to likely future embargos and the power implications of the oil revenues. Moreover, this negative risk premium is not reflected in the price. To the contrary, United States policy is to hold down domestic oil prices, paying a premium price for imported oil. One economic solution might be to impose taxes on oil products to incorporate the true costs of dependence into the price thereby raising the price of oil relative to other forms of energy.(2) The Cheap Non-Oil Scenario shows the effect of oil product prices rising with inflation while all other energy prices escalate at a 2 per cent slower rate.

---

1) See Griffin op. cit., Ch. VIII.
2) Even better than a general tax on oil products would be a tariff on imported oil equal to the negative externalities a barrel of imported oil entails.

Aggregate energy consumption growth is 3.8 per cent in the Cheap Non-Oil Scenario as compared to 3.2 per cent in the Baseline in which all fuels escalate with inflation. Compared to the Cheap Energy Scenario in which all fuel prices escalate at 2 per cent less than the inflation rate, the growth rate declines from 4.1 per cent to 3.8 per cent.

The Cheap Non-Oil Scenario emphasizes the considerable possibilities for long-run interfuel substitution. Compared to the Baseline Scenario, consumption rates increase sharply for coal (5.4 per cent vs 3.5 per cent), gas (3.8 per cent vs 2.3 per cent) and electricity (5.3 per cent vs 4.3 per cent). The oil consumption growth rate declines from 2.9 per cent to 2.5 per cent. In view of the relatively larger percentage increases in coal, gas, and electricity growth, a decline in oil consumption of only .4 per cent might seem surprising. The explanation rests with the fact that the Cheap Non-Oil Scenario exhibits cheaper aggregate energy prices than the Baseline Scenario since in the former, coal, gas, and electricity prices are escalating at 2 per cent slower rate than inflation. Lower aggregate energy prices stimulate aggregate energy consumption, thereby tending to raise the consumption rates for all fuels, including oil products. At a second stage, interfuel substitution effects occur holding aggregate energy consumption constant. These effects tend to further raise coal, gas and electricity consumption. Conversely, oil consumption is lowered due to the substitution effects. The net effect of these two forces is that for oil, the interfuel substitution effect dominates the energy/non-energy substitution leading to a net decline in oil consumption.

To obtain a better idea of the magnitude of the interfuel substitution effects, compare the Cheap Non-Oil Scenario with the Cheap Energy Scenario. Recall that the only difference in the two scenarios is that in the former oil product prices escalate at a 2 per cent faster rate than the prices for other fuels. The higher oil prices lead to a slight decline in aggregate energy consumption - from 4.1 per cent to 3.8 per cent. In this comparison, both the effects of reduced aggregate energy consumption and interfuel substitution work together to accomplish a major reduction in oil consumption from 3.8 per cent to 2.5 per cent.

The welfare loss of a policy to restrain oil demand growth to 2.5 per cent instead of 3.8 per cent is approximately $100 billion over the 25-year period. Note that this welfare loss is less than half that obtained by a policy aimed at conserving all fuels. Moreover, the degree of conservation obtained for oil exceeds that in the general energy conservation policy (the Baseline Scenario) as oil demand growth is 2.5 per cent annually as compared to a 2.9 per cent growth in the Baseline Scenario. If indeed oil should be conserved and not energy in general, then the tax on oil products is more desirable since the welfare loss is much smaller and the degree of oil conservation greater.

High Nuclear Scenario:

The High Nuclear Scenario is particularly interesting in that it emphasizes the effects of increased reliance on nuclear power generation. Since the impact of nuclear power occurs in electricity generation, the effects on demand relationships in the residential and industrial sectors are small. Consequently, aggregate energy consumption growth is affected only slightly (3.4 per cent vs 3.2 per cent).[1]

The effects of individual fuel demand are more pronounced. Coal, which is used primarily for electricity generation, experiences a decline in its consumption growth rate from 3.5 per cent to 2.5 per cent. The effects on other fuels such as gas and oil are negligible. Moreover, the demand for electricity generation is not affected as it depends on electricity demand in the residential and industrial sectors.

An interesting policy implication of the High Nuclear Scenario is that nuclear power will displace coal, but have no appreciable direct effect on oil consumption. If our model included supply side responses, it is conceivable that increased nuclear power might have indirect effects on oil consumption by lowering the price of coal and inducing oil/coal substitution. While we cannot entirely dismiss these indirect effects, they seem unlikely to alter the conclusion that nuclear power and oil are only weak substitutes.

SCENARIO RESULTS BY SECTOR

1. The Transportation Sector

Compared to aggregate energy growth of 3.2 per cent, the transportation sector promises to be a relatively slow growth sector averaging 2.8 per cent per year to 2000. Since gasoline, aviation fuel, and motor diesel fuel constitute the bulk of transportation fuel usage, other minor fuels are disregarded in Figure 2. The growth rate of 2.8 per cent can be attributed primarily to motor gasoline consumption. In part, the slow growth can be attributed to an income elasticity for gasoline which declines with per capita income. In the auto stock equation, cars per capita

---

1) Actually, the growth rates would be identical if the same conversion efficiency of fossil fuels to electricity were applied for each respective country.

income. In the auto stock equation, cars per capita approached a saturation level. For the high per capita income countries, which constitute
the bulk of the consumption, the income elasticity is below unity. This
effect is apparent from the Slow Growth Scenario in which oil consumption
growth declines by .9 per cent while real income declined by 1 per cent.

Figure 2
TRANSPORTATION SECTOR
CONSUMPTION OF OIL PRODUCTS

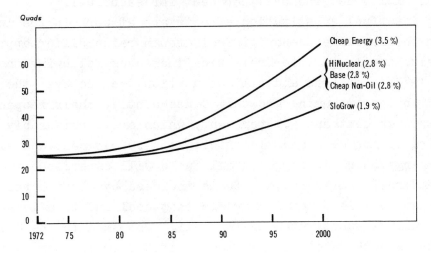

A second and more important factor contributing to the slower growth
is the gasoline price increases occurring following the Arab Oil Embargo
of 1973. As indicated in Figure 2, consumption is virtually unchanged over
the period 1972 to 1980 with a resumption of growth in the early 1980s.
The importance of gasoline price is also evidenced in the Cheap Energy
Scenario. If gasoline prices decline 2 per cent annually in real terms,
oil consumption rises at a 3.5 per cent rate instead of the 2.8 per cent
in the Base Scenario. This difference can have an appreciable effect on
oil consumption over time.

The policy implications of these results are two-fold. First, gasoline taxes are an effective method of reducing the consumption of oil products. Second, gasoline taxes should be imposed gradually with prior-
announced schedules for implementation. In the short run, higher gasoline
prices have little effect on gasoline consumption per car. The response
to higher prices occurs only over a much longer time horizon in which
motorists can adapt the weight and utilisation of the auto fleet. Given
that the response is lagged, income distribution effects can be minimised
through phasing in gasoline taxes, while having similar efficiency effects.

2. The Industrial Sector

Compared to primary energy growth, energy consumption in the industrial sector is relatively fast-growing. Figure 3 shows that the Baseline Scenario results in a 3.9 per cent growth rate. Aggregate energy consumption is affected both by price and economic output. With a decline in economic growth of 1 per cent, energy input drops by 1 per cent, falling from 3.9 per cent to 2.9 per cent. The effects of slow growth on fuel consumption are essentially identical. The growth rates for coal, gas, oil and electricity decline by 1 per cent for each fuel.

Strong interfuel substitution effects are at work in the industrial sector. In the Baseline Scenario, fuel growth rates differ appreciably between fuels, due in part to fuel price increases following the Arab Oil Embargo of 1973. Both gas and oil consumption decline over the period 1972 to 1980 before resuming growth. Consequently, their respective growth rates are 2.5 per cent and 4.2 per cent, which are considerably less than the growth rates for coal and electricity of 4.9 per cent and 4.7 per cent. The Cheap Energy Scenario suggests all fuels will experience increased growth. The growth rates for all fuels increase by about 5 per cent.

The Cheap Non-Oil Scenario reveals that coal and to a lesser extent gas, are principal substitutes against oil. Compared to the Cheap Energy Scenario, the growth rate for coal accelerates from 5.3 per cent to 5.9 per cent while the growth rate for gas increases from 3.0 per cent to 3.6 per cent. In contrast, the growth rate for oil declines from 4.7 per cent to 3.6 per cent.

Both the Cheap Energy Scenario and the Cheap Non-Oil Scenario have important policy implications. There may be a presumption to attempt fuel conservation only in the transportation sector where the potential for conservation is more apparent. These scenarios suggest that in the industrial sector there is also considerable potential for conservation, either for total energy or oil.

3. The Residential Sector

Figure 4 depicts the response of the residential sector under the five scenarios. A reduction of 1 per cent/year in the growth of real per capita income leads to a decline in the energy growth rate from 3.2 per cent to 2.3 per cent. Not only is energy consumption quite income responsive, but it is also quite price responsive. The Cheap Energy Scenario indicates an acceleration in consumption growth from 3.2 per cent to 4.4 per cent. Compared to the industrial sector, the residential sector shows even greater potential for energy/non-energy substitution. Recall

Figure 3

SCENARIO RESULTS FOR INDUSTRIAL SECTORS

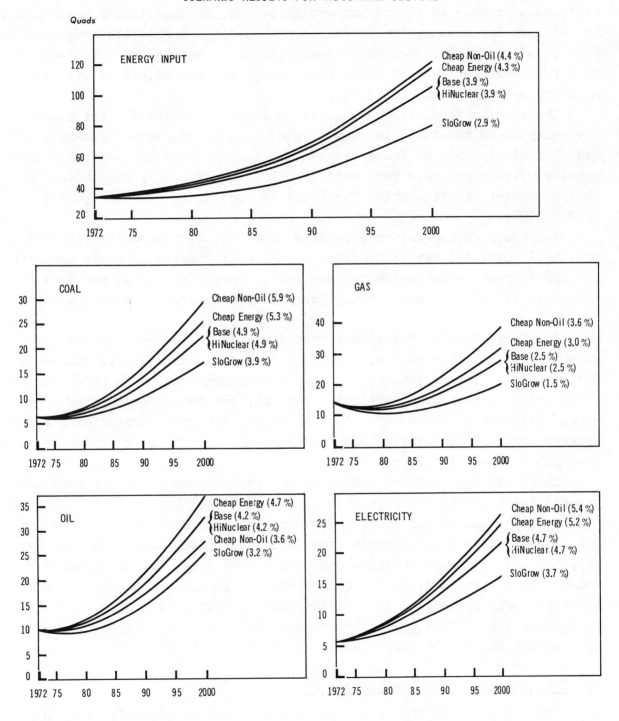

that primary energy consumption increased by .4 per cent in the industrial sector compared to the 1.2 per cent increase in the residential sector.

Even the Cheap Non-Oil Scenario projects energy growth at a 4.3 per cent annual rate. As in the industrial sector, the relatively rapid growth in total energy consumption in the Cheap Non-Oil Scenario occurs partially because of the cheaper aggregate energy and also because of substitution of a fuel with low thermal efficiency (coal) for one with a relatively high thermal efficiency (oil).

The results for individual fuels in Figure 4 indicate considerable diversity among the fuels in growth rates. In the Baseline Scenario, coal and electricity grow at faster rates (4.0 per cent and 3.7 per cent, respectively) than gas (2.8 per cent) or oil (3.1 per cent), due primarily to the greater relative price increases in the latter over the period 1973 to 1975.

The Cheap Energy Scenario results in rather even increases in fuel growth of about 1.3 per cent. Even though coal remains of minor importance, its growth rate increases to 5.2 per cent. Gas grows at 4.1 per cent and oil grows at 4.4 per cent. Electricity records a growth rate of 5.0 per cent.

The Cheap Non-Oil Scenario suggests that coal is the principal substitute against oil products. Coal's growth rate jumps to 7.9 per cent, but even then, its importance to total sector consumption in 2000 is minor. Electricity and gas are substitutes for oil, but the substitutions are weak. Compared to the Cheap Energy Scenario, the growth rate for gas increases from 4.1 per cent to 4.3 per cent while electricity growth increases from 5.0 per cent to 5.2 per cent.

These findings suggest that the residential sector has considerable potential for conservation of both energy and oil. The decrease in energy consumption from 4.4 per cent to 3.2 per cent comparing the Cheap Energy Scenario with the Base Scenario attests to this potential. Likewise for oil, the Cheap Non-Oil Scenario compared to the Cheap Energy Scenario shows that a 2 per cent relative increase in oil product prices relative to other fuels leads to a drop in the oil's growth rate from 4.7 per cent to 3.6 per cent. Such conservation potential occurring largely in home heating and cooling is appreciable and should not be overlooked.

4. The Electricity Generation Sector

The top graph in Figure 5 depicts the growth in electricity demand in the five scenarios. The subsequent four graphs indicate by what fuels those electricity generation requirements are to be met. First of all,

Figure 4
SCENARIO RESULTS FOR RESIDENTIAL SECTOR

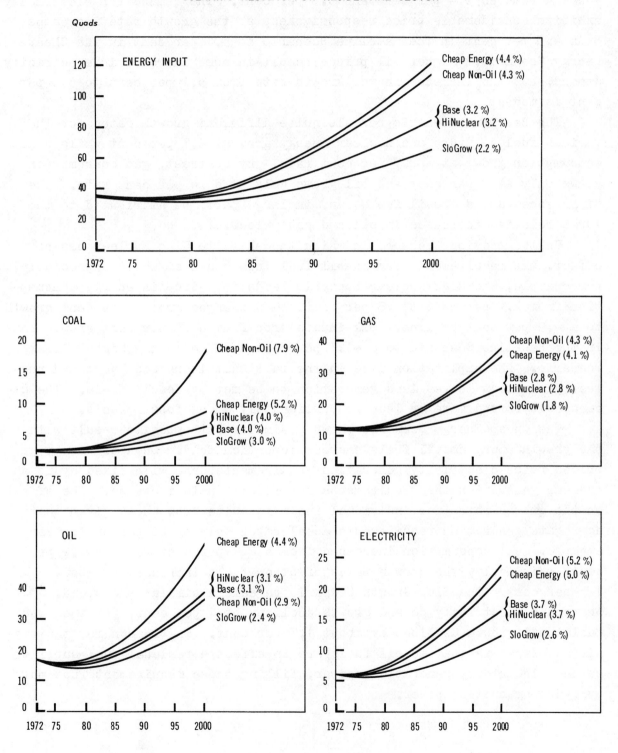

it should be noted that the trend towards increased electrical energy will continue. In the Baseline Scenario and other cases, electricity will claim an increasing fraction of the future energy mix with growth rates ranging between 3.4 per cent and 5.3 per cent. The demand for electricity exhibits considerable price responsiveness as the growth rate increases from 4.3 per cent in the Baseline Scenario to 5.1 per cent in the Cheap Energy Scenario. Higher oil prices result in some increase in electricity demand, but the increase in the growth rate from 5.1 per cent to 5.3 per cent is modest.

The Baseline Scenario reveals quite different growth rates for the various fuels. Hydro and nuclear energy grow at 5.7 per cent while coal consumption grows at a 2.7 per cent rate. By contrast, gas consumption grows only at 1 per cent and oil grows only at a .4 per cent rate. The differences among fossil fuels can, in large part, be attributed to the large relative increases in oil and gas prices.

Unlike previous sectors in which the shift to High Nuclear has no effect, the requirements for fossil fuel inputs are affected appreciably. For example, the High Nuclear Scenario leads to a drop in energy consumption from 2.7 per cent to .8 per cent. Gas changes from 1 per cent growth to a -.9 per cent decline. Oil inputs drop from a .4 per cent growth rate in the Baseline Scenario to a -1.1 per cent decline in the High Nuclear Scenario. The explanation lies in the fact that increased hydro and nuclear generation leaves less generation to be met by fossil fuels. The decrease is shared more or less proportionately among fossil fuels.

The Cheap Energy Scenario shows a similar but opposite result with the growth among fossil fuels shared proportionately. Cheaper energy prices raise electricity generation requirements, and with fixed hydro and nuclear generation the requirements for fossil fuels increase. The effect on the growth rates for individual fuels is quite pronounced given that aggregate generation requirements accelerated only by .8 per cent. For example, coal consumption increased from a 2.7 per cent rate to 4.2 per cent. Similarly, the growth rates of other fuels increased by about 1.5 per cent. The Slow Growth Scenario reveals a similar phenomena. Aggregate electricity demand growth declines by 1 per cent, yet the fossil fuel input growth declines by about 1.7 per cent. The reason for the magnified effect is that fossil fuels are in effect a residual fuel supplier to the electricity generation sector, filling those requirements not met by hydro or nuclear power.

Figure 5
SCENARIO RESULTS FOR ELECTRICITY GENERATION

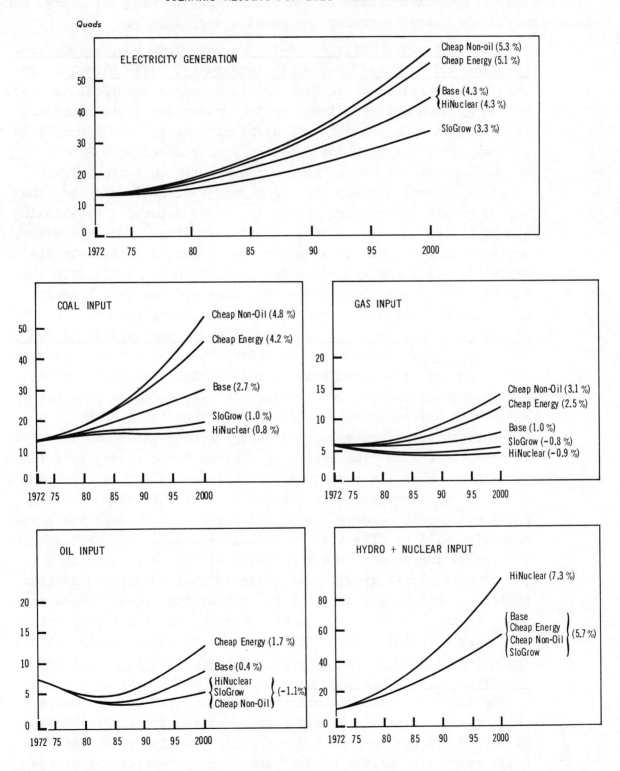

POLICY IMPLICATIONS AND CONCLUSIONS

The preceding statistical estimations and scenarios lead to six major policy conclusions which relate to the controversy between the energy conservationists and those favouring increased energy supply.

1. <u>Both in the very short run and in the very long run, energy consumption and economic growth are intertwined quite closely.</u> The Baseline Scenario and the Slow Growth Scenario reinforce the role of energy as a critical input in the production process and an important item in the modern household. Energy growth parallels economic growth. Both the Baseline Scenario and Slow Growth Scenario predicted energy prices rising with inflation, i.e., constant in real terms. The Cheap Energy Scenario did show that energy growth and economic activity can be decoupled by allowing energy prices to rise or fall relative to other goods. However, for very long periods of a century or longer, continuously rising or falling real energy prices are probably unrealistic. To the extent that constant, real energy prices are the very long-run norm, energy growth will closely track economic growth.

2. <u>Over the intermediate and long run, energy conservation is technically feasible.</u> The Cheap Energy Scenario showed that higher relative energy prices can lead to substantial reductions in the growth rate of energy consumption. In effect, it is possible to alter both the energy/GDP ratio and the fuel input in Btu's to GDP through using price-induced energy conservation and interfuel substitution. The time period to achieve these adjustments tends to coincide with the gestation and replacement periods for capital since energy consumption is closely tied to the stock of energy using equipment. Energy conservation achieved through the price mechanism will be effective in reducing demand. The fact that a 2 per cent increase in the real price of energy results in a .9 per cent slower growth rate suggests that fuel price and tax policies can play a vital role in controlling demand. Thus contrary to the pessimism of the WAES Report, we view energy prices as a very powerful vehicle through which to shape energy policy.

3. <u>While technically feasible, energy conservation is costly and practical only within limits.</u> The Cheap Energy Scenario compared to the Baseline Scenario indicated that a 2 per cent increase in the annual increase in energy prices reduced the energy growth rate from 4.1 per cent to 3.2 per cent. We find these results quite supportive of the effectiveness and potential for energy

conservation. It should be overlooked that conservation involves
real sacrifices in living standards; the present value of welfare
losses under this policy would be about $220 billion over the
period 1976 to 2000. For greater degrees of desired conservation,
the welfare losses rise exponentially, suggesting there are limits
to the potential for conservation. For example, even though zero
energy growth is technically feasible, the welfare losses would be
large and price increases necessary to achieve it would probably
be politically unrealistic. Overly ambitious energy conservation
may be limited for other reasons not examined in this study. The
macro-economic effects of the much higher energy prices necessary
to achieve a conservation goal could exacerbate inflation and lead
to unacceptably slow rates of economic growth. Non-price methods
of achieving conservation could also lead to shortages of energy
intensive goods, higher prices, and even plant shutdown to those
plants not receiving a sufficient allocation of energy. While this
study does not attempt to establish the linkage from energy back to
macro-economic activity, it should be clear from the worldwide recession following the Arab Oil Embargo of 1973 that these limits
are real.

4. <u>Conservation policies should be adopted economy wide and not restricted to one or a few sectors.</u> Among politicians, there is a
tendency to call for higher energy prices and conservation by industry leaving the residential sector to purchase energy at the
lower prices. While such policies may be good politics, the sectoral scenario results in Figures 2 and 5 suggest that energy conservation is achievable in all sectors. In fact, the residential
sector tends to be somewhat more responsive to incentives than the
industrial sector. To force one sector to achieve a degree of conservation which could be spread among the various sectors is to
lead to much greater welfare losses and economic dislocations.

5. <u>If the principal concern is national security, conservation policies would be more effectively redirected from energy to oil consumption.</u> If we accept the view that energy supplies are large
and virtually limitless at higher prices, then we may conclude that
the problem lies in the national security implications of rising
dependence on imported oil. A comparison of the Cheap Non-Oil
Scenario with the Cheap Energy Scenario showed that an escalating
tax on oil products could induce significant interfuel substitution responses causing the growth in oil demand to drop from
3.8 per cent to 2.5 per cent. Moreover, by focussing only on oil

products, the welfare losses are less than half those occurring with the energy conservation policy and yet oil conservation is greater. While this discussion does not suggest that the national security problem is synonymous with the energy problem, it does suggest that more selectivity is desirable in identifying those fuels for which conservation is desirable.

6. <u>Balance is needed between policies promoting conservation and those stimulating energy production.</u> While this study does not address the potential for increased energy production, we tacitly assume that the potential is significant. Like conservation, we suspect that within limits, increased supply can make an important contribution towards improving the world's energy future. If both conservation and increased supply are practical only within limits, it is essential that policies be balanced to utilise both supply and demand forces. To date, United States energy policy through price controls restrains production and stimulates consumption. Somehow the message must have been reversed!

ENERGY DEMAND TO 1990 IN THE EUROPEAN COMMUNITY

COAL - POSSIBLE DEVELOPMENTS AND ITS ROLE IN THE

ENERGY ECONOMY

by

C. Waeterloos

BACKGROUND

In December, 1974, the Council of Ministers of the European Community
decided on a coherent set of objectives for the "Community of the Nine"
to achieve by 1985 in order to whittle down the share of imported energy
needed to cover primary energy requirements from 63 to less than 50 per
cent.(1)   The plans called for a major energy saving effort to hold primary energy requirements at no more than 1,475 million toe in 1985 and,
maximum development and diversification of the Community's own energy
resources as follows:

- a halt to the run-down in coalmining with output maintained at
  about 210 million toe;
- considerable growth in the production of oil and gas based mainly
  on the result of offshore prospecting and possibly yielding
  350 million toe in 1985; and
- a resolute and ambitious nuclear development policy whereby installed nuclear power would climb to at least 160 GWe by 1985.

The twofold background to these jointly-agreed objectives was the
1973/74 oil crisis and the very satisfactory "economic performance" so far
achieved. In the five years that have gone by since then, many of the
factors governing the energy market have altered radically, to such an
extent that it has become necessary to consider whether certain changes
observed both in the economy and in the energy market are likely to be
short-lived or, on the contrary, more fundamental and lasting. These
factors - whose effects may incidentally, counteract each other - include:

---

1) These objectives are summarised in Annex 1.

- the protracted slowdown in economic growth bound up, among other
  things, with the substantial drop in output from certain old-
  established European industries;
- increasingly strong and well-organised public opposition to the
  building of nuclear power stations;
- growing awareness of the need to give preference to the energy
  cycles most likely to avoid loss and wastage;
- possible changes in the policy of certain industrialised countries
  towards a greater effort to conserve resources; and
- the waning pressure that high energy prices were exerting on
  demand growth.

In these circumstances the need was felt in 1977 to reconsider the
1985 targets, at least in part, in the light of the new developments that
had taken place. Fresh targets were therefore proposed by the Commission
and adopted by the European Council (Heads of State and Governments) at
its meeting in Bremen on 6th and 7th July, 1978. Alongside these revised
targets for 1985 there seemed a need to widen the field of research by
undertaking a first study, using the scenario method, with 1990 as the
selected horizon. In view of its exploratory nature, a simulation model
at a relatively high level of aggregation was felt to be adequate, the
economic implications of energy constraints being taken into account on
an exogenous basis.

FIVE ENERGY SCENARIOS FOR THE EEC

A first study of possible trends in the EEC energy market based on
the scenario method was carried out by the Commission in 1977. The study
covered a set of five different scenarios based on various combinations
of assumptions relating primarily to the economic growth rate, the success or otherwise of energy savings policies, the development of indigenous energy resources and the FOB price of imported crude oil.

Two first "limit" scenarios were prepared in order to define the
extremes within which energy consumption and supply conditions might
fluctuate over the next 10-15 years. The combinations of assumptions
made for these were designed to give the highest and lowest possible
levels of oil demand and energy dependence:

Scenario 1: upper limit
High GDP, stable crude oil prices, weak conservation policy,
weak nuclear programme.

Scenario 2: lower limit

Moderate GDP, rising crude oil prices, strong conservation policy, stronger nuclear programme.

Two other scenarios were variants on these limit scenarios and were designed to show the implications for the energy market of scoio-economic options in the direction of resource conservation policy or reflecting the ideas of the "new energy policy strategy".(1)

Scenario 3: resource conservation

(Variant of scenario 2) but:

Stable crude prices and weak nuclear programme.

Scenario 4: 1974 strategy trends

(Variant of scenario 1) but: stronger nuclear programme with more use of electricity for heating purposes.

A fifth scenario corresponding to an intermediate situation is based on a different set of assumptions from the others. As far as the economy is concerned a "high GDP" was adopted in view of the social constraints bound up with the "moderate GDP" objective.

Scenario 5: "medium"

High GDP, rising crude oil prices, weak energy conservation policy, stronger nuclear programme with greater use of electricity for heating purposes.

ENERGY DEMAND IN 1990

In the light of the assumptions described above, gross domestic consumption of energy in the Community could increase over the period 1976 t 1990 at an average growth rate for the whole period ranging from a maximum of 4 per cent to a minimum of 1.8 per cent. In 1990, therefore, the absolute figure for total energy demand(2) in the Community should therefore lie between 1,160 and 1,610 million toe. These estimates should be compared with the first data obtained for 1977 suggesting a gross consumption of just under 950 million toe.

In fact, this total energy demand envelope is arrived at by adding together estimates for each of the consumption sectors, enabling the limits and constraints as regards substitution between different forms of energy to be taken into account at the level of the final consumer.

---

1) See Communication by the Commission to the Council: "Towards a new energy policy strategy for the European community", April 1974.
2) Gross domestic consumption (including non-energy uses) plus bunkers.

The point is that the existence of certain types of specific uses
and the relative inertia of some areas of consumption clearly rule out
interchangeability in these cases between energies capable of meeting
demand. What might be called the captive part of the energy market,
therefore, can be identified only at a sufficiently disaggregated level
of that market. As a guide, energy demand in the Community for 1975 and
1990 has been broken down as follows for the two limit scenarios(1):

	1975	1990	
		Scenario 1 Upper limit	Scenario 2 Lower limit
Industry	343	670	470
of which non-energy uses	56	120	80
Transports	123	210	150
Household sector, ...	309	520	390
Final consumption(*)	775	1,400	1,010

*) Excluding consumption and losses in the energy industries themselves.

CONSTRAINTS AT THE LEVEL OF FINAL DEMAND

Even in 1990, at the level of final fuel consumption, a number of
requirements will still have to be met preferentially by a pre-determined
type of energy, unless, in other words, there is a fundamental technological revolution, which would in any case be unlikely to have any mass
effect because of the relative nearness of the horizon concerned:

- the steel industry will continue to rely largely on coal and its
  derivates;
- non-energy uses will still be largely based on hydrocarbons;
- private transport will continue to be a practically exclusive
  market for oil products; and
- the household sector will continue to depend, to a large extent,
  on oil and gas, especially because of the rate of renewal of
  housing stock. In this sector, the eventual use of coal will
  primarily depend on the rate of development of district heating
  which is slow to establish itself because of the scale of work
  and considerable capital cost involved.

1) In actual fact, some of the sectors have been reaggregated in this
   table for simplicity's sake.

In the light of the above limits - but disregarding any constraints in terms of oil supplies - the coverage of final energy consumption in 1990 could take the following pattern:

(million toe)	1975	1990	
		Scenario 1 Upper limit	Scenario 2 lower limit
Solid fuels	67	65	55
Oil products	381	710	375
Gas (natural and derived)	121	185	255
Electricity(*)	206	440	325
Total "Final consumption" of energy	775	1,400	1,010

*) The conversion of electricity to toe is based on the specific average consumption of primary energy in conventional thermal power stations.

COVERING THE ELECTRICITY DEMAND

Power generation is one of the most fruitful sectors as regards substituting one form of energy for another because of the ability of some power stations to switch from one fuel to another but also because of the flexibility of their use, enabling the load to be varied and the inter-connection between supply networks.

Nevertheless, although the scope for substitution is wider than in other user sectors, it is limited by the characteristics of existing power stations and their fairly slow rate of renewal and also because of the economic advantages attaching to certain types of electricity produc- tion (power stations using lignite and by-product gas, and hydro and nuclear power stations).

If these factors are taken into account, the pattern of electricity generation could be as follows (nuclear energy providing 30 - 50 per cent by 1990):

(million toe)	1975	1990	
		Scenario 1 Upper limit	Scenario 2 Lower limit
Hydro and geothermal	27	30	30
Nuclear	20	150	190
Conventional thermal i.e.	193	335	160
- solid fuels	94	175	100
- natural gas	30	50	20
- oil	60	100	30
- other products(*)	9	10	10
Total "production"	240	510	380

*) Including derived gases.

POTENTIAL ENERGY DEMAND

To sum up(1), with the energy industries' own requirements and losses taken into account, the Community's potential demand for primary energy in 1990 could, in theory, lie between the following limits:

(million toe)	1975	1990	
		Scenario 1 Upper limit	Scenario 2 Lower limit
Solid fuels	195	290	190
Oil	508	920	480
Natural gas	143	215	265
Primary electricity and other	52	185	225
Total "Gross energy consumption"(*)	898	1,610	1,160

*) Internal consumption (including non-energy uses) + bunkers.

PREFERENTIAL SUPPLY OF INDIGENOUS ENERGY

Both for purely economic reasons and as the result of active policy measures, initial priority will be given - for the coverage of part of the energy requirement - to indigenous primary energy products, availabilities of which will be completely taken up by the market.

1) For a more detailed description of possible trends in energy demand in the Community see - for example - the description of the scenario given in Annex 3.

The products to be included under this "preferential supply" heading and likely to make an appreciable contribution to supplies in 1990, include:

- lignite, the supply of which may be estimated at about 30 million toe;
- by-product gas (from coking plants and blast furnaces) and "other primary products" (refuse, wood ...) for burning in power stations, about 10 million toe;
- hydro and geothermal electricity totalling 30 million toe;
- nuclear energy (mainly nuclear electricity) which should give the equivalent of 150 - 190 million toe;
- certain new forms or sources of energy, such as geothermal energy, solar energy, wind power, gasified coal (SNG), etc. Their contribution however, though it might become substantial towards the end of the century, will presumably be only minimal or marginal by 1990 and is mentioned purely for completeness.

OTHER INDIGENOUS ENERGY RESOURCES

In addition to the preferential forms of energy listed in the preceding paragraph, the Community will also be continuing its production of various other energy sources and it has been assumed that, by 1990, the Community will have:

- coal production capacity for mining 145 - 160 million toe a year in economically satisfactory conditions; and
- oil and gas production facilities for 110 - 165 million toe of oil and 135 - 170 million toe of natural gas a year.

In view of the serious dangers of too great a dependence on foreign supply sources, particularly for oil - from the twofold standpoint of security of supply and pressure on the balance of payments - it has been assumed that concerted policy at the national and Community levels would enable at least the minimum level shown for indigenous production to be achieved in all the scenarios considered.

TRENDS IN THE VARIOUS IMPORT MARKETS

On the basis of the quantities of natural gas covered by current firm contracts with other countries falling due after 1990 and contracts now being negotiated which may reasonably be expected to be signed,

413

the Community's imports of natural gas should lie somewhere between
80 and 95 million toe in 1990.

As for coal imports, it is more than likely - in view of the present
situation (30 million toe in 1975) - that the 1985 target (40 million toe)
will be exceeded. For 1990 coal imports of the order of 50 million toe
would appear to be the very minimum for all the scenarios concerned except one, namely scenario 2 (lower limit) for which there is practically
no import requirement. This would be due to the general low level in the
demand for energy and electricity in conjunction with surplus availabilities of imported oil.

Conversely, in scenario 1 (upper limit), the minimum amount of coal
required, purely to balance fuel supply and demand, is 115 million toe.

Because of the decline in physical availabilities and the possible
effects of political and economic decisions, recent world-level estimates
for oil supplies suggest that the Community's imports of crude are likely
to level off as early as 1985 at about 480 million toe a year. This
ceiling has been established by splitting up the deficit that, at first
approximation there is likely to be between possible supply and potential
world demand among the various economically developed countries.

"Possible supply" compared with "potential demand"

A comparison between the above energy supply and demand figures
makes it fairly clear that in neither case have they been established
completely independently of one another and that a first set of constraints
has been taken into account in establishing what, for simplicity's sake
(though rather arbitrarily it is true), is called "possible supply" and
"potential demand". This first set of constraints may be summed up as
follows:

- the figure given for solid fuel imports is limited to what is
  necessary to cover requirements established in the demand calculation. In the event of further policy measures or new technological developments leading to an increased demand for solid fuels
  the supply of coal could therefore be increased - particularly
  imported coal;
- gas demand has been kept to the maximum level of available resources on the basis that the relative price of natural gas should
  remain favourable in the long-term and that, in those circumstances,
  any availabilities would be taken up by consumers.

Summarised, the "possible supply" of primary energy for the Community in
1990 should lie between the following high-low limits:

	Community production	Net imports	Total "possible supply"
Solid fuels	175-190	115-0	290-190
Oil	110-165	480	590-645
Natural Gas	135-170	80-95	215-265
Nuclear Energy	150-190	–	150-190
Hydro, and geothermal energy and other primary products	35	–	35
Total	605-750	675-575	1,280-1,325

If the "possible supply" of energy (as defined above) and the level of "potential demand" established for the two limit scenarios (scenario 1, upper limit, 1,610 million toe - scenario 2, lower limit, 1,160 million toe) are compared, it can be seen that in relation to both energy and oil demand, scenario 1 shows a very large supply shortfall whereas in scenario 2 supply exceeds demand.

It may at first sight, of course, be objected that the possible bracket is not as wide as it could be and that it could amount to, say, 1,180 - 1,425 million toe (instead of 1,280-1,325). In fact, to be coherent, the basic criteria used for defining demand in scenarios 1 and 2 also need to be applied to the supply assumptions for the same scenarios, so the supply bracket arrived at is based on these assumptions, particularly those concerning the rate of exploitation of indigenous resources and the level of world energy prices.

In the case of the limit scenarios (1 and 2), the comparison between energy supply and demand works out as follows for each of the primary energy sources:

Community 1990 (million toe)	Solid Fuels	Oil	Natural Gas	Primary Electricity + other	Total
Scenario 1: upper limit					
Demand (see para. 9)	290	920	215	185	1,610
Supply (see para. 15)	290	590	215	185	1,280
Shortfall (-)(*)	(*)	-330	(*)	(*)	-330
Scenario 2: lower limit					
Demand (see para. 9)	190	480	265	225	1,160
Supply (see para. 15)	190	645	265	225	1,325
Surplus (+)(*)	(*)	+165	(*)	(*)	+165

*) Taking into account the points made at the beginning of page 414.

<u>Specific conclusions regarding scenarios 1 and 2</u> (described above)

From scenario 1 the main deduction is that there would probably be energy supply difficulties as early as 1985. The only ways to solve the specific problem of the shortfall are:

- a more vigorous energy savings policy including, for example, the setting of binding sectoral targets;
- the immediate intensification of prospecting and technological research to bring about a substantial increase in indigenous energy availabilities (particularly the new energies);
- greater use of imported steam coal in power stations and of natural gas (negotiation of new LNG contracts, etc.) in industry and in the household sector.

This latter condition would seem realistic from the standpoint of the existence of reserves but it raises the problem of the need to base practically all conventional thermal power generation of the non-preferential kind (in other words using fuels other than lignite and by-product gas) on steam coal, an extremely improbable eventuality. In addition, coal would have to regain favour in industry so as to cover, as far as possible, at least the equivalent of the increase in industrial demand and this is hardly feasible unless the use of coal gasification and fluidised bed technology begins to spread widely.

Unlike scenario 1, scenario 2 shows an energy surplus instead of a shortfall and does not, therefore, present any problems from the energy supply standpoint. On the other hand, in view of the socio-economic problems that could arise, the scenario is hardly "viable" in the medium term.

GENERAL CONCLUSIONS APPLYING TO ALL THE FIVE SCENARIOS STUDIED

Scenarios 1 and 2 described above are merely intended to define the possible extremes of development for the Community's energy market by 1990 and for this reason they contain with themselves the factors that make them improbable. Intermediate - and complementary - scenarios were therefore devised.(1) These helped to refine the first general conclusions to be drawn from the exercise although they did not change their general tenor. Very briefly these conclusions may be summed up as follows:

---

1) See synthesis of results in Annex 2.

- the assumption of a low growth rate for GDP mitigates and even resolves energy supply difficulties but transfers the problem from the energy level to the socio-economic level;
- the price of energy is one of the decisive factors in practically all the cases studied: if too low, it does little to encourage energy savings (or investment) and if too high it seriously mortgages the economic future;
- there are serious doubts whether the Community would be able to import the quantities of oil that the market could well want and be able to absorb in 1990;
- by 1990, nuclear energy, coal and energy conservation will be the only decisive factors capable of restoring a certain flexibility to energy supplies. It seems doubtful, however, whether - in the light of present progress - nuclear energy will be in a position to make a major contribution in 1990 unless there is a sharp acceleration in nuclear programmes over the next four or five years;
- the gradual substitution of oil and gas by solid fuel in conventional thermal power stations is clearly a basic policy to be encouraged;
- any strategy designed to increase the share of electricity in heating uses must inevitably be associated with an effective policy for the recovery of residual heat and the generalised introduction of heat pumps;
- the coal market could grow considerably in the long term but, in the medium term, this will depend on the extent to which power stations use more coal. In the longer term expansion in the market will partly depend on the rate of economic growth. If the latter is slow it could extend by several years the temporary period of supply facility currently being experienced on the international oil market.

ADDITIONAL CONCLUSIONS REGARDING THE SPECIFIC FIELD OF COAL

In the more specific case of coal, the future level of consumption in Europe would seem, therefore, to depend largely on three internal factors:

- policy decisions to promote the use of coal in power stations;
- the acceptance by European countries of a high volume of imports (on the assumption that present relative prices are maintained);

- the rapid introduction of new consumption technologies (mainly fluidised bed and gasification).

In 1976, Europe consumed 300 million tonnes (tce) of solid fuel (22 per cent of total energy demand). This consumption was split up among power stations (53 per cent or 158 million tonnes), coking plants (33 per cent or 100 million tonnes) and the household sector (12 per cent or 35 million tonnes). The balance, 2 per cent, went to various industrial uses.

In 1990 it may reasonably be supposed that there will be no growth in requirements for coking (low level of activity in the steel industry) or for the domestic sector (inconvenience of use).

In the power station sector, however, any slowdown in nuclear programmes will have to be offset by growth in coal consumption. In scenario 1 this could well amount to 190 million tonnes (tce) by 1985 (150 million tonnes of coal plus 40 million tce of lignite) and even 250 million tonnes by 1990 (equivalent to 3.5 million barrels of oil a day). The annual average increase in consumption would therefore be 2 per cent up to 1985 and 5.5 per cent thereafter. The 2 per cent rate for the period 1976-85 may seem low and even negligible but in fact - taking into account the number of obsolete plants to be shut down - it implies building a further 30 GWe capacity by the year 1985 (total nuclear power station capacity for the same period would increase by about 50 GWe).

The use of coal in industry (for making cement, bricks, etc.) would necessarily mean the introduction of new technologies since modern production processes require the use of fluid fuels, particularly because of the need to regulate heat in industrial furnaces. In the case of coal, therefore, this means using fluidised bed techniques or the development of coal gasification on a vast basis (processes for producing "low btu gas"). In view of the probable cost of coal gasification ($90-120/tce, fuel included, according to process) this technology would have to profit by economic support. If this may be assumed, it might cover 1-2 per cent of the Community's energy requirements in 1990.

Coal's possible contribution in this way has not, in fact, been taken into account in any of the scenarios considered and is referred to here purely pro mem. and in the interests of completeness.

+

+   +

418

Annex 1

RESOLUTION OF THE COUNCIL OF 17TH DECEMBER, 1974

(O.J. No. C 153 of 9th July, 1975)

<u>Synthesis:</u> Community energy policy - 1985 Objectives (final).

I. <u>Total energy requirements:</u> 1,475 million toe
II. <u>Coverage of requirements:</u>
     Common nuclear objective: 160 - 200 GWe installed.
     Natural gas objective: internal production 175 - 225 million toe.

		Minimum	Maximum
a)	<u>Internal production:</u>		
	Solid fuels	210 mtoe	210 mtoe
	Natural gas	175 mtoe	225 mtoe
	Hydro, geothermal and heat	45 mtoe	45 mtoe
	Crude oil	180 mtoe	180 mtoe
	Nuclear energy	190 mtoe(*)	240 mtoe(**)
	TOTAL	800 mtoe	900 mtoe
b)	<u>Imports:</u>		
	Solid fuels	40 mtoe	40 mtoe
	Natural gas	95 mtoe	115 mtoe
	Oil (crude and products)	540 mtoe	420 mtoe
	TOTAL	675 mtoe	575 mtoe
	c) <u>Degree of energy dependence:</u>	46%	±40%
III.	<u>Structure of supply</u>		
	Solid fuels	17%	17%
	Oil	49%	41%
	Natural gas	18%	23%
	Hydro and geothermal energy	3%	3%
	Nuclear energy	13%	16%
	TOTAL	100%	100%

*) 865 TWh (**) 1,100 TWh.

SUMMARY OF SCENARIOS STUDIED

---

1. Assumptions

Scenario 1: upper limit

GDP		Price of crude	RUE(*)	Internal production		Nuclear		Share of electricity	
(1976–1990):	+ 4.2%	stable	medium	(1990):	610 mtoe	(1990):	121 GW	(1990):	31.5%

Scenario 2: lower limit

(1976–1990):	+ 3.1%	rising	strong	(1990):	770 mtoe	(1990):	166 GW	(1990):	32.5%

Scenario 3: resource conservation

(1976–1990):	+ 3.1%	stable	strong	(1990):	610 mtoe	(1990):	121 GW	(1990):	32.5%

Scenario 4: 1974 strategy trends

(1976–1990):	+ 4.2%	stable	medium	(1990):	770 mtoe	(1990):	166 GW	(1990):	34%

Scenario 5: a medium scenario

(1976–1990):	+ 4.2%	rising	medium	(1990):	770 mtoe	(1990):	166 GW	(1990):	34%

2. Results

Scenario 1

	Energy demand	Energy imports	Consumption in power stations
1990:	1,610 mtoe	1,000 mtoe	515 mtoe
	of which: oil 915	of which: oil 800	of which: oil 100
	coal 295	coal 115	coal 175

Oil imports considerably more than 450 mtoe

Scenario 2

1990:	1,160 mtoe	390 mtoe	380 mtoe
	of which: oil 460	of which: oil 295	of which: oil 20
	coal 190	coal 0	coal 100

Economic difficulties

Scenario 3

1990:	1,190 mtoe	580 mtoe	390 mtoe
	of which: oil 565	of which: oil 450	of which: oil 45
	coal 220	coal 45	coal 125

Active government policies and sense of responsibility from consumers

---

Scenario 4

1990:	1,630 mtoe	860 mtoe	560 mtoe
	of which: oil 835	of which: oil 670	of which: oil 100
	coal 285	coal 95	coal 170

Acceptable strategy for 1985 but would require an additional effort in 1990

Scenario 5

1990:	1,520 mtoe	750 mtoe	520 mtoe
	of which: oil 755	of which: oil 590	of which: oil 95
	coal 255	coal 65	coal 145

Acceptable degree of energy dependence; conversion of all power stations from oil to coal wherever possible (necessary in order not to exceed the limit set for oil imports).

*) Degree of energy savings.

Example: Outline energy scenario for Europe.

Apart from the five scenarios described in this paper, the following outline scenario would be possible - just as an example - for the European Community. No degree of probability should be attached to the likelihood of this sixth scenario materalising.

	1976-1980	1981-1985	1986-1990
1. Basic assumptions:			
GDP (Δ% per year)	4.0%	3.9%	3.2%
Price of crude (Δ%)	-10%	+20%	+30%
Conservation (as compared with unit consumption in 1975)	-3.5%	-8.5%	-15%
or toe/10^3 EUA 1975(*)	0.850	0.805	0.750
Nuclear (end of period)	46.5 GWe	80-90 GWe	118-148 GWe
2. Results			
Energy demand (Δ% a year)	+3.9%	3.1%	1.8%(**)
of which oil	+2.9%	1.9%	-0.1%
of which electricity	6.2%	5.1%	3.7%

In this same scenario, consumption sectors should grow at the rate shown below between period 1976-1980 and period 1986-1990 (annual average):

	Δ% a year
Steel industry	2.5 to 1.2
Chemicals	6.5 to 2.2
Other industries	4.5 to 2.0
Transport	3.5 to 1.3
Households, ...	3.5 to 1.8
Energy sector	4.0 to 2.1

For demand as a whole, possible trends for each primary source of energy may be summed up as follows (in million toe):

	Solid Fuels	Hydrocarbons	Primary Electricity	TOTAL
1975	194	650	52	896
1990	225-245	915-930	230-195	1,370

*) 1 EUA (European unit of account) = US$1.2 at 1975 prices and exchange rates.

**) i.e. 1,370 million toe in 1990.

ENERGY DEMAND PROSPECTS IN NON-OPEC DEVELOPING COUNTRIES

by

B.J. Choe(1)

1) The views expressed in this paper are those of the author and do not necessarily reflect those of the World Bank.

INTRODUCTION

This paper attempts to project the demand for primary energy in non-OPEC developing countries on the basis of energy demand functions estimated from historical data of 35 sample panel countries.(1) It is concerned only with inland consumption of total primary energy. The demand equations take the usual form in which per capita consumption of energy is expressed as a function of per capita gross domestic product (GDP), price of energy, and lagged per capita consumption of energy. This type of specification has been widely used in energy demand studies for developed countries and, to some extent, developing countries. In applications to developing countries, however, estimates often suffer from pseudo energy price variable (for example, OPEC export price of crude oil) used in estimating the equation. The emphasis in this paper is on improving estimates of price elasticity - and income elasticity at the same time - by using the appropriate energy price variable, that is, the prices to the consumers.

The next section briefly reviews historical trends in energy consumption. It is followed by a discussion of estimation methods and results. The final section shows the demand projections to 1985.

---

1) For a list of the 35 sample panel and other countries included in this study, see Annex 1. Throughout this paper, the non-OPEC developing countries are divided into four groups by the level of per capita income. Primary energy is defined so as to include petroleum, coal, natural gas, and primary electricity (hydro, geothermal, and nuclear electricity). Non-commercial energy (firewood, animal dung, agricultural wastes) is not included in this study.

HISTORICAL TRENDS IN PRIMARY ENERGY CONSUMPTION

Primary energy consumption in all non-OPEC developing countries increased at 6.8 per cent annually between 1960 and 1975, from 3.6 million barrels per day (b/d) of oil equivalent to 9.7 million b/d. For the 35 sample panel countries, the consumption growth rate was practically the same as that of all non-OPEC developing countries. Table 1 summarizes the energy consumption and GNP trends by income groups.

Several interesting aspects emerge from Table 1. If one compares the energy consumption growth rates to those of GDP of the sample panel countries, the lower middle income group shows the highest ratio (1.65); followed by the low income group (1.44), the intermediate middle income group (1.19), and the upper middle income group (1.11). This would seemingly suggest a link between income elasticity and level of per capita income. The other important aspect is that energy consumption growth rates substantially slowed down in the two years after the oil price hike in 1973-74, whereas GDP growth rates were reduced only slightly. Furthermore, the slowdown is more dramatic in the upper and immediate middle income groups than in the lower middle income group. The same panel low income group shows a perversity of sorts in that energy consumption growth rate increased substantially in the 1974-75 period whereas GDP growth rate did not.(1)

Natural gas was the fastest growing among the primary energy sources during the 1960-1975 period, followed by primary electricity, oil, and solid fuels (see Appendix Table 1). After the oil price hike, however, consumption of solid fuels and primary electricity increased faster than before, whereas consumption growth rates of oil and natural gas declined dramatically.

Energy consumption trends are also closely related to energy prices. Table 2 shows the historical movement of the average price of energy in real terms to the consumers.(2) It shows that the average price of energy declined steadily through 1970 in most of the countries and even through 1973 in the lower middle and low income countries. It also shows that the quadrupling of international market price of crude oil in 1973/74 resulted in sudden increases in the average price of energy in almost all of the countries. The percentage increase in the average price of energy, however, was far less than that of crude oil - only 42 per cent increase

---

1) Rapid increases in consumption in India explains most of the perverse behaviour.

2) For the details and nature of this price data, see the section on data.

Table 1

ENERGY CONSUMPTION AND GDP IN

NON-OPEC DEVELOPING COUNTRIES BY INCOME GROUPS

Income groups	1960	1970	1973	1975	Average annual growth rate (%)		
					1961-73	1974-75	1961-75
Energy consumption, '000 b/d							
All non-OPEC Developing countries							
Upper middle income	1188	2183	2671	2803	6.4	2.4	5.9
Intermediate Middle inc.	1069	2337	3083	3437	8.5	5.6	8.1
Lower middle income	286	680	801	918	8.2	7.1	8.1
Low income	1080	2005	2245	2509	5.8	5.7	5.8
Total	3623	7205	8800	9667	7.1	4.8	6.8
35 Sample panel countries							
Upper middle income	931	1744	2078	2246	6.4	4.0	6.0
Intermediate middle inc.	851	1889	2424	2746	8.4	6.4	8.1
Lower middle income	242	551	677	775	8.2	7.0	8.1
Low income	931	1582	1798	2101	5.2	8.1	5.6
Total	2955	5766	6977	7868	6.8	6.2	6.7
GDP, billions of constant 1970 US$							
35 Sample panel countries							
Upper middle income	44.0	76.2	88.6	96.8	5.5	4.5	5.4
Intermediate middle inc.	56.0	100.2	131.2	149.7	6.8	6.6	6.8
Lower middle income	20.7	33.3	38.4	42.3	4.9	5.0	4.9
Low income	48.6	75.7	80.3	86.7	3.9	3.9	3.9
Total	169.3	285.4	339.1	375.5	5.5	5.2	5.5

Source: World Bank, Economic Analysis and Projections Department, and Appendix Table 1.

compared to 253 per cent increase in constant terms for crude oil. The average price of energy increased more in the upper middle and intermediate middle income countries than in the other two income groups. This partially explains the difference in energy consumption growth rates in Table 1 for the 1974-1975 period.

Table 2

ENERGY PRICE INDEXES IN REAL TERMS BY INCOME GROUPS

(1970 = 100)

Income groups	1960	1973	1974	1975	Percentage change	
					1961-73	1974-75
Upper middle income	107.8	109.6	142.5	150.8	1.7	37.6
Intermediate middle income	105.5	108.2	146.7	166.8	2.6	54.2
Lower middle income	124.9	94.0	119.9	125.8	-24.7	33.8
Low income	98.5	82.0	101.9	108.0	-16.3	31.7
Average	115.4	100.3	132.5	142.8	-13.1	42.4

METHODOLOGY

The energy demand function to be estimated takes the following form:
$$\ln(EN/POP) = \alpha + \beta * \ln(GDP/POP) + \gamma * \ln(ENP) + \delta * \ln(EN_{-1}/POP_{-1}) + U,$$
where U is the error term and

EN: total consumption of primary energy expressed in oil equivalents,

ENP: index of average price in real terms to consumers of primary energy,

GDP: gross domestic product in constant 1970 US$,

POP: mid-year population.

The subscript (-1) denotes one-year lagged values of the respective variables. The equation assumes partial adjustment of energy consumption to changes in GDP and prices, where $(1-\delta)$ is the adjustment coefficient. It also assumes constant income and price elasticities over time.[1]

The equation is estimated using pooled time-series of cross-sections data for the 35 sample panel countries over the 1960-1975 period. The estimation techniques is ordinary least squares. The equation is estimated

---

1) As is well known, β and γ are the short-run income and price elasticities respectively and $\beta/(1-\delta)$ and $\gamma/(1-\delta)$ are the corresponding long-run elasticities.

for each income group separately. In regressions for each income group, dummy variables are introduced for each country and each observation is weighted by the average energy consumption of the country in the period.

Country dummies help eliminate between-country variations in energy consumption from estimates of the coefficients. This is necessary because the estimates will be used for projections over time on the one hand and because between-country variations reflect a number of factors other than GDP and prices on the other. The coefficients, therefore, will be estimated primarily from between-year variations. Also note that we have already eliminated between-country variations in energy prices by expressing them as indexes. The weighted regression adopted here simply recognises the importance of large countries in terms of energy consumption.

It should be recognised at this point that the simple approach outlined above is not expected to capture all the underlying factors that determined the historical energy consumption trends. One such factor is the role of non-commercial energy in developing countries, which is believed to have been undergoing significant changes over the years. If the significance of non-commercial energy has been declining, as presumably is the case in most of the countries,(1) an estimate of income elasticity of commercial energy alone would overstate the true overall income elasticity. When the estimates are used to project demand for commercial energy, an implicit assumption is that the role of non-commercial energy will keep declining in the future at about the same rate as it did in the past.

Another important factor inadequately treated in this approach is the changes in the efficiency of energy use, particularly those relating to the shifts in the fuel mix from coal to more efficient oil, natural gas, and primary electricity. In such a case, an aggregation of the fuels based on their inherent heat values rather than usable energy would result in an underestimate of the income elasticity. The bias, however, is not expected to be substantial except for countries where a lot of transportation coal was replaced by oil during the period.(2)

1) Available evidence on this point is only scanty. The UN statistics on fire-food – perhaps the most important non-commercial energy source – show that developing countries' production increased only at 1.3 per cent annually between 1971 and 1975. Brazilian statistics show that the share of firewood and bagasse in total primary energy consumption decreased from 44 per cent in 1966 to 26 per cent in 1976. Furthermore, the Brazilian government projects the share to decline to 14 per cent by 1986. See, "National Energy Balance", Ministry of Mines and Energy, Federal Republic of Brazil, 1977.

2) A prime example of this would be India. See P.D. Henderson, "India: The Energy Sector", Oxford University Press, 1975, p. 30.

Finally, an argument often raised is that historical experience is not likely to be relevant to projections for the future because of the recent dramatic increase in prices of oil and other energy sources. Implicit in this argument is the suggestion that income elasticity in the future is likely to be lower than in the past because of structural (output mix) changes and price elasticity higher because, at substantially higher prices, consumers would take prices more seriously and try harder to save energy. Although the point appears to be a valid one, the net effect of it is not likely to be significant enough to invalidate the whole approach. Because output structure also depends on a number of other important factors - for example, demand for final goods, technology, costs of other inputs etc. - it is hard to imagine that developing countries will significantly change their historical pattern of structural developments, influenced only by higher prices of energy. The argument for higher price elasticity also could be moderated. First of all, the percentage increases in the consumer price of energy shown in Table 2 are far less than those of OPEC crude oil, and thus would not carry us to the upper end of the demand curve. Secondly, indications are that the bulk of energy consumed in the developing countries goes to production processes rather than being consumed directly, in which case much of the increases in energy costs can be transferred to higher output prices. Thirdly, because energy saving often requires capital investments and the cost of capital is high in developing countries, substitution of capital for energy may not prove economical in many areas. As a whole, for the medium-term projections to 1985, we feel that projections based on historical experience can be fairly sound.

THE DATA

Energy consumption (EN) is the total inland consumption of energy and non-energy petroleum products, solid fuels, natural gas, and primary electricity, all converted into b/d of crude oil equivalents on the basis of the inherent heat values. Oil equivalent of primary electricity, however, represent the fossil fuel requirements to generate that much of electricity in conventional thermal power plants.

For most of the sample panel countries, we were able to collect data on consumer prices of energy products. National statistical yearbooks provided the principal source for energy product prices and/or indexes at either the wholesale or retail level. These were supplemented by: a) retail energy product prices in "Preise und Preisindizes im Ausland", Statistisches Bundesamt, Germany; b) retail prices of petroleum products in

"International Petroleum Annual", United States Bureau of Mines for the
1970-1975 period. The price of petroleum is the weighted average of petroleum product prices with weights given by the consumption level of each
product. The average price of energy is the weighted average of the four
primary energy prices with weights given by the consumption level of each.
For several (small) countries for which no price data were available, we
used the international market prices of petroleum products (FOB prices of
motor gasoline, kerosene, and fuel oil from Netherland Antilles) and coal
(average import price of bituminous coal, CIF, Japan). The average prices/
indexes of energy were then converted into prices/indexes in real terms by
dividing them by IMF's wholesale or consumer price indexes.

REGRESSION RESULTS

The weighted ordinary least squares results are as follows:

Upper-middle income countries

$$LN(EN/POP) = -1.568 + 0.509 \; LN(GDP/POP) - 0.122 \; LN(ENP) + 0.626 \; LN(EN_{-1}/POP_{-1})$$
$$\quad\quad\quad (-5.37) \quad (8.34) \quad\quad\quad\quad (-4.19) \quad\quad\quad\quad (12.1)$$

$\bar{R}^2 = 0.999$ D.W. = 1.999   + Dummies

d.f. = 67   S.E.E. = 9.881

Intermediate middle income countries

$$LN(EN/POP) = -1.347 + 0.398 \; LN(GDP/POP) - 0.085 \; LN(ENP) + 0.695 \; LN(EN_{-1}/POP_{-1})$$
$$\quad\quad\quad (-6.95) \quad (9.04) \quad\quad\quad\quad (-4.56) \quad\quad\quad\quad (18.0)$$

$\bar{R}^2 = 0.999$ D.W. = 2.217   + Dummies

d.f. = 147   S.E.E. = 8.361

Lower middle income countries

$$LN(\;N/POP) = -2.481 + 0.680 \; LN(GDP/POP) - 0.134 \; LN(ENP) + 0.649 \; LN(EN_{-1}/POP_{-1})$$
$$\quad\quad\quad (-3.10) \quad (4.61) \quad\quad\quad\quad (-1.83) \quad\quad\quad\quad (11.44)$$

$\bar{R}^2 = 0.991$ D.W. = 2.335   + Dummies

d.f. = 151   S.E.E. = 6.373

Low income countries

$$LN(EN/POP) = -2.084 + 0.621 \; EN(GDP/POP) - 0.152 \; LN(ENP) + 0.458 \; LN(EN_{-1}/POP_{-1})$$
$$\quad\quad\quad (-4.27) \quad (6.76) \quad\quad\quad\quad (-1.89) \quad\quad\quad\quad (5.44)$$

$\bar{R}^2 = 0.998$ D.W. = 1.816   + Dummies

d.f. = 95   S.E.E. = 15.477

Estimates of the coefficients are all significant at better than
0.01 level of significance except those of ENP for the lower middle and
low income groups, which are significant at 0.05 level (see the t-ratios
in the parentheses). The adjusted R-squares ($\bar{R}^2$) are high. Estimates of
the price coefficient do show significant improvement. They are all significant at 0.05 or better and the estimated price elasticities are substantially higher than those obtained from pseudo energy price variables.

The estimates imply the following income and price elasticities.

Table 3

ESTIMATES OF INCOME AND PRICE ELASTICITIES

Income groups	Income elasticities		Price elasticities	
	Short-run	Long-run	Short-run	Long-run
Upper middle income	.509	1.361	-.122	-.326
Intermediate middle income	.398	1.305	-.085	-.279
Lower middle income	.680	1.937	-.134	-.382
Low income	.621	1.146	-.152	-.280

PROJECTIONS TO 1985

Before going into projections to 1985, the predictive ability of the
estimated regression equations are tested against the historical data.
Figures 1-4 compare the projected energy consumption using the regression
estimates with the actual historical figures for the 35 sample panel countries. Simulation for the historical 1960-1975 period starts from the
actual energy consumption in 1960 and uses the actual GDP, POP, and ENP in
1961 to compute the predicted energy consumption (EN) for 1961. Projections for the later years repeat the same calculation, but use the predicted
rather than the actual energy consumption of the previous year. Except
perhaps for the low income group, Figures 1-4 strongly demonstrate that
the regression equations are able to simulate the historical trend closely.
The problem with the low income group is caused by the sudden jump in energy
consumption in 1975 not fully attributable to income and price changes.

For projections to 1985, we take the recent World Bank's GDP and population growth projections for the sample panel countries.(1) Table 4 shows
the assumed growth rates.

_____

1) See, "Prospects for Developing Countries", Development Policy Staff,
World Bank, November 1977.

Table 4

GDP AND POPULATION GROWTH ASSUMPTIONS FOR
THE 35 SAMPLE COUNTRIES

	1980	1985	Average annual growth rates (%)		
			1976–80	1981–85	1976–85
GDP in billions of constant 1970 US$					
Upper middle income	118.7	158.7	4.2	6.0	5.1
Intermediate middle income	205.6	298.2	6.6	7.7	7.1
Lower middle income	57.7	78.5	6.4	6.4	6.4
Low income	105.3	130.1	4.0	4.3	4.1
Total	487.3	665.5	5.4	6.4	5.9
Per capital GDP in constant 1970 US$					
Upper middle income	948	1129	1.8,	3.6	2.7
Intermediate middle income	672	855	3.8	4.9	4.4
Lower middle income	265	313	3.4	3.4	3.4
Low income	111	121	1.3	1.7	1.5
Average	305	367	2.6	3.8	3.2

We assume two alternative cases for the average price of energy:

Case A: The average price of energy remains unchanged in constant terms at its 1975 level through 1985.

Case B: The average price of energy remains unchanged in constant terms at its 1975 level through 1980, but increases by 15 per cent in 1981 and remains at that level through 1985.

The estimates of the constant terms in our regressions are those of the country left out in the set of country dummies. The constant terms, therefore, are replaced by the estimated average of residuals for each income group, the residuals being defined by:

$$e = \ln(GDP/POP) - \hat{\beta}*\ln(GDP/POP) - \hat{\gamma}* \ln(ENP) .- \hat{\delta}*\ln(EN_{-1}/POP_{-1}),$$

where the parameters with hats are the least squares estimates.

For the 35 sample panel countries, Table 5 shows the resulting demand projections. Figure 1-4 plot the projected energy demand - case A and case B - against the historical trend. For the low income group, the case A projected growth rate for 1976-1985 is substantially lower than that of the historical 1961-1975, despite the fact that GDP is projected to grow more rapidly than in the past. This is mostly due to the fact that 1975 actual energy consumption was higher than those predicted by the regression equations (see Figure 4). For other income groups, the projections are generally in line with the historical trend.

Figure 1

UPPER MIDDLE INCOME COUNTRIES, HISTORICAL AND PROJECTED ENERGY CONSUMPTION

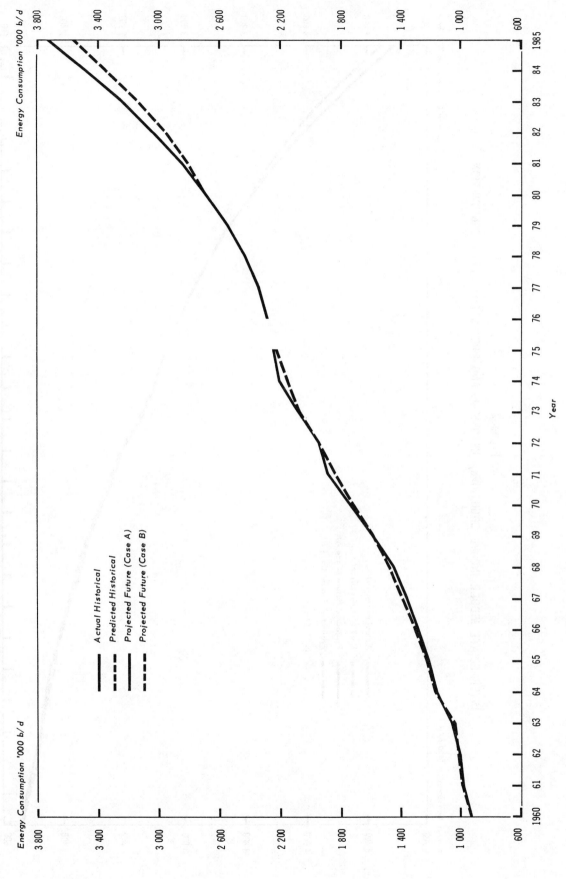

Figure 2

INTERMEDIATE MIDDLE INCOME COUNTRIES, HISTORICAL AND PROJECTED ENERGY CONSUMPTION

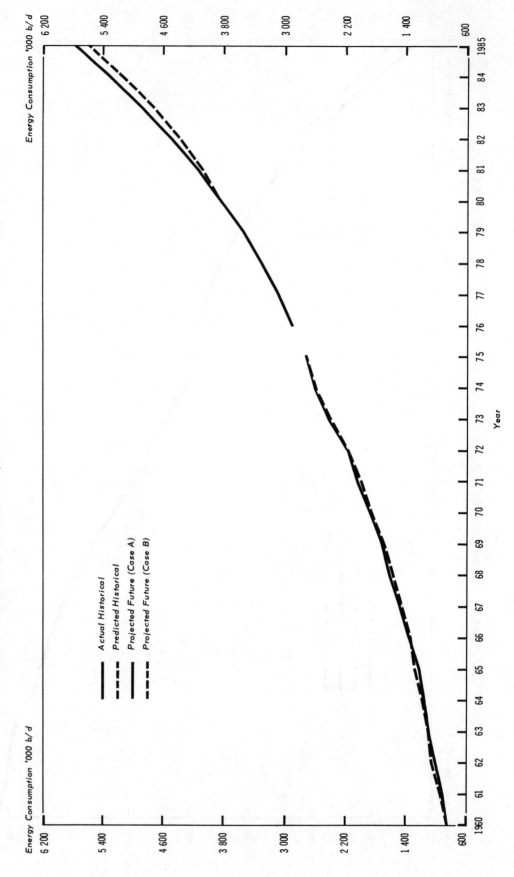

Figure 3

LOWER MIDDLE INCOME COUNTRIES, HISTORICAL AND PROJECTED ENERGY CONSUMPTION

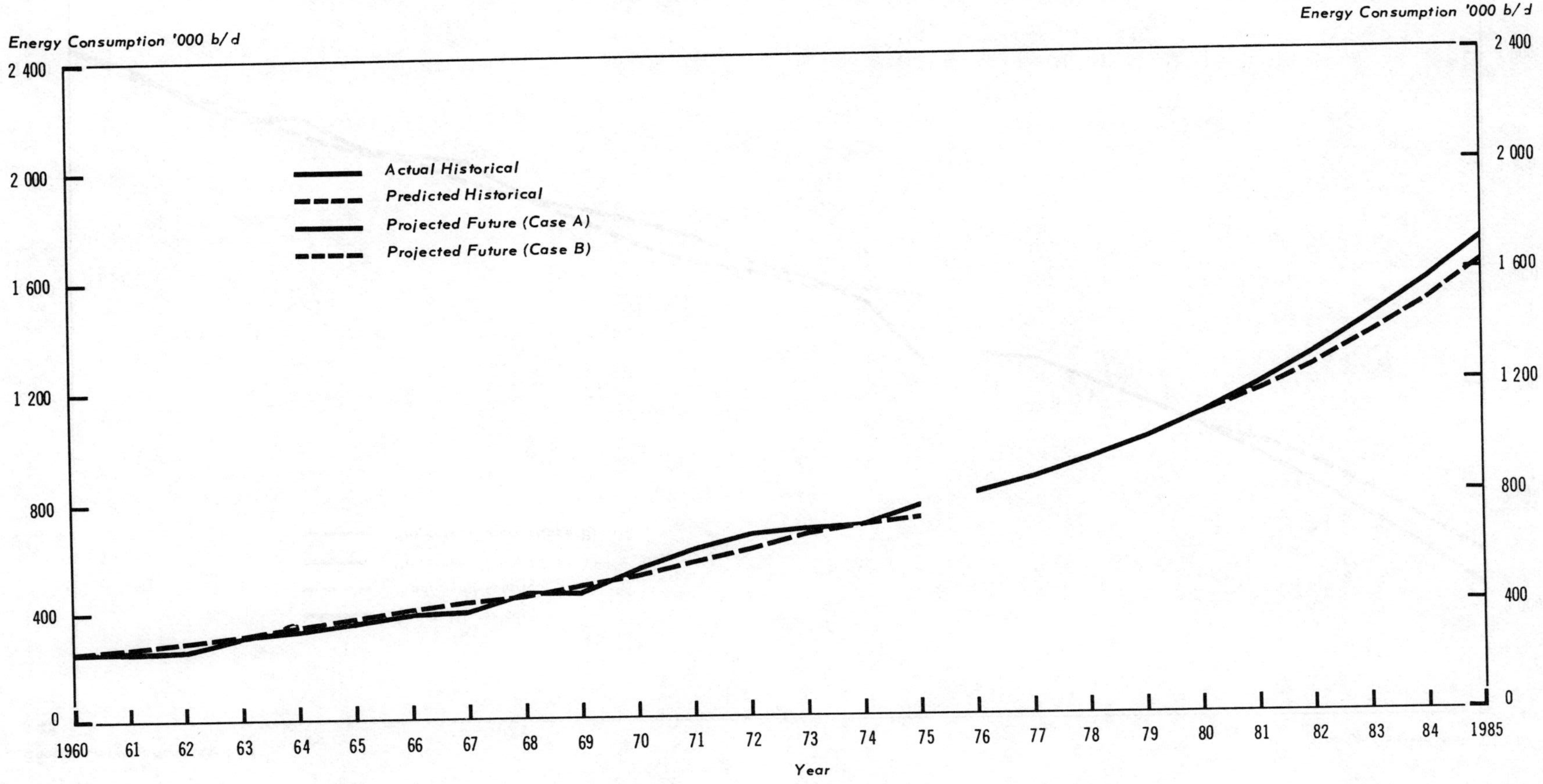

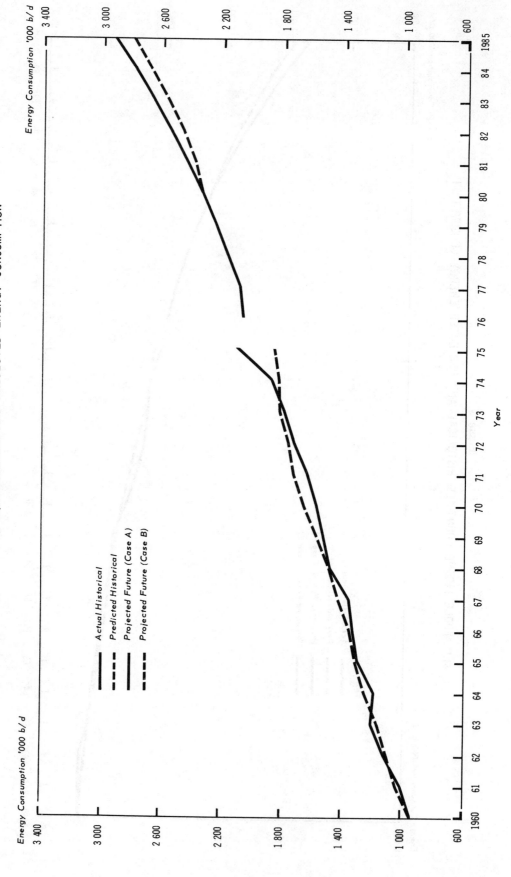

Figure 4

LOW INCOME COUNTRIES, HISTORICAL AND PROJECTED ENERGY CONSUMPTION

Table 5

PROJECTED DEMAND FOR PRIMARY ENERGY TO 1985 FOR SAMPLE PANEL COUNTRIES BY INCOME GROUPS

| | Projected | | | | Average annual growth rate (%) | | | | |
| | Actual 1975 | 1980 | 1985 | | 1976–80 | 1981–85 | | 1976–85 | |
			Case A	Case B		Case A	Case B	Case A	Case B
'000 b/d of oil equivalent									
Upper middle income	2 246	2 694	3 727	3 577	3.7	6.7	5.8	5.2	4.8
Intermediate middle income	2 746	3 830	5 792	5 606	6.9	8.6	7.9	7.7	7.4
Lower middle income	775	1 092	1 710	1 632	7.1	9.4	8.4	8.2	7.7
Low income	2 101	2 347	2 924	2 814	2.2	4.5	3.7	3.4	3.0
Total	7 868	9 963	14 153	13 629	4.8	7.3	6.5	6.0	5.6
Energy consumption/GDP ratio in b/d per million $									
Upper middle income	23.2	22.7	23.5	22.5	.4	.7	.2	.1	.3
Intermediate middle income	18.3	18.6	19.4	18.8	.3	.8	.2	.6	.3
Lower middle income	18.3	18.9	21.8	20.8	.6	2.9	1.9	1.8	1.3
Low income	24.2	22.3	22.5	21.6	-1.6	.2	-.6	-.7	-1.1
Average	20.9	20.4	21.3	20.5	-.5	.9	.1	.2	-.2

For the non-OPEC developing countries as a whole, the case A projected consumption growth rate for 1976-1980 is substantially lower than that for 1981-1985 - 4.8 per cent per annum compared with 7.3 per cent. The projected acceleration is partly due to the assumption that GDP growth rate will be higher in 1981-1985 than in 1976-1980.

Another factor responsible for the projected acceleration is the delayed impact of the oil/energy price hike in 1973/74, which continues to suppress demand through 1980 and early 1980s. The average energy intensity for the sample panel countries as a whole is projected to decline at .5 per cent annually between 1976 and 1980. It is projected to resume slightly lower than historical rate of increase between 1981 and 1985.

The case B demand projections for 1985, which assumes a 15 per cent increase in the real price of energy in 1981, are lower than the case A projections by about 4 per cent on the average. The implied medium-term price elasticity is about -.25.

The 35 sample panel countries accounted for about 81 per cent of the total primary energy consumption of the non-OPEC developing countries in 1975. For projections for all non-OPEC developing countries, we assume the same GDP growth rates and demand structure as those of the sample panel countries and, hence, the same energy demand growth rates. Under these assumptions, Table 6 shows the projections for all non-OPEC developing countries.

Table 6

PROJECTED INLAND DEMAND FOR PRIMARY ENERGY IN
NON-OPEC DEVELOPING COUNTRIES BY INCOME GROUPS

Income group	Actual 1975	1980	Projected 1985	
			Case A	Case B
'000 b/d of oil equivalent				
Upper middle income	2 803	3 362	4 651	4 464
Intermediate middle income	3 437	4 794	7 250	7 017
Lower middle income	918	1 293	2 025	1 932
Low income	2 509	2 803	3 492	3 361
Total	9 667	12 252	17 418	16 774

In smaller countries not included in the sample panel, energy consumption has usually been increasing at faster rates than in the larger sample panel countries. For this reason, the assumptions made above could well result in an underestimate of energy demand for the whole non-OPEC developing group.

CONCLUSION

Despite the recent oil price hike, primary energy consumption in non-OPEC developing countries is expected to increase at 6 per cent per annum between 1976 and 1985 compared with 6.8 per cent growth between 1961 and 1975. Demand is expected to reach 17.4 million b/d in 1985 from 9.7 million b/d in 1975. A 15 per cent increase in 1981 in the real price of energy in all countries could lead to a total reduction in consumption of about .6 million b/d by 1985.

Annex I

NON-OPEC DEVELOPING COUNTRIES INCLUDED IN THE STUDY

I.   LOW INCOME COUNTRIES

Rwanda
Burundi
Ethiopia(*)
Somalia
Malawi
Zaire
Tanzania(*)
Lesotho
Madagascar(*)
Kenya(*)
Uganda
Comoro Islands
Mali
Upper Volta
Chad
Guinea
Niger
Benin
Gambia

Sierra Leone
Central Afr. Rep.
Yemen Arab Rep.
Yemen P.D.R.
Cambodia
Laos
Bhutan
Sikkim
Bangladesh(*)
Burma
Maldive Is.
Nepal
Afghanistan
Pakistan(*)
India(*)
Timor
Sri Lanka(*)
Viet-Nam
Haiti

II.   LOWER MIDDLE INCOME COUNTRIES

Sudan(*)
Mozambique
Botswana
Swailand
Seychelles
Cameroon(*)
Togo
Mauritania
Equatorial Guinea
Senegal(*)
Liberia(*)
Guinea-Bissau
Ghana(*)
Congo, P.R.
Ivory Coast(*)
Cape Verde

Egypt(*)
Jordan
Morocco(*)
Macao
Thailand(*)
Philippines(*)
Bolivia(*)
Honduras
Grenada
St. Vincent
Dominica
El Salvador
Western Samoa
British Solomon
Tonga
Papua New Guinea

III. INTERMEDIATE MIDDLE INCOME COUNTRIES

Rhodesia
Zambia(*)
Mauritius
Angola
Namibia
Sao Tome(*) Principe
Ceuta and Melilla
Syria(*)
Tunisia(*)
Korea(*)
Malaysia(*)
China, Rep. of
Colombia(*)
Guyana
Paraguay
Chile(*)
Peru(*)

Brazil
St. Kitts-Nevis
St. Lucia
Antigua
Guatemala(*)
Dominican Rep.
Nicaragua
Belize
Costa Rica
Panama
Turkey(*)
New Hebrides
Trust Territory of
    Pacific Islands
Gilbert & Ellice Is.
Fiji

IV. UPPER MIDDLE INCOME COUNTRIES

Reunion
Afars and Isaas
Lebanon
Oman
Bahrain
Hong Kong
Surinam
Uruguay(*)
French Guiana
Argentina(*)

Mexico(*)
Barbados
Jamaica(*)
Guadeloupe
Neth. Antilles
Martinique
Trinidad & Tobago
Cyprus
Malta
Yugoslavia(*)

*) Denotes the countries included in the sample panel.

NON-OPEC DEVELOPING COUNTRIES' CONSUMPTION OF PRIMARY ENERGY, 1955-1976

	1955	1960	1965	1970	1973	1974	1975
	1000 b/d oil equivalent						
Total Non-OPEC Developing Countries							
Oil	1438	1910	2597	4175	5275	5388	5466
Solid fuels	963	1291	1660	1773	1919	2067	2301
Natural gas	49	119	306	464	615	659	692
Primary electricity	166	303	488	793	991	1117	1208
Total	2616	3623	5033	7205	8800	9231	9667
Upper middle income							
Oil	609	797	968	1387	1715	1786	1714
Solid fuels	162	229	278	284	327	354	373
Natural gas	44	96	242	340	434	462	485
Primary electricity	37	66	103	172	195	235	231
Total	852	1188	1591	2183	2671	2837	2803
Intermediate middle income							
Oil	460	580	823	1454	2027	2123	2159
Solid fuels	255	318	422	474	502	531	573
Natural gas	4	12	30	54	79	88	92
Primary electricity	91	159	238	355	475	551	613
Total	810	1069	1513	2337	3083	3293	3437
Lower middle income							
Oil	175	248	361	572	675	684	760
Solid fuels	15	18	19	24	23	29	42
Natural gas	0	0	2	4	5	5	5
Primary electricity	8	20	38	80	98	108	111
Total	198	286	420	680	801	826	918
Low income							
Oil	194	285	427	762	858	795	833
Solid fuels	531	726	941	991	1067	1153	1313
Natural gas	1	11	32	66	97	104	110
Primary electricity	30	58	109	186	223	223	253
Total	756	1080	1509	2005	2245	2275	2509
Upper middle income							
Sub-Sahara Africa	0	1	1	2	3	3	4
Asia & Pacific	10	17	27	32	21	41	43
North Africa & Middle East	20	22	39	53	104	112	101
Latin America	668	903	1170	1622	1919	2038	2004
Southern Europe	154	245	354	474	624	643	651
Total	852	1188	1591	2183	2671	2837	2803
Intermediate middle income							
Sub-Sahara Africa	57	72	86	108	139	149	153
Asia & Pacific	173	205	348	639	793	826	880
North Africa & Middle East	19	29	38	70	80	94	87
Latin America	488	669	886	1283	1732	1877	1958
Southern Europe	73	94	155	237	339	347	359
Total	810	1069	1513	2337	3083	3293	3437
Lower middle income							
Sub-Sahara Africa	23	47	70	126	152	149	171
Asia & Pacific	61	84	147	295	362	374	370
North Africa & Middle East	100	137	179	221	238	252	322
Latin America	14	18	24	38	49	51	55
Total	198	285	420	680	801	826	918
Low income							
Sub-Sahara Africa	52	64	82	124	148	149	153
East Asia & Pacific	8	14	37	140	115	62	56
North Africa & Middle East	8	10	9	19	15	10	12
Latin America	2	2	2	2	3	3	3
South Asia	686	990	1379	1720	1964	2051	2285
Total	756	1080	1509	2005	2245	2275	2509
13 non-OPEC oil exporting developing countries							
Oil	321	489	590	845	1023	1079	1135
Solid fuels	2	2	3	4	6	7	8
Natural gas	34	77	174	238	303	326	343
Primary electricity	27	43	72	126	142	144	145
Total	384	611	839	1213	1474	1556	1631

Source: "World Energy Supplies", UN Series J.

LATIN AMERICAN ENERGY AND OIL: PRESENT SITUATION AND PROSPECTS

AN OVERVIEW

by

L.A. da Silva(1)

Latin America's commercial energy consumption has grown at an annual
average cumulative rate of 5.9 per cent during the 1960-1975 period. This
rate of growth exceeded that of the developed countries as a whole and of
the developing countries as a group. Compared to the world's average (including the centrally planned economies), the rate of growth of energy
consumption in Latin America was higher than that of the group of all other
countries by 2 percentage points.

As in other regions of the world, energy consumption in Latin America
was closely tied to the area's economic expansion. The high rates of GDP
growth in the 1960s and early 1970s (6 per cent a year in the period 1960-
1973), and the relative decline in 1974 and 1975 were paralleled by similar growth pattern of energy consumption. In spite of its rapid growth,
by the end of 1975 Latin America energy consumption represented only
3.7 per cent of the world's total and 36 per cent of the overall consumption of the developing countries.

On a per capita basis, energy consumption in Latin America amounted
to only 632 kilograms of oil equivalent (kg.o.e.), which represents 45 per
cent of the world's average and only 15 per cent of the level reached by
the developed countries in 1975 per capita. Latin American energy consumption, however, is more than double that of the developing countries as a
group, reflecting the region's relatively higher level of development.

---

1) Szabolcs Szekeres and Jorge Marcano from Informations for Investment
    Decisions Inc. (Washington, D.C.), were responsible for the design
    and implementation of the forecasting model used in this study.
    Christian Gomez collaborated in the revision of the forecasts and up-
    dating of the study for a forthcoming version covering historical data
    up to 1977. The views expressed do not necessarily reflect the posi-
    tion of the Inter-American Development Bank.

Within Latin America, rates of growth of energy consumption have varied greatly from country to country. More than half of the countries, however, showed rates of growth that exceeded the region's average. Four countries of the region - Argentina, Brazil, Mexico and Venezuela - accounted for almost 80 per cent of the region's total energy consumption.

Table 1

WORLD DISTRIBUTION OF ENERGY CONSUMPTION IN 1960 and 1975
(Millions of Tons of Oil Equivalent)

	1960	%	1975	%	Rate of Growth 1960-1975 %
Developed countries	1,773.5	59.8	3,165.2	60.7	4.0
U.S. & Canada	1,084.4	36.6	1,772.2	34.0	3.3
Europe	584.9	19.7	1,009.7	19.4	3.7
Japan	75.9	2.6	276.5	5.3	9.0
Others	28.3	1.0	106.8	2.0	9.3
Developing countries	242.7	8.2	535.2	10.3	5.4
Latin America(1)	78.3	2.6	190.9	3.7	5.9
Others	164.4	5.6	344.3	6.6	5.0
Centrally-planned Economies	950.1	32.0	1,515.0	29.0	3.2
Total	2,966.3	100.0	5,215.4	100.0	3.8

Source: IDB estimates based on United Nations, World Energy Supplies, Series J., Nos. 18 and 20.

1) IDB members.

Unlike the rest of the world, the Latin American countries witnessed relatively less inter-fuel substitution. The role of oil in the consumption pattern of the region has not varied significantly throughout the 1960-1975 period. The Latin American countries as a group are more dependent on oil than any other region of the world. In 1975 oil supplied 70 per cent of the total energy requirements. In many countries of the area oil accounts for more than 90 per cent of the total energy consumption and in only four countries (Trinidad and Tobago, Venezuela, Mexico and Colombia), was its share of the total below the region's average. In 1960 and 1975, the structure of energy consumption by sources of fuel in Latin America and in the world was as follows:

Table 2

LATIN AMERICA AND WORLD ENERGY USE BY SOURCE
(Per cent)

| | Structure | | | | Rate of Growth | |
| | Latin America | | World | | Latin America | World |
	1960	1975	1960	1975	1960-75	1960-75
Solid fuels	9.0	6.1	52.0	32.8	3.4	1.2
Liquid fuels	72.7	69.7	32.0	44.1	5.7	6.6
Natural gas	14.5	18.5	14.0	20.4	7.5	7.0
Primary electricity	3.7	5.8	2.0	2.7	9.0	6.7
Total	100.0	100.0	100.0	100.0	5.9	3.8

Source: IDB estimates based on United Nations, World Energy Supplies, Series J., Nos. 18 and 20.

In reviewing the region's energy balance by sources of fuel, the more striking features are that in spite of the efforts to increase production, the oil importing countries have not been able to reduce their dependency on external energy supplies. Energy exports from the region have decreased by 10 per cent between 1960 and 1975 and, expressed as percentage of the region's indigenous supply, they fell from 73 per cent in 1960 to 50 per cent in 1975. On a geographical basis, Latin American oil importing countries have significantly increased their purchases of oil from the Middle East (estimated to have reached 65 per cent of total imports in 1975 as against 25 per cent in 1960), while decreasing those originating in the region itself.

Energy consumption per unit of output (expressed in terms of GDP in constant U.S. Dollars), increased between 1960 and 1970, from an average of 764 kg.o.e. to 829 kg.o.e. Since 1970, however, this ratio has been declining, partly associated with the drop in refining activities in Venezuela and Trinidad and Tobago and the slackening in the rate of growth of some of the energy-intensive industries. Nonetheless, there is no clear indication that the rising trend of the region's energy/output ratio has reached its high mark in 1970. Further research and more detailed analysis is needed to explain recent energy/output trends. On a per capita basis, the region's average conceals wide differences at the country level. Only three countries of the region (Argentina, Trinidad and Tobago and Venezuela), showed per capita energy consumption in excess of one ton of oil equivalent (t.o.e.). Haiti showed the region's lowest level - 20 kg.o.e. in 1975.

Energy consumption in the region has been increasingly taken in the form of electricity. The region's electrification coefficient (kwh/kg.o.e.), rose from 0.834 in 1960 to 1.148 in 1975. In the aggregate, the electrification coefficient in Latin America is only slightly lower than that of the world as a whole, but is still significantly less than that of the developed countries. The secular growth of the electricity coefficient in the region has been exacerbated by the rapid population growth, the shift from rural to urban-based societies and the extremely high speed of the urbanisation process. In spite of the rapid growth of electricity generation, it is estimated that, on the average, less than 60 per cent of the region's population is supplied with electric energy.

It should be noted that the energy data referred to previously, understate the region's real level of energy consumption as they do not take into account consumption of energy in the form of direct uses of firewood, bagasse and other non-commercial sources of energy. These are relatively important even in the more developed countries of the region, but far more substantial in the less developed ones, especially in Haiti.

Considering the utilisation of energy by the different economic sectors, the region shows a comparatively higher concentration in the productive sectors, with residential and services sectors accounting for only a small proportion of the total. It is estimated that the industrial sector accounted for approximately 40 per cent of the total fossil fuels consumption, transportation 34 per cent, electricity generation 15 per cent and the remaining 10 per cent was used by residential and other sectors. Industry also accounts for the greater proportion of electricity generation.

The region has expanded considerably its refining capacity, to an extent that it presently supplies almost all the internal demand of refined products. Present levels of imports of refined petroleum are restricted to a few products not locally produced or to complement temporary production deficiencies. Venezuela and Trinidad and Tobago have refining capacity for exports, mostly of residual fuel oil. Notwithstanding, with some few notable exceptions, refining processes in the region are relatively simple, being mainly concentrated in the production of the first-line of products, with catalytic cracking and reforming process accounting for only a small proportion of the total.

Total capital and exploration expenditures in the petroleum and gas industry in Latin America increased from an estimated annual average of US$833 million in 1965-1967 to US$4.3 billion in 1975 (data in current values). The bulk of these expenditures, however, has been concentrated in production, transportation and marketing, with outlays for exploration representing only a small fraction of the total. On the whole, the region has been able to mobilise a larger share of the total world resources allocated to this sector up to 1970. Since 1970, however, Latin America's participation in the total has decreased in relative terms, although expenditures on exploration activities have increased slightly up to 1974. Venezuela's share of the region's total has declined steadily, from 21.2 per cent in 1965-1967 to 8.7 per cent in 1976. Nonetheless, its share in the region's total expenditures on exploration has been maintained at a relatively high level ranging from 17 per cent to 25 per cent of the total. The overall investment pattern in the petroleum and natural gas industry, where major efforts have been directed to production and refining, has been based on the unquestioned assumption of ample and readily available supplies.

Reviewing the Latin American energy reserve base is fraught with difficulties due to the lack of published data. Besides, insufficient exploratory activity in the past limits the knowledge of the region's energy potential. Present proven reserves of petroleum in Latin America amount to 39.7 billion barrels, almost half of which corresponds to Venezuela. Recent discoveries in Mexico have greatly enhanced this country's reserves, which have been estimated at 14 billion barrels by mid-1977. In comparison, Brazil's proven reserves stood at 877 million barrels. As a whole, the region's "life expectancy" of present proven reserves (reserves divided by total production in 1976) are sufficient to maintain production for 24 years. Reserves in Venezuela and Ecuador can maintain production for almost 22 years. In Mexico, they are sufficient for approximately 43 years, although presently increased planned exports could reduce their "life expectancy" significantly. In Argentina and Colombia, reserves are only sufficient for 16 and 14 years respectively, while in Brazil the production potential is only equivalent to 10 years. Outlook for natural gas is more promising for many countries of the region (as an average, proven reserves can maintain production for 40 years), while coal reserves are small and with a high level of impurities. Colombian coal, however, seems to be available in great quantities and is of good quality. The region is particularly well endowed with hydropower, which has not been sufficiently utilised, except in a few countries. Uranium is also available in some quantities in Brazil, Argentina and Mexico and nuclear power is expected to play an important role in electricity generation in these countries.

Latin America's economic growth has required increased levels of imports of capital and intermediate goods for the expansion and continuing modernisation of productive capacity. Although exports have grown in the post-war period at significant rates, they have been outpaced by imports and, as a result, the region has been characterised by current account deficits of important magnitudes. As Latin America increased its rate of growth, current account deficits increased from about US$1.3 billion in 1960-1969, to US$3.8 billion in 1973. Long-term capital inflows helped the region cover these current account deficits, maintain high levels of investments, and at the same time increase its international reserve position. The oil price increases since 1973 added to the difficulties facing the world economy and especially the developing countries from the severe recession and inflationary process in the industrialised countries and declining real prices for their main primary export commodities. Excluding Venezuela, the region's consolidated current account deficit increased from US$4.1 billion in 1973 to US$12.8 billion in 1974 and to US$ 16 billion in 1975, in spite of substantial increases in the real value of the region's main primary export commodities in 1973 and 1974. Evidently, the impact of the oil price increases has affected the countries of the region with different magnitude and force. The "energy burden" (as expressed by the relationship between the total value of oil imports and the total export bill) of the net oil-importing countries rose from an average of 8 per cent in 1972 to approximately 24 per cent in 1976. In Brazil, it reached 37.8 per cent in 1976, in Panama 69.9 per cent, in Uruguay 30.1 per cent, in Barbados 30.0 per cent, and in Jamaica 30.2 per cent. These figures are indicative of the claim on these countries' foreign exchange earnings, a situation of concern, particularly in view of the significant gains in exports registered in 1973-74 and the subsequent deterioration of the region's export prices.

Latin American countries have been able to maintain their level of oil imports and, in certain cases relatively high levels of economic growth, by reducing their levels of international reserves and by obtaining unprecedented high levels of external financing, made possible by the successful recycling of financial resources to the oil importing countries of the region. However, the increased recourse to external financing has been more unfavourable, in terms of the debt's structure and cost. It is estimated that, while the portion of the region's (excluding Venezuela) public outstanding debt (public and publicly guaranteed), at the end of 1973 with a maturity period of less than five years was 43 per cent of the total, in 1975 this proportion reached 50 per cent. The average interest rate to be paid on this external debt, on the other hand, increased from about 5.3 per cent per year in 1972 to 6.0 per cent in 1973 and to 7.4 per cent in 1975. (To the public debt indicated should be added the private debt without government's guarantee, for which complete and reliable information is not yet available.) Preliminary data for 1976 indicate an increase in the region's outstanding debt of approximately 26 per cent. Loans from the

private sector increased their participation in the total, from 55 per
cent in 1975 to 59 per cent in 1976, which has resulted in still higher
costs for servicing the area's foreign debt.

An analysis of the economic impact of the petroleum price increases
on several Latin American countries led to the conclusion that there is
a two-way relationship between liquid fuels demand and national income.
This effect takes place: (a) through the impact of higher prices of oil
on the capacity to import raw materials and capital goods needed to support economic expansion, in both oil importing and exporting countries,
and (b) through the direct relationship between growth and the demand for
oil. To model this relationship,(1) a system of simultaneous equations
was set up relating liquid fuel demand, capacity to import capital goods
and raw materials other than oil, investment and GNP. The model's exogenous variables were defined to include total capacity to import, domestic
production of liquid fuels, prices of liquid fuels and population. The
simultaneous system of equations was solved by numerical methods since
the non-linear nature of the demand equation precludes an explicit analytical solution. Regression techniques were used to estimate behavioural
equations of the system and to project the exogenous variables. To avoid
the use of "scenarios", probability distributions for the exogenous variables were used to construct unconditional confidence intervals for the
forecasted endogenous variables. The estimated errors of the model itself
(based on a simulation of the model over a historic period), and the likely
errors of its exogenous variables, were combined by the use of the Monte
Carlo simulation technique. This process defined frequency distributions
and confidence intervals for each endogenous variable (an 80 per cent confidence interval was chosen to present the results of the model). Since
the model yields solutions to a system of simultaneous equations, the forecasted values of liquid fuels consumption, liquid fuels imports, energy
demand and GNP for each country of the region, are compatible with each
other. Forecasted low, medium and high values were calculated up to 1990;
however, because of the higher degree of uncertainty associated with the
forecasts for later years, the results are more reliable through 1985.
Therefore, this is the period for which the main results are presented.

To evaluate the overall performance of the model, a simulation run
was made using data from 1962-1975 for each country. With few exceptions,(2)
the model developed for this study fits the historical data well, and its

---

1) See Appendix.
2) Barbados, Chile and Haiti.

results should be interpreted as statistically significant, within the hypothesis formulated for each of the individual equations.

Table 3 summarizes the projected (mid estimates) quantities of liquid fuels consumption, production and imports and commercial energy consumption in each of the IDB member countries. The corresponding income elasticities of liquid fuels and energy demand are also included in this Table.

In 1975-1985, liquid fuels consumption for the region as a whole is estimated to almost double, reaching by the end of the period almost 265 million t.o.e. Of this total, the net oil importing countries are expected to account for approximately 172 million t.o.e. and the oil exporting countries for the remaining 93 million t.o.e. The corresponding overall rate of growth is 6.7 per cent per year, with consumption of oil exporting countries growing at a slightly higher rate than that of oil importing countries, but both groups registering virtually the same growth as that of the region (see Table 4). Given the projected increases in oil production, liquid fuels imports by the net oil importing countries will increase from 48 million t.o.e. in 1975 to 89 million t.o.e. in 1985. Net oil exporting countries, on the other hand, are expected to generate exportable surplus amounting to 220 million t.o.e. in 1985, as compared to 139 million t.o.e. in 1975. Under the assumption of the model, as can be seen in Table 4, the region's overall rate of economic growth, endogenously determined by the model, is expected to be 5.6 per cent per year between 1975-1985, with the oil importing countries registering a projected growth slightly higher (5.9 per cent) than that of the oil exporting countries (4.8 per cent), mainly as a result of the rapid increase estimated for Brazil (7.4 per cent).

Comparison of the expected growth between 1975 and 1985 with past economic performance, indicates that several countries of the region will not be able to maintain the high rates registered in 1960-1975. In other countries, however, the projected increases in exports and net capital inflow will be translated into high investment rates and accelerated GNP growth. Among the net oil exporting countries, liquid fuels consumption in Venezuela and Bolivia are expected to increase at a faster rate than production, thus decreasing, in relative terms, these countries' future levels of oil exports. Under these assumptions, rates of growth of GNP projected for the 1975-1985 period are likely to be lower than those for the 1960-1975 period. In the case of Mexico, it should be borne in mind that the high rates of growth of oil exports (25.4 per cent in 1975-1985), reflects in effect the low level in the initial year of the forecast period. Although production is projected to increase at a significantly faster rate than internal consumption, oil exportable surplus will not be

Table 3

LATIN AMERICAN LIQUID FUELS CONSUMPTION, PRODUCTION, IMPORTS,
ENERGY CONSUMPTION AND INCOME ELASTICITIES OF ENERGY DEMAND

(1975-1985)

(Million Tons of Oil Equivalent)

Countries	Liquid Fuels Consumption		Liquid Fuels Production		Liquid Fuels Imports		Energy Consumption		Income Elasticities(3)	
	1975(1)	1985(2)	1975(1)	1985(2)	1975(1)	1985(2)	1975(1)	1985(2)	Liquid fuels Consumption	Energy Consumption
Argentina	22.284	28.965	21.135	23.411	1.149	5.555	31.767	41.978	1.00	1.08
Barbados	0.214	0.338	0.017	0.019	0.196	0.390	0.182	2.077	2.15	2.10
Bolivia	0.886	1.891	1.881	4.346	-0.994	-2.455	1.131	2.359	1.32	1.27
Brazil	39.460	90.559	8.692	40.555	30.768	50.004	50.076	112.080	1.08	1.01
Chile	4.463	7.106	1.692	1.183	2.771	5.923	7.084	9.527	1.50	1.00
Colombia	5.893	9.377	8.280	7.419	-2.387	1.958	10.878	17.209	0.96	0.96
Costa Rica	0.580	1.257	-	-	0.580	1.257	0.707	1.382	1.59	1.37
Dominican Rep.	1.400	4.978	-	-	1.400	4.978	1.549	4.377	1.91	1.59
Ecuador	1.975	6.410	8.168	18.026	-6.193	-11.616	2.140	5.809	1.69	1.41
El Salvador	0.672	1.222	-	-	0.672	1.222	0.673	1.147	1.59	1.38
Guatemala	0.947	1.526	-	-	0.947	1.526	0.986	1.585	1.02	1.04
Guyana	0.581	1.525	-	-	0.581	1.525	0.617	1.369	2.35	1.91
Haiti	0.105	0.143	-	-	0.105	0.143	0.110	0.164	0.34	0.20
Honduras	0.481	0.969	-	-	0.481	0.969	0.494	0.984	1.27	1.23
Jamaica	2.005	5.087	-	-	2.005	5.087	2.060	4.726	1.74	1.51
Mexico	32.946	67.806	40.135	135.347	-7.189	-67.541	50.673	99.223	1.29	1.22
Nicaragua	0.676	1.516	-	-	0.676	1.516	0.657	1.336	1.64	1.40
Panama	0.966	2.020	-	-	0.966	2.020	0.937	1.866	1.27	1.15
Paraguay	0.267	0.443	-	-	0.267	0.443	0.290	0.490	1.16	1.20
Peru	5.904	9.185	3.650	9.998	2.253	-0.813	7.134	10.947	1.05	1.07
Trinidad & Tobago	1.192	1.978	11.217	24.490	-10.025	-22.513	3.069	4.515	2.17	1.74
Uruguay	1.784	2.662	-	-	1.784	2.662	1.920	2.673	2.50	2.13
Venezuela	10.453	14.759	125.240	131.013	-114.787	-116.254	22.423	32.455	1.03	1.09
Latin America	138.571	264.262	230.106	395.808	-91.536	-131.546	198.797	359.669	1.16	1.05
Net Oil Importers	91.118	171.418	43.466	82.585	47.652	88.833	119.361	215.307	1.09	1.05
Net Oil Exporters(4)	47.453	92.844	186.641	313.223	-139.188	-220.379	79.436	144.362	1.33	1.10

1) Source: U.N. Energy Supplies, Series J, Nos. 18 and 20.
2) Source: IDB mean values of projection in simulation model.
3) Defined as percentage change in consumption over percentage change in GNP.
4) Excludes Colombia.

sufficiently high to spur economic growth at rates exceeding the 6.2 per cent average registered between 1960-1975.

Given the observed relationship between liquid fuels consumption, energy consumption and GNP, overall energy requirement is estimated to reach 360 million t.o.e. in 1985, from 200 million t.o.e. in 1975. Energy consumption of the oil importing countries is expected to increase at a faster rate (6.2 per cent) than that of the group of the oil exporting countries (5.3 per cent) and the region as a whole (5.8 per cent). Income elasticity of liquid fuels demand and energy demand is calculated to be 1.16 and 1.05, respectively, for the region as a whole, with the group of the oil exporting countries showing higher values as expected.

Values shown in Tables 3 and 4 are based on the most likely projections. Nevertheless, the simulation projections actually provide a range of future values each of which has a probability of occurrence. Using this range for 1985 projections, Latin American liquid fuels consumption has an 80 per cent probability of being between 173 and 377 million t.o.e. It is assumed that liquid fuels production will range between 360 and 433 million t.o.e. Energy consumption is expected to range between 262 and 478 t.o.e. in 1985 for the region as a whole.

An unconstrained forecast, not taking into account the effect of the higher levels of oil prices on the balance of payments and the corresponding income effect among the model's variables, was also developed as a benchmark against which the results of the constrained forecast could be compared. Under this hypothesis, consumption of liquid fuels was forecasted using only the deterministic values of GNP and oil prices. It represents a possible behavioural pattern of each of the variables and economic growth under the assumption that past observed relationships and performances will be repeated in the future. Therefore, the differences between the rates of growth of the constrained and unconstrained forecasts are an indication of the possible direction and magnitude of the income effect, that is, the impact of oil price increases on liquid fuels demand, investment and GNP growth. It is clear, however, that a highly aggregated and simplified model to depict the relationships between economic growth and energy requirements is inadequate in many respects. Notwithstanding, it highlights in many cases the importance of the income effect originating from the higher prices of petroleum and therefore points out a line of energy modelling worth pursuing.

Table 4 presents in a summary form the four most important variables of the model. The mean (or most likely) values of the constrained forecasts are compared to the deterministic values of the unconstrained forecasts. For each of these variables, the Table indicates two sets of values:

Table 4

COMPARISON OF GROWTH RATES(1) BETWEEN THE CONSTRAINED AND THE UNCONSTRAINED FORECASTS

(Per cent)

Mid estimate, 1975 to 1985

Countries	Consumption of Liquid Fuels		Imports of Liquid Fuels(2)		Gross National Product		Total Energy Consumption	
	Con-strained	Uncon-strained	Con-strained	Uncon-strained	Con-strained	Uncon-strained	Con-strained	Uncon-strained
Argentina	2.6	4.6	17.1	24.4	2.6	4.4	2.8	4.9
Barbados	6.4	6.4	6.8	6.8	3.3	3.3	5.8	5.8
Bolivia	7.9	7.2	(9.5)	(10.0)	6.0	5.4	7.6	6.9
Brazil	8.7	7.9	5.0	4.0	7.4	7.6	7.5	7.8
Chile	5.0	5.0	8.1	8.1	3.3	3.3	3.0	3.0
Colombia	4.8	5.4	*	*	5.0	5.6	4.8	5.3
Costa Rica	8.0	10.3	8.0	10.3	4.9	6.6	6.7	9.1
Dominican Rep.	13.5	12.8	13.5	12.8	6.6	6.5	10.5	10.5
Ecuador	12.5	10.7	(6.5)	(7.3)	7.1	6.5	10.0	9.3
El Salvador	6.2	8.4	6.2	8.4	3.9	5.2	5.4	7.3
Guatemala	4.9	5.9	4.9	5.9	4.7	5.7	4.9	5.7
Guyana	10.1	8.6	10.1	8.6	4.3	3.7	8.2	7.1
Haiti	1.1	1.1	1.1	1.1	1.8	1.8	1.2	1.2
Honduras	7.3	5.3	7.3	5.3	5.6	4.2	6.9	5.4
Jamaica	9.8	8.1	9.8	8.1	5.7	4.7	8.6	7.3
Mexico	7.1	7.9	(25.4)	(24.1)	5.5	6.7	6.7	7.7
Nicaragua	8.4	9.2	8.8	9.2	5.0	5.8	7.0	8.3
Panama	7.7	8.7	7.7	8.7	6.1	6.9	7.0	8.1
Paraguay	5.2	5.4	5.2	5.4	4.5	4.7	5.4	5.7
Peru	4.5	5.5	*	(-17.7)	4.2	5.2	4.5	5.4
Trinidad & Tobago	5.2	9.6	(8.4)	(8.4)	2.3	3.9	4.0	6.9
Uruguay	4.1	2.5	4.1	2.5	1.6	1.0	3.4	2.1
Venezuela	3.5	5.3	(0.1)	(-0.2)	3.3	5.3	3.6	5.7
Net Oil Importers	6.5	7.4	6.4	6.4	5.9	6.9	6.2	7.4
Net Oil Exporters	6.9	6.3	(4.7)	(4.3)	4.8	4.8	5.3	5.7
Total	6.7	7.0	3.7	3.1	5.6	6.2	5.8	6.7

1) Growth rates are calculated on the basis of extreme points.

2) Data in parenthesis denote growth rates of net exports.

*) Not defined because of changes from net-exporting to net-importing position and vice versa in the forecasting period (1975-1985).

451

(a) their cumulative rate of growth over the forecast period under the constrained projection, and (b) their cumulative rate of growth under the unconstrained projection, over the same period.

It can be observed that, for the region as a whole, the average rate of growth of liquid fuels consumption is 6.7 per cent under the constrained forecast as compared to 7.0 per cent under the unconstrained one. These results reflect the combined income effects of oil importing and exporting countries, as described below.

Most of the oil importing countries show a slower rate of growth of liquid fuels consumption in the constrained forecast. The most affected, however, are Costa Rica, Argentina, Panama and Nicaragua, where there is either no domestic oil production, or in spite of the expected increase in indigenous output of liquid fuels (as is the case of Argentina), limited future availability of foreign exchange for imports would constrain purchases abroad of these products.

The ability to maintain imports of oil among the other oil importing countries varies considerably. In the case of Peru, for instance, consumption of liquid fuels is expected to outpace local production between 1975 and 1983, but importing capacity will be restrained by balance-of-payments difficulties, resulting in lower rates of growth than those corresponding to the unconstrained forecast. This country's effort to increase oil production, however, is expected to yield increased exportable surplus beginning under the median projection in 1984. Colombia, on the other hand, would change from a net exporting to a net importing position(1) as early as 1978. As in the case of the other oil importing countries, the annual rate of growth of liquid fuels consumption will reach only 4.8 per cent under the constrained forecast, against 5.4 per cent in the absence of the impact of the oil price increases in the country's capacity to import.

Although Brazil has been and will continue to be a net oil importer, growth of liquid fuel consumption is expected to be similar under the constrained and unconstrained forecasts. This can be explained by the fact that up to 1975 Brazil was able to maintain high rates of growth of GNP and liquid fuels consumption in spite of the income effect created by higher prices of oil since 1973. Even allowing for projected rapid increases of oil production, Brazil is expected to continue to be a net oil importer, but the relative balance of payments constraining effect of these imports will not be significantly different from the historical one.

_____

1) More recent data indicate that beginning in 1976 this country had to import oil, as production was insufficient to meet internal requirements.

Among the remaining oil importing countries, only Guyana, Honduras, Jamaica and Uruguay will be in a position to accelerate liquid fuels consumption, assuming evidently, that the very favourable past improvements in their capacity to import would be maintained in the future.

Among the oil exporting countries, Bolivia and Ecuador will be able to increase their importing capacity substantially during the forecasting period through their higher volume of oil exports. As can be seen in Table 4, the simple extrapolation of historical levels of liquid fuels consumption in the unconstrained model produces lower forecasts than those in which the full effects of petroleum exports begin to show in the constrained forecast.

Liquid fuels consumption in other oil exporting countries is expected to slow down throughout the 1975-1985 period. This is particularly true in the case of Venezuela and Trinidad and Tobago, where the rate of growth of liquid fuels consumption is expected to be 3.5 per cent and 5.2 per cent respectively, in the constrained model, as compared to 5.3 and 9.6 in the unconstrained forecast. These seemingly paradoxical results follow from the fact that the historical rates of growth of liquid fuels production and exports on which the unconstrained model is based are high as compared to those of the constrained forecast. The latter assumes a more gradual rate of growth of oil production, foreign exchange earnings and GNP in the forecasting period and therefore a moderation in the rate of growth of liquid fuels consumption.

As was previously noted, in the case of Mexico, the slowdown in the rates of growth of liquid fuels consumption reflects the fact that higher levels of oil exports affect the balance of payments more significantly toward the end of the forecasting period, although production is projected using a relatively high rate of growth (13 per cent per year between 1975 and 1985). Even at the highest expected level of oil output (1985), only 70 million t.o.e. out of 136 million t.o.e. produced would be left available for exports and thus could not be expected to significantly affect the balance of payments or GNP growth.

Assuming the past relationship between liquid fuels consumption, overall energy consumption and economic growth, it can be observed that with a few exceptions the region will not be able to maintain the high rate of growth of energy consumption (6.7 per cent) that would have been associated with an economic performance without the effects of the higher prices of oil. Energy consumption for the region as a whole, assuming a rate of growth of GNP of 5.6 per cent (constrained forecast) is projected to increase by only 5.8 per cent per year between 1975 and 1985, as compared to 6.7 per cent implied by an overall growth of economic activity of 6.2 per cent if past performance were to be maintained in the future.(1)

In interpreting these results, the highly aggregative nature of the model and the simplified assumptions adopted should be taken into account; notwithstanding, as indicated, it reflects the energy problem faced by most of the countries of Latin America. Since the model is heavily based on the future performance of each country's capacity to import (exports plus net capital inflows), three additional elements should be considered: first, the recent financing of the region's current account deficit with less favourable terms may reduce its ability to secure increased levels of external loans necessary for the financing of new investments and purchases of raw materials; second, economic recuperation in the developed countries is taking place at a slower rate than expected, and this would affect the projected rate of growth of the region's exports; and third, restrictive balance of payments policies adopted by the industrial countries to reduce their current account deficits will inevitably affect the region's export performance in the future. Although further research is needed to determine more precisely the region's and each of the Latin American countries' export and capital inflow outlook, it would seem, from the results shown, that unless the oil importing countries of the region are able to significantly increase their capacity to import, they will experience a reduction in the rate of economic growth due to the impact of the oil prices in the balance of payments.

In the light of changes in cost conditions derived from the growing depletion of world's oil reserves, the high mark of oil consumption in relative terms seems to have been reached in Latin America and elsewhere. The contribution of oil to incremental energy supplies in the years ahead will be considerably less important than in the past decade or two. For the Latin American countries, the problem associated with the transformation from an economic and social structure highly dependent on oil to one in which other and more abundant sources of fuels prevail, lies in the short-run in three main factors: (a) the degree of success of energy conservation practices and fuel substitution by the developed countries, with a view to decreasing significantly their rate of growth of oil consumption; (b) the speed and cost at which new sources of energy are developed in the industrialised countries and the technology transferred to the countries

---

1) It is unlikely that both energy input per unit of production and the substitution of other sources of energy for liquid fuels as a result of the higher prices of oil, will significantly affect the projected energy demand, liquid fuels consumption and GNP relationships up to year 1985.

of the area, and (c) the continued and increased access by the countries
of the region to sizeable financial and technical resources to meet development goals.

The increased competition of the industrial countries for dwindling
supplies of fuels in the world market will undoubtedly be reflected in
higher costs for these scarce resources. The Latin American countries,
with a much weaker and less flexible economic structure will, in these
circumstances, be denied, under normal market conditions, access to these
resources and their growth prospects will be severely constrained. Given
the growing interdependence between developed and developing countries,
it is to their mutual interest to find global and specific country solutions which are compatible with a steady growth of the world economy and
to avoid making more difficult the attainment of the developing countries
socio-economic goals.

On the basis of the present pattern of demand-supply relationships
and the region's energy reserves, it will be difficult for the net oil importing countries to develop their energy resource bases and to decrease
their external dependence on imported oil in the short and medium term.
Acceleration of the region's economic activities will be reflected in
higher levels of energy requirements, given the observed levels of income
elasticity of energy demand and the rigidities in the structure of energy
demand in Latin America. High rates of population growth and the dramatic
increases in the urbanisation process with their concomitant demand on
electrification and transportation requirements, will add to that effect.
There is a certain latitude for energy conservation and for increasing
the efficiency in the utilisation of available energy resources in transportation and industry, but especially in the electric power sector. However, potential reductions in the rate of growth of energy consumption deriving from these measures will not significantly affect the region's energy
picture.

Changing the region's energy mix requires high investments and long
lead times. Considerable development is taking place in the sector, both
to increase the supply of fossil fuels as well as of other sources of
energy, but because of the long time required for the implementation of
these projects and for the transition to an energy economy less dependent
on oil, the structure of the region's overall and sectoral energy consumption will not vary significantly before the mid-1980s.

Development of the region's energy resources and continued access to
foreign energy supplies are closely linked with the future growth of the
Latin American economies. As indicated, successes in increasing and diversifying exports, improving terms of trade and obtaining foreign capital on

adequate terms will be necessary for the region to maintain its capacity
to import and to improve economic growth. This will be especially true
during the transition period until present world market desequilibria are
corrected and the region's exports find greater access in the developed
countries. To meet the region's high level of energy requirements while
diversifying from the present reliance on scarce fossil fuels will indeed
require an extraordinary effort to mobilise both internal and external
financial and technical resources, which will make all that has been accomplished to date by the countries of the region appear small by comparison.

Over the last decade or so, changes have occurred in the traditional
patterns of financing petroleum exploration and development in many countries of the area. Socio-economic and political considerations in the host
countries, the increased national control of the natural resources and
nationalisation policies, changing investment policies of the capital-
exporting countries, just to mention a few, have affected and in some cases
precluded the flow of private foreign capital and technology to the energy
sector of the Latin American countries. More recently, however, many countries have revised their foreign investment policies and have adopted a
series of incentives and other instruments (such as risk-contracts,
production-sharing agreements, national investment guarantees and insurance
schemes), to attract the necessary foreign private capital for the exploration and mobilisation of their petroleum resources. Many organisations and
countries are increasing their efforts in the search for effective financial mechanisms to provide adequate conditions for an increased flow of
external financial resources into the developing countries in the form of
equity or debt financing.

Latin America cannot avoid sharing the consequences of the rapid depletion of oil reserves and the energy crisis facing the developed and developing countries alike. A frontal attack on energy problems is required
in each of the countries of the region. A strong commitment to implement
rational energy policies, to undertake conservation measures (including
changing the product mix and promoting the utilisation of less energy-
intensive technologies) and to mobilise technical and financial resources,
both internal and external to develop conventional and non-conventional
energy resources, is of paramount importance if past high rates of growth
and employment are to be approximated. Only through an active awareness
of the energy crisis and a determined effort to deal with the energy problems will the Latin American countries be able to diversify and build
energy systems which are less dependent on oil.

APPENDIX

DESCRIPTION OF THE MODEL, ESTIMATION AND FORECASTING METHODS

This Appendix defines the variables and equations utilised, presents goodness of fit statistics for the model as a whole, and explains the forecasting techniques employed.

A. THE PRIMARY (CONSTRAINED) SIMULTANEOUS EQUATION SYSTEM

The following model was estimated and solved for twenty-three Latin American IDB member countries:

$$\text{Log } M_t = \beta_{11} + \beta_{12} T \tag{1}$$

$$I_t = \beta_{21} + \beta_{22} N_t + \beta_{23} Y_{t-1} \tag{2}$$

$$Y_t = \beta_{31} + \beta_{32} I_t + \beta_{33} I_{t-1} + \beta_{34} I_{t-2} + \beta_{35} H_t \tag{3}$$

$$\text{Log } D_t = \beta_{41} + \beta_{42} \text{ Log } Y_t + \beta_{43} \text{ Log } P_t + \beta_{44} \text{ Log } D_{t-1} \tag{4}$$

$$Eh_t = \beta_{51} + \beta_{52} Y_t \tag{5}$$

$$El_t = \beta_{61} + \beta_{62} D_t \tag{6}$$

$$N_t = M_t - (D_t - L_t) P_t \tag{7}$$

where:

M	=	total imports
T	=	time
I	=	total investment
Y	=	GNP
H	=	population
D	=	liquid fuels consumption
E	=	total energy consumption (subscripts h = high prediction, and l = low prediction)(1)
L	=	domestic liquid fuels production
P	=	liquid fuels price (CIF)
N	=	non oil imports
β_{ij}	=	regression coefficients
t	=	time subscript

---

1) The same series for E is used in the estimation of regressions (5) and (6).

M, I, N and Y are measured in millions of 1973 US dollars, while D, E and L are in million tons of oil equivalent. The P used are Saudi Arabian 31° light crude FOB prices, adjusted by the International Monetary Fund's FOB-CIF factors for each country.(1) H is measured in millions of inhabitants.

The system contains seven endogenous variables (M, I, Y, D, Eh, E1 and N); four exogenous variables (T, L, P and H), and uses lagged values of I, Y and D as predetermined variables.

The regressions were estimated on the basis of 1960-1975 yearly data, using the ordinary least squares procedure. Equations (5) and (6) use data prior to 1973 only, to avoid any distortions that may have been introduced in the following years as a result of higher oil prices.

The system is block-recursive, i.e. it can be broken into three sub-systems: equation (1) establishes values of M to be used in the second sub-system - equations (2), (3), (4) and identity (7) - a block of four simultaneous equations and four unknowns. Finally, the results of this sub-system are fed into the third block - equations (5) and (6) - in order to arrive at the high and low forecast values of E, repsectively.

Simplifying assumptions: In order for the model to hold, and to capture the income effect in a direct fashion, the following simplifying assumptions were made:

a) Total importing capacity is given by actual imports:

$$M_t = X_t + K_t + \Delta^- R \tag{8}$$

where:

X = total exports

K = net capital inflows from abroad

$\Delta^- R$ = decrease in the level of international reserves

b) Investment levels depend in part on the capacity to import capital goods, and in part on the availability of domestic resources. N and Y are used as proxies for these two phenomena, respectively.

c) The level of GNP is directly related to investment outlays in the current and past two years.(2)

---

1) Although for obvious reasons the desirability of these prices is dubious, they were used since no other reliable and consistent information was universally and readily available. For most of those countries where domestic prices could be found, the elasticity results were not found to be significantly different, but they were not used due to the difficulty of transforming the figures into border prices as required in identity (7).

2) H is used in equation (3) to establish a trend since its rate of growth is practically constant, and in order to increase the regression's predictive power, should the link between Y and I prove to be weak.

d) Liquid fuels demand is not only affected by the current year's income and price, but also by their levels in previous years. The Koyk transformation can be used to show that this relationship is captured by the inclusion of the lagged value of D in equation (4).

Model for oil exporting countries: For Venezuela and Trinidad and Tobago, where oil constitutes a propulsive rather than a restrictive force in the economy, the model was modified so as to include this effect on growth. The changes affect both identity (7) and equation (2) as explained below. Assuming net capital inflows and changes in reserves are zero or balance out, the balance-of-payments identity is:

$$X = M = X_1 + X_2 = M_1 + N$$

where:

X_1 = exports of liquid fuels
X_2 = non-liquid fuels exports
M_1 = imports of liquid fuels

we derive therefore that:

$$N = X - X_1 - M_1$$

but

$$M_1 = (D - L)P$$

therefore

$$N = X - X_1 - (D - L)P$$

Equation (2) becomes:

$$I_t = \beta_{21} + \beta_{22}X_t + \beta_{23}Y_{t-1} \tag{2a}$$

Since Trinidad and Tobago imports crude for both its refineries and for exports, equation (7) becomes:

$$N_t = X_t - X_1 - M_1 - \left\{D_t - (L_t + M_1)\right\} P_t \tag{7b}$$

B. REGRESSION RESULTS

Due to space limitations the estimated coefficients and other statistics are not presented here. Instead, goodness of fit statistics for the model as a whole in each country are presented below.

The results were, for the most part encouraging. The model behaved well for all countries with the exception of Barbados, Chile and Haiti, where the convergence conditions for the simultaneous equation system as determined by $\beta_{22} > 0$, $\beta_{32} > 0$, $\beta_{33} > 0$, $\beta_{34} > 0$, $\beta_{35} > 0$, and $\beta_{43} > 0$, did not hold.

In order to capture the goodness of fit of the model as a whole, a historic simulation was performed for the period 1962-1975 and the resulting values of I, Y and D were compared with their actual values. The statistic used to measure the goodness of fit was a measure of the deviation between fitted and actual values of endogenous variables(1) to determine what part of the total forecast error is explained by the equation in consideration. Results for all the countries are presented in Table 1 below.

Table 1

GOODNESS OF FIT HISTORIC SIMULATION
(1962-1975)

Countries	Measures of deviation(1)			Weighted(a) Measure of deviation
	Demand	Income	Investment	
Argentina	0.9806	0.9820	0.8705	0.9738
Barbados	0.6976	0.6704	0.4542	0.6511
Bolivia	0.9744	0.9927	0.8979	0.9882
Brazil	0.9626	0.9530	0.8798	0.9439
Colombia	0.9396	0.9811	0.7722	0.9742
Chile	0.8782	0.8182	0.3175	0.8082
Costa Rica	0.9592	0.9966	0.9565	0.9940
Dominican Rep.	0.9046	0.9380	0.8953	0.9326
Ecuador	0.9478	0.9476	0.9029	0.9425
El Salvador	0.9357	0.9939	0.6975	0.9835
Guatemala	0.9719	0.9901	0.8022	0.9856
Guyana	0.9053	0.9610	0.6780	0.9154
Haiti	0.9553	0.9165	0.8465	0.9090
Honduras	0.9438	0.9827	0.7365	0.9644
Jamaica	0.7715	0.8353	0.6258	0.8110
Mexico	0.9862	0.9963	0.9898	0.9959
Nicaragua	0.9476	0.9713	0.4955	0.9235
Panama	0.9587	0.9491	0.7704	0.9228
Paraguay	0.8888	0.9695	0.7870	0.9507
Peru	0.9329	0.9823	0.5491	0.9569
Trinidad & Tobago	0.8991	0.9661	0.4477	0.9388
Uruguay	0.8261	0.9338	0.3668	0.9102
Venezuela	0.9782	0.9812	0.9445	0.9748

a) Computed as an average, using errors explained by each equation as weights.

---

1) This statistic uses the R^2 definition of deviation with the simulated substituted for the fitted value.

The majority of the countries studied showed a good fit of the predicted values of D, I and Y when related to actual values. The Demand and Income functions /equations (3) and (4)/ tend to predict better, however, than the investment equation. For those countries where this equation presents poor measures of deviation, this means that the structure of the model does not apply rigorously to the country's particular historic behaviour. In fact, these countries show a poor (or even negative in some cases) relationship between N and I, and therefore other ways of modelling their economies even at this highly aggregated level must be found; further research and new model specifications should therefore be tried for these countries.

C. FORECASTS

Rather than making a deterministic projection of future values of the endogenous variables, a range of values for each future year were predicted, each being assigned a particular probability of occurrence. It should be pointed out that only the mean value out of this range is comparable to the value that would be produced in a deterministic fashion.

When forecasting on the basis of simultaneous equation models, two types of errors are common: (1) errors in the forecasting of exogenous variables and (2) errors of the model itself (hereafter referred to as errors type 1 and type 2).

Errors type 1 frequently appear in forecasts in the energy field. In order to remedy problems arising from them, "scenarios" are often constructed considering alternative value of input parameters that cannot be foretold with certainty.

An alternative approach to the use of "scenarios" (which generally reduce the information content of the forecast), is to construct probability distributions for the exogenous variables and use these to define unconditional confidence intervals for the forecasted endogenous variables. In addition, the effects of the likely error of the model itself - error type 2 - can also be included in these confidence intervals. Since this approach yields a forecast that is not conditional on arbitrarily chosen "scenarios", it was adopted in this study.

The following paragraphs describe: (a) how the probability distributions for the exogenous variables were defined, (b) what simulation techniques were used to introduce errors type (1) and (2) in the forecast, and (c) how the two sources of error were combined.

a) Probability distributions for exogenous variables.

The variables for which probability distributions were constructed were M, P and L. It was assumed that H (population) can be forecasted well enough to consider it a deterministic variable.

i) M (_imports_) was forecasted by way of a regression against time. Assuming a correct equation specification, the standard error of that regression defines a normal distribution of the error, which was used to establish the probability distribution of future values of M. The mean of the distribution for any given year is the value computed from the regression equation, and its standard deviation is given by the standard error of estimate.(1)

ii) Since variable P (_petroleum prices_) is uncertain and depends on many unmeasurable factors, there is no way of assigning it a probability distribution. Hence values of the distribution were assigned by judgement.

It was assumed that the expected value of future prices would remain constant in real terms and that with 80 per cent probability the price would not increase or fall by more than 10 per cent in real terms.(2) That is, a 10 per cent change in price defines an 80 per cent confidence interval. A normal distribution was used to describe this assumption, hence allowing the probability, albeit small, of vastly larger changes. A strong serial correlation between prices of successive years was assumed, so a 0.9 correlation coefficient was established between them.

iii) The probability distributions of L (_domestic production_) were also judgementally assigned. After careful scrutiny of all available information regarding oil related investment programmes of the oil producing countries, triangular (skewed upwards and downwards), trapezoidal, uniform and normal probability distributions were used to estimate future levels of oil output for each year in the forecasting period. These were also serially correlated with a 0.9 correlation coefficient.

b) Simulation technique to introduce errors type (1) and (2).

For each forecast year and exogenous variable a probability distribution is defined:(3)

---

1) This result is an underestimate of the standard deviation of the probability distribution, for it is held constant, regardless of the year for which the forecast is made, while in fact the standard deviations ought to depend on the difference between the values of the independent variables used in the forecasts and the means of the observations used in estimating the equations.

2) Since a normal distribution was used, this implies a standard deviation of 12.8 per cent.

3) For explanation purposes only, M and the normal distribution will be used. However, the same analysis applies to P and L and their particular distributions.

i.e. P_{rt} = f (M_t)

where:

P_r = probability of outcome of M

M_t = value of imports at time t

In this function each possible value of M is assigned a probability out-come. In the case of a normal distribution, its shape is completely de-fined at each point in time if values are given for its mean and standard deviation,

i.e. $M \sim N (\hat{M_t}\sigma)$

where M_t is the mean of the distribution and is calculated thus

$$M_t = \hat{\beta}_{11} + \hat{\beta}_{12} T$$

$\hat{\beta}_{11}$ and $\hat{\beta}_{12}$ are the estimated coefficients of regression (1), T is the year for which the distribution is being calculated, and σ is the standard deviation of M, calculated thus

$$\sigma = \sqrt{\frac{\sum_{t = 1962}^{1975} (M_t - \hat{M_t})^2}{n}}$$

where:

M_t = the actual values of M

$\hat{M}$ = fitted values of M

n = number of observations or periods

Once the probability function is defined in this way, a particular shape is arrived at for each time period, as seen in Figure 1.

Figure 1

P_r = f(M_t)

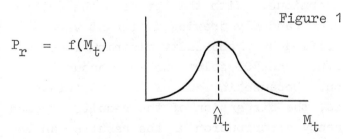

The horizontal axis shows dif-ferent values that M_t can take in period t, and the vertical axis, the probability that each value has of occurring.

In order to be able to randomly generate values of M from this distribution, it is necessary to integrate the function and the result will be the cumu-lative probability function. Its shape resembles Figure 2, where

$$F(M_t) = \int_0^{M_t} f(M_t) d M_t$$

Figure 2

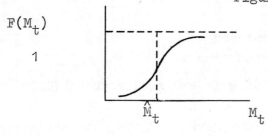

F(M_t)

1

Once the cumulative F is arrived at[1] 500 values between zero and one are introduced into the function F and values of M are arrived at from the implicit form

$$M_t = F^{-1}(M_t)$$

The number of occurrences of these values will also have a normal distribution, i.e. more values will be chosen near the mean $\hat{M}_t$ (value with highest probability) and vice versa.

Errors type (2) were estimated using the results of the historic simulation (see section B). From the deviations between the forecast and observed values in the historic period a standard error of estimate was calculated for equations 2, 3 and 4. These standard errors were used as standard deviations of the probability distributions of the errors in each equation.

While the confidence intervals derived from these distributions apply to each equation independently, they nevertheless describe errors in the entire system, i.e. errors due to a less than perfect fit for each equation, and errors in variables computed by other equations (the standard errors computed in this fashion are therefore larger than the regression standard errors of estimate for each equation).

c) Combining the errors of the model and the errors of its exogenous variables.

The task of combining both kinds of forecasting error was accomplished using the Monte Carlo simulation technique. With this process the entire model was solved 500 times, each time randomly drawing different values for those variables which were specified by probability distributions, and for the error terms of each equation. The 500 solutions were accumulated and formed a frequency distribution. Because in each of the 500 trials both types of errors were simulated, the dispersion of the results stemmed from their combination. The frequency distribution of the results can be used to define any confidence interval. An 80 per cent confidence interval was chosen to present the results of the model.

---

1) In the case of the normal distribution, integration is not possible so linear approximations to the function were used.

D. THE UNCONSTRAINED FORECAST

The main thrust of the unconstrained model is to detect the presence of income effects in the primary forecast when compared with this benchmark.

The model specification in this case is:

$$\text{Log } Y_t = -\beta_{71} + \beta_{72}T \tag{9}$$

$$\text{Log } D_t = \beta_{81} + \beta_{82} \log Y_t + \beta_{83} \log P_t + \beta_{84} \log D_{t-1} \tag{10}$$

$$Eh_t = \beta_{91} + \beta_{92} Y_t \tag{11}$$

$$E1_t = \beta_{101} + \beta_{102} D_t \tag{12}$$

The model differs from the primary one in that it is completely recursive and values for D and E result from the simple logarithmic extrapolation of Y and the projection of deterministic values for P. Since equation (9) was estimated for the period 1960-1975, the resulting trend line is frequently some distance away from the last observation. To remedy this, the trend line was shifted by changing intercept β_{71} so as to make the line pass through the last available datum on GNP. Thus, sudden discontinuities between historic and forecast periods are avoided while preserving the benefit of a larger number of observations in the regression.

E. SOURCES OF DATA

Data related to energy and liquid fuels were taken from: United Nations, World Energy Supplies (Series J), Nos. 18 and 20. Balance-of-payment data are from the IMF computer tapes of the Balance of Payments Yearbook. National accounts (GNP and I) and other data are IDB figures based on information from national statistical offices.

CANADIAN ENERGY DEMAND OUTLOOK, 1985 AND 2000

by

W.A. Bain

This paper addresses three main subjects: recent changes in methods used by Imperial Oil Limited in preparing energy projections; the Canadian energy demand outlook in the medium term to 1985; and some preliminary thinking about how to approach the very long range demand outlook to 2000 and beyond.

ENERGY PROJECTIONS

Methods used prior to 1973

During the 10 years before 1973, our company's projections of energy requirements were based on historic relationships between energy consumption and a number of demographic and economic variables. We had up to 20 years of historical data on energy demand for the normal end-use sectors in the five major regions of Canada. A forecast of Canada's economic performance was used in combination with historical energy/economic correlations to develop the energy projections. To reduce manual effort and ensure consistency between assumptions and economic/energy output, several computer models were developed. All energy costs were assumed to remain constant in real terms and the main areas of judgement or uncertainty, aside from the economic projections themselves, were the pace at which coal would lose market share and natural gas would capture market share. Supply constraints were not seen as a problem.

These old relationships and underlying assumptions were made obsolete by the dis-equilibriums beginning in 1973. Forecasts have necessarily been very judgmental in recent years. By the early 1980s, the data base may again be sufficient to permit at least a partial use of econometric modelling techniques. Information collected in the years prior to 1973 is unlikely to be useful in projecting what might happen in the future. Real energy price increases, new technology, possible supply constraints, lower economic growth and the emergence of energy into the political arena all serve to reduce the value of pre-1973 data.

Current methods

Imperial's current methods begin with an analysis of the end-use energy demand in each major consuming sector based on such underlying factors as population, economic activity, energy prices and conservation initiatives. This demand is divided among competing fuels according to historical market share, as modified by the relative prices and availability. Finally, the primary energy required to supply these fuels is calculated.

The residential sector provides a good example of the reasons for changing the method used to project energy demand. The energy forecast for this sector has a significant element of uncertainty, even in the short and medium term.

When energy was cheap, Canadian houses were insulated to minimum levels. The 1971 building code called for 10 centimeters of insulation in the ceiling, for an insulation rating of R-7. The upper curve on Figure 1 represents a normalized projection of residential energy use, assuming a continuation of past relationships with economic and other parameters. This "trends continue" case shows a growth rate of 2.5 per cent annually reflecting an average building rate of approximately 200,000 units per year and 0.5 per cent annual increase in energy use in existing houses.

There is a very large potential to increase efficiency versus past trends. The national Building Code was raised in 1975 (to R-20 in the ceiling for example) and is in the process of being raised again to a range of R-28 to R-33, depending on climate, an economic level according to the Government's perception of energy costs. The new code could improve the heating efficiency of new houses by 50 per cent over the 1971 standards. The first adjustment on Figure 1, labelled "New Code" assumes that this opportunity is fully implemented.

The second adjustment, labelled "Retrofit" assumes that existing houses achieve two-thirds of this improvement. There are indications that about 5 per cent of today's houses are being brought up to this level each year. In theory, the above two steps could almost eliminate growth in space heating needs in this sector for some time to come.

The last two adjustments indicate the potential reductions from improved furnace efficiency, an assumed shift to more electric heating, heat pumps, and the use of solar energy.

Figure 1

PROJECTION RANGE FOR RESIDENTIAL ENERGY CONSUMPTION
(1977 = 100)

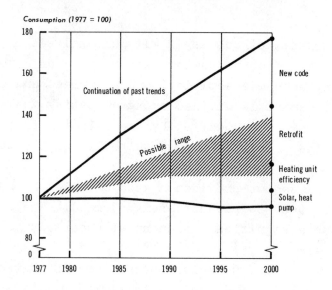

It is expected that the efficiency steps described above will be implemented in the order discussed. As the heat lost by a house is reduced
by better insulation and sealing, the internal sources of heat - people,
appliances, water heaters and passive solar, will provide an increasing
proportion of the space heating requirements. Solar heating will face an
uphill economic battle, since the other improvements will significantly
reduce the amount of energy purchased for space heating, thus reducing the
justification for an active solar system.

There is still a lot to learn in the residential sector, but it is
reasonable to assume that most of the potential in the first two categories
(New Code, Retrofit) will be achieved. The cross-hatched area on Figure 1
represents the range used by Imperial in its planning and analysis.

PROJECTION OF CANADIAN ENERGY DEMAND

In total

As shown on Figure 2, real GNP grew at 5.4 per cent annually in
Canada and energy consumption grew slightly faster at 5.7 per cent over
the 1965-1973 period. This ratio of 1.06 to 1 was actually increasing and
looked like it would level out at 1.1 to 1, which is the ratio that we have
used for our trends continue projection.

Figure 2

PROJECTION OF CANADIAN ENERGY DEMAND TO 1985

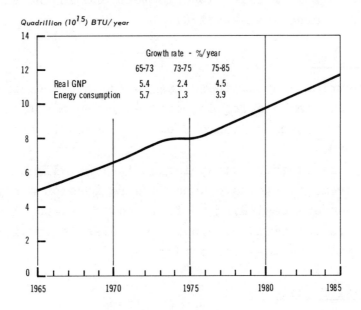

During 1974 and 1975, real GNP grew at only about 2½ per cent per year and energy consumption averaged only 1.3 per cent. We still do not have final data for 1976 but it looks as though the economy grew at about 4.9 per cent and energy use increased by about 3 per cent. For 1977 results to-date indicate that both real GNP and energy may grow about 2 per cent.

In the future, real GNP is assumed to grow at 4.5 per cent per year to 1985. Current energy consumption of about 8.5 quadrillion BTUs (Q) is expected to reach 11.7 Q in 1985. The 3.9 per cent annual growth rate in energy divided by 4.5 per cent economic growth rate yields a ratio of 0.87 versus the historical ratio of 1.1.

Figure 3 illustrates the breakdown of total demand by end-use sector. Each of these sectors is described below:

Residential/Commercial sector

The energy efficiency activities described in the previous residential example also apply to the commercial sector. One of the major differences between the two is that institutional and technical factors are expected to slow the pace at which the efficiency of existing commercial buildings is improved.

Transportation sector

Major changes are projected for the Transportation Sector. More than half of the savings will result from the Canadian Government's adoption of

469

automotive efficiency targets equivalent to those adopted by the United States. Improvements in airline load factors and in the efficiency of commercial trucking are also expected.

Industrial sector

In the Industrial Sector, there are 14 Industry/Government task forces studying and setting targets for improved energy-use efficiency in key industries. Individual targets range from 3 per cent to over 20 per cent improvement by 1980.

We believe a reasonable extrapolation to 1985 will show an overall 18 per cent improvement. We have done some regression analyses including recent data, and we are beginning to feel that a demand elasticity of -0.25 is reasonable for manufacturing. This elasticity yields efficiencies consistent with the task force targets.

Petrochemical feedstocks

The projection of total primary energy requirements in Canada must also give specific consideration to the energy resources required for non-energy uses. A major example is petrochemical feedstock requirements. This sector is expanding very rapidly with requirements projected to triple between 1975 and 1985. This outlook is based primarily on a specific construction schedule for new petrochemical facilities.

Energy industries

Electricity generation is the largest component of energy industry use. Other uses include oil refining and gas transmission. On an input basis (one kilowatt hour is equivalent to 10,000 BTU), electricity generation accounted for 35 per cent of Canada's total primary energy requirement in 1976. This dependence on electricity could increase to 45 per cent in 1985, as electricity gains market share from the direct consumption of fossil fuels. Thus there will be a rapid buildup in the energy resources required to supply electricity.

Oil refineries are expected to make significant improvements in efficiency - some 25 per cent over the period 1972 to 1980.

Although not significant in this time frame, the following two factors, among others, will cause Canadian energy industry requirements to expand significantly in the late 1980s and 1990s. Synthetic crude oil plants, whether of the strip mining type like GCOs and Syncrude, or the in-situ type like Cold Lake, will use the oil equivalent of 0.5 million barrels per day by the year 2000. There are, of course, almost no similar

Figure 3
PROJECTION OF CANADIAN ENERGY DEMAND TO 1985

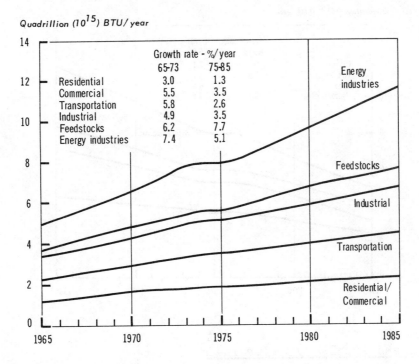

requirements today. Secondly, new natural gas and hydraulic supply will come from remote northern areas; transmission requirements per unit delivered to market will increase significantly.

Summary - All sectors

Figure 4 presents a summary of the changes in energy usage expected to occur by 1985. Approximately 5 points of the 14 per cent overall improvement have already been realised.

Figure 4
CHANGES IN ENERGY USAGE BY 1985
(expressed as % deviation from continuation of trends to 1975)

due to use efficiency (constant real prices for world oil)	CHANGE	
	% of sector	% of total energy
Residential/Commercial	17	
Transportation	21	
Industrial	18	14
Energy Industries	6	

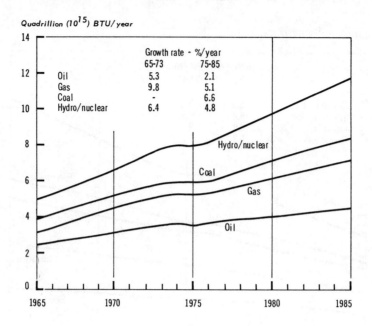

Figure 5

PROJECTION OF CANADIAN ENERGY DEMAND TO 1985

	Growth rate - %/year	
	65-73	75-85
Oil	5.3	2.1
Gas	9.8	5.1
Coal	-	6.6
Hydro/nuclear	6.4	4.8

Energy demand and supply by major fuel

Figure 5 shows how we expect the energy market to be divided among the various fuels. Growth will be supplied largely from hydraulic, nuclear and coal resources used to generate the increasing requirements for electricity.

Natural gas will also be a larger factor in the supply picture than it is today. Gas is generally in plentiful supply and is currently very competitive relative to oil. The projection assumes a strong perception of supply continuity on the part of large customers and distribution utilities. This supply is expected from Alberta and from northern areas during the 1980s.

Oil is expected to retain its position as the only supply source for the transportation market, but will continue to lose market share in the residential, commercial and industrial markets. The result is that oil requirements probably will grow at no more than 2 per cent a year through 1985 (a large part of this growth is accounted for by three new electric generating plants and a new petrochemical refinery).

There is considerable variability around this oil demand as oil is the swing fuel in this time frame. Consumption of oil is sensitive to a number of variables including: the rate of economic growth, the pace of improving energy use efficiency, the attractiveness of electricity for space heating purposes, and the availability of natural gas.

472

THE YEAR 2000

How do we get there from here?

For the next decade, most of the equipment and structures that will
consume energy are now in place. The same is true for the supply side.
Long lead times will prevent rapid change in the total energy picture.
The main unknowns are: the pace and type of economic growth; the economics
of implementing efficiency improvement; and Government actions on the conservation front. It is expected that policy makers will continue to recognise that the energy-economic linkage cannot be uncoupled to any significant extent in less than 10 years. Any mandating of lower energy consumption would result in reduced economic performance with harmful effects
on employment and living standards.

For the very long range, we believe that the passage of time will permit the development and implementation of new technology and the acceptance
of some lifestyle changes. By the turn of the century, a majority of
equipment and buildings then in use will be new and will incorporate the
latest consumption processes. Those of today's plants and equipment still
operating will have been retrofitted to the maximum economic level. The
passage of time should permit us to overcome major roadblocks - but we must
admit to uncertainty as to the pace of change. Some examples of the possible long-range improvements in the efficiency of energy use appear on
Figure 6.

Figure 6

EXAMPLES OF POSSIBLE LONG-RANGE EFFICIENCY IMPROVEMENTS
IN CANADIAN ENERGY USE

Industrial - with today's modern processes

Paper	40
Chemicals	40
Petroleum refining	30

Residential/Commercial

Latest residential building code	50
Commercial construction	50-60

Transportation

Automotive (with latest mileage targets)	50

Scenarios for Canadian energy demand in the year 2000

The possibility of large changes in energy efficiency as suggested on Figure 6 suggest that extrapolation of past trends, or even extrapolation of one's perception of current trends may not produce a useful framework for analysis of long-term energy demand.

One possible alternate technique for projecting long-term demand is to examine scenarios for demand in the year 2000 itself. The scenarios described on Figure 7 are only two out of the many possible, but they are expected to provide a reasonable range for analysis. The demand range in the year 2000 defined by these two scenarios is the oil equivalent of 8 million barrels per day (Scenario A) and 5 to 6 million barrels per day (Scenario B). Simple extrapolation of either the current 4 per cent annual trend, or the historic 5.7 per cent trend would yield a demand range of 11 to 15 million barrels per day in the year 2000.

Figure 7

POSSIBLE SCENARIOS FOR CANADIAN ENERGY DEMAND IN YEAR 2000

Assume real G.N.P. growth averages about 4% per year,
per capital wealth almost doubles

Scenario A

- Towards the upper range of energy use.
- Use efficiency (conservation) grows from 14% in 1985 to 25% in 2000. (The limit of likely improvement with today's world energy cost.)
- Energy consumption grows at an average annual rate of 2.8%, (3.9% to 1985 and 2.2% thereafter).

Scenario B

- Towards the lower range of feasibility with the assumed level of economic growth.
- Conservation reaches 50% assuming that greater economic and government stimuli accelerate technological improvements and the pace of implementation.
- To achieve this, energy growth rate would approach zero in the 1985-2000 period.

A range of 5 to 8 million barrels per day indicated by Scenarios A and B is dramatically different from the 11 to 15 million barrel range generated by projecting past trends. This difference emphasizes the necessity to get away from the use of trend projections alone when examining the very long range.

Despite the difficulties involved, long-range energy projections, for the year 2000 for instance, are becoming increasingly necessary. The long lead times required to institute major changes in the structure of either energy supply or demand mean that decisions taken in the next few years will have their major impact during the 1990s. The International Energy Agency could perhaps assist in the task of planning for the 1990s by encouraging the study of the year 2000 on a world-wide basis.

The impact on energy demand of Scenarios A and B in the year 2000 allow another important observation. Given the appropriate economic stimuli and government initiatives, the market place is capable of maintaining a reasonable energy balance without massive disruptions in life-style or economic growth.

DEVELOPMENT OF ENERGY CONSUMPTION IN WESTERN GERMANY

by

H. Kraft

The worldwide slow down in demand will lead to a lower rate of growth in the Gross National Product (GNP). In addition, the population figure for West Germany is stationary or even falling slightly. Also, productivity in all sectors is rising more slowly than before. Furthermore, the effects of greater rationalisation in the use of energy are likely to be felt. All these factors will limit the future increase of energy consumption and thus also the consumption of mineral oil.

The following data on the overall development of economy and on primary energy consumption (PEC) have been taken as a basis for the period up to 1990:

Period	Average growth-rate p.a. (%)		PEC in Mill t SKE (HCE)
	GNP	PECC	
1975/80	3.5	3.3*)	348 / 415
1980/85	3.2	3.0	415 / 480
1985/90	2.8	2.0	480 / 530

(*) Without allowance for influences of temperature in 1975, the PEC-rate is 3.6 per cent.

The development of demand for energy forecast on the basis of these assumptions is well below previous estimates. However, the demand for energy during the period under review is unlikely to be even lower than the figures given above, as expected substantially reduced growth rates of GNP, of overall primary energy consumption and also of electricity consumption are already taken into consideration in the forecast of demand.

While the contributions of the various energy sources to the meeting of demand can largely be estimated with a fair degree of certainty up to about 1980, they appear to be relatively uncertain for the following decade, at least in some cases. This applies in particular to the contribution they will make towards meeting the demand for electricity. Calculations A and B on meeting demand set out in the attachments take the above into consideration.

Calculation B on Meeting Demand.

Calculation B is based on the following assumptions as regards electricity consumptions.

- Available Power Plant Capacity.
    - Nuclear Energy
        1980: 13,000 MW
        (Power plants - existing and under construction)
        1985: 30,000 MW
        (Including nuclear power plants for which initial partial approval has been granted)
        1990: 45,000 MW
        (In line with the Federal government's energy-policy targets, but delayed by five years)

In the case of nuclear power plants, the load factor has been put at 6,000 hours per annum.

- Hard Coal (Including Dual-Fired Power Plants).
    The presently existing capacity of about 29,000 MW will also be in existence in 1990. Additional capacity constructed as a result of plant expansion projects will be offset by the shutdown of plant. The load factor has been assumed to be about 3,700 hours per annum up to 1985 and about 4,000 hours per annum up to 1990.

- Fuel Oil and Natural Gas.
    In the case of fuel oil and natural gas, it has been assumed that, because of available capacity, an input will be possible that is as high as or even slightly higher than the limit laid down by the electricity-from-coal law.

- The average utilisation of all power plant capacities is put
    for 1985 at 5,000 hours p.a.
    for 1990 at 5,200 hours p.a.

Because of the load distribution conditions these figures constitute the upper limit of what is possible.

Calculation B on the meeting of demand is based on the goals set by present energy policy. It therefore assumes that the plan for nuclear energy will be realised, although with a certain amount of delay.

Calculation A on the Meeting of Demand.

For Calculation A it has been assumed that, outside the sphere covered by the electricity-from-coal law, the various energies will make the same contributions to meeting demand as has been assumed for Calculation B.

The following lower nuclear power plant capacities have been assumed in respect of electricity generation:

1985: 18,000 MW
(No further plant construction in addition to the existing or approved plants)
1990: 30,000 MW (Construction - with five years' delay - of the plants for which initial partial approval has been granted)
This reduced availability of nuclear power plant capacities means that there will be:

in 1985 a shortage of 12,000 MW
= 23 Mill tons HCE
in 1990 a shortage of 15,000 MW
= 30 Mill tons HCE

taking as a basis the forecast electricity consumption and the load distribution conditions.

Because of the already high average utilisation and the load distribution problems, it will not be possible to balance the shortages by higher utilisation of the assumed availability of power plant capacity.

Therefore, in order to fully meet the demand for electricity, it would be necessary to use additional fossil energies for the generation of electricity. In order to assess these possibilities, the following criteria should be borne in mind in respect of the various energies:

- Lignite Coal.

Limits on the supply side; problems of electricity distribution due to binding of the power plants to the extraction area; no additional contribution possible.
- Hard Coal.
12,000 MW (net) up to 1985 and 15,000 MW up to 1990 would be necessary in addition to the planned 6,000 MW. The construction of coal-fired power plants to this extent does not seem realisable (location problems, approval needed for construction, environmental protection, etc.). These coal-fired power plants, to be operated in the base-load range, would have input requirements amounting to 23 mill. tons HCE in 1985 and 30 mill. tons HCE in 1990. To meet these requirements the domestic coal production would have to be raised from 110

to 120 mill. tons. Such an increase in the period concerned does not seem possible.

Regarding the alternative of substantially increasing the use of imported coal, it is questionable whether this would be possible regarding the availability or acceptable from the point of view of energy policy.

Following this it seems improbable that the shortage can be met fully by hard coal alone.

- Fuel Oil and Natural Gas.

A greater contribution by these energies seems possible, from the point of view both of present availability and of existing refinery and power plant capacity. However, this would lead to higher percentages of imported energies being used for electricity generation. Because of the limited oil and gas reserves the question of adequate availability in the long term has to be raised.

If the nuclear power plant capacities available up to 1990 are only those for which partial approval has already been given, there will be an electricity shortage up to 15 per cent of the total needed hard coal, which will have to make a decisive contribution, will not be able to fully close the gap.

Because of the existing power plant capacities, a certain amount of flexibility can be seen with regard to the utilisation of fuel oil and natural gas. If advantage were taken of the opportunities in this respect, the shortage could be reduced temporarily, but not fully eliminated. An increase in the fuel oil input would, in addition, improve the utilisation rate of the domestic refining capacities, regarding the present refinery structure and the supply situation on the international crude oil market.

In view of the fact that the energy industry's decisions on its investments for the eighties have to be taken now and regarding the limitations on the use of other energy sources, the existing impediments to the construction of coal-fired power stations and nuclear power stations will have to be removed as soon as possible. The government's energy policy should create the basic requirements for all energies to make that contribution to energy supplies that is possible for them, bearing in mind existing temporary flexibilities.

Annexes

DEVELOPMENT OF PRIMARY ENERGY CONSUMPTION
1973 - 1977, 1980, 1985, 1990

PRIMARY ENERGY CONSUMPTION BY SOURCES OF ENERGY

in million tons coal equivalents (mtce)

Energy Sources Year	1973	1974	1975	1976	1977	1980	1985 A	1985 B	1990 A	1990 B
Hard coal	84	83	67	71	65	70	(71)	71	(72)	72
Lignite	33	35	34	38	35	34	(34)	34	(34)	34
Petroleum	209	188	181	196	193	210	(224)	224	(238)	238
Natural Gas	39	47	49	51	55	69	(83)	83	(87)	87
Hydroelectricity/ Net imports of elecriticity	8	7	8	4	7	8	(8)	8	(8)	8
Nuclear Energy	4	4	7	8	12	22	35	58	58	88
Others	2	2	2	2	2	2	(2)	2	(3)	3
Total	379	366	348	370	369	415	(457)	480	(500)	530

Nuclear-Energy-Deficit: $\frac{23}{480}$ $\frac{30}{530}$

Hard coal	22.2	22.6	19.1	19.1	17.7	16.9	(14.8)	14.8	(13.6)	13.6
Lignite	8.7	9.6	9.9	10.2	9.5	8.2	(7.1)	7.1	(6.4)	6.4
Petroleum	55.2	51.5	52.1	52.9	52.1	50.6	(46.6)	46.6	(44.9)	44.9
Natural Gas	10.2	12.7	14.0	13.9	14.9	16.6	(17.3)	17.3	(16.4)	16.4
Hydroelectricity/ Net imports of electricity	2.2	2.0	2.2	1.2	2.0	1.9	(1.7)	1.7	(1.5)	1.5
Nuclear Energy	1.0	1.1	2.0	2.1	3.2	5.3	7.3	12.1	10.9	16.6
Others	0.5	0.5	0.7	0.6	0.6	0.5	(0.4)	0.4	(0.6)	0.6
Total	100.0	100.0	100.0	100.0	100.0	100.0	(95.2)	100.0	(94.3)	100.0

Nuclear-Energy-Deficit:

							4.8		5.7	
							100.0		100.0	

PRIMARY ENERGY CONSUMPTION BY SECTORS

in mtce

Sectors/Year	1973	1974	1975	1976	1977	1980	1985	1990
Households, handicraft, agriculture etc. (incl. military)	112	104	104	112	111	123	136	150
Industry	96	96	84	88	86	95	102	110
Transportation	46	44	46	49	51	53	60	65
Final Energy Consumption	254	244	234	249	248	271	298	325
Transformations	95	93	90	96	93	109	140	157
Non-Energy Consumption	30	29	24	25	28	35	42	48
Primary Energy Consumption	379	366	348	370	369	415	480	530

Final Energy Consumption	67.1	66.7	67.2	67.3	67.2	65.3	62.1	61.3
Transformations	25.0	25.3	25.9	25.9	25.2	26.3	29.2	29.6
Non-energy Consumption	7.9	8.0	6.9	6.8	7.6	8.4	8.7	9.1
Primary Energy Consumption	100.0	100.0	100.0	100.0	100.0	100.0	100.0	100.0

FINAL ENERGY CONSUMPTION OF SECTORS BY ENERGY SOURCES

in mtce

Energy Sources/Year	1973	1974	1975	1976	1977	1980	1985	1990
Households, handicraft, agriculture etc. (incl. military)								
Hardcoal	9	9	7	6	6	4	2	2
Lignite	4	4	3	3	3	2	2	1
Fuel Oil	67	57	58	65	61	65	68	70
Gas	12	13	14	16	16	23	29	32
Electricity	13	14	15	15	17	20	25	31
Others	7	7	7	7	8	9	10	14
Total	112	104	104	112	111	123	136	150
Industry								
Hard coal	16	19	15	15	14	12	12	12
Lignite	1	1	1	1	1	-	-	-
Fuel Oil	36	31	28	29	28	27	24	23
Gas	24	26	22	25	24	32	37	39
Electricity	16	17	15	17	17	21	25	30
Others	3	2	3	1	2	3	4	6
Total	96	96	84	88	86	95	102	110
Transportation								
Hard coal	1	1	1	0	0	0	0	0
Motor gasoline	27	27	29	31	33	33	36	37
Diesel fuel	13	12	12	13	14	15	18	21
Electricity	1	1	1	1	1	1	2	2
Others	4	3	3	4	3	4	4	5
Total	46	44	46	49	51	53	60	65

ENERGY INPUT FOR PRODUCTION OF ELECTRICITY BY ENERGY SOURCES

in mtce

Energy Sources/Year	1973	1974	1975	1976	1977	1980	1985 A	1985 B	1990 A	1990 B
Hard coal	34	32	25	31	28	34	(34)	34	(37)	37
Lignite	28	30	30	33	31	31	(31)	31	(31)	31
Nuclear Energy	4	4	7	8	12	22	35	58	58	88
Fuel oil	13	9	9	11	9	11	(11)	11	(12)	12
Gas	11	16	18	17	18	20	(20)	20	(20)	20
Hydroelectricity/Imports	8	8	8	5	7	8	(8)	8	(8)	8
Others	4	5	4	4	5	4	(4)	4	(4)	4
Energy Input	102	104	101	109	110	130	(143)	166	(170)	200
Total available electricity in TWh	309	318	310	334	341	410	(453)	525	(540)	635

Nuclear Energy-Deficit: mtce $\dfrac{23}{166}$ $\dfrac{30}{200}$

TWh $\dfrac{72}{525}$ $\dfrac{95}{635}$

Electricity Deficit: in per cent 13.7 15.0

PETROLEUM CONSUMPTION

in million tons

Sectors/Year	1973	1974	1975	1976	1977	1980	1985	1990
I. Inland Sales								
Naphtha (net) (outside refineries)	5.6	5.0	3.5	4.9	5.3	6.1	8.0	10.0
Motor gasoline	18.5	18.0	19.7	20.6	21.8	22.6	24.0	25.0
Diesel fuel	10.8	10.0	10.3	10.9	11.7	12.3	14.0	16.0
Fuel oil, light	52.0	44.8	44.8	49.0	47.1	50.5	52.0	53.0
Fuel oil, medium and heavy	29.7	24.8	22.4	24.3	22.1	24.0	24.0	24.0
Other products	18.0	17.8	15.4	16.0	15.9	18.7	21.5	24.0
Total	134.6	120.4	116.1	125.7	123.9	134.2	143.5	152.0
II. Military	2.0	1.8	1.6	1.5	1.5	1.9	2.0	2.0
III. Refinery own consumption (incl. losses)	9.4	9.0	8.1	8.9	8.7	10.0	10.5	11.0
IV. Total I. - III. in oil equivalents	146.0	131.2	125.8	136.1	134.1	146.1	156.0	165.0
in coal equivalents	209	188	181	197	193	210	224	238
V. Bunker deliveries	3.7	3.1	2.9	2.8	3.0	3.3	3.5	3.7

ENERGY PROSPECTS FOR THE FEDERAL REPUBLIC OF GERMANY

by

D. Schmitt

After the profound changes on the world energy markets and a deep recession in 1974 and 1975 from which the German economy seems to recover
slower than expected, it is more and more doubted, whether the development
of the energy sector still can stick to guidelines set up in the past.
The views on the direction and necessary size of a change are nevertheless
controversial: On one side the "business as usual" hypothesis, i.e. the
belief that the trend earlier to 1973 will continue has been given up more
or less, because this approach does not reflect the real factors influencing energy demand and supply; on the other side the "doomsday approach"
i.e. the belief that energy consumption cannot and should continue to
grow, because otherwise serious disturbances in meeting energy demand in
the long run, the risks of destroying the ecological equilibrium, and a
threat of human life as such would be unavoidable is dissatisfying, too,
because it excludes adjusting policies in the long run.

Therefore we are not in favour of one of the extreme viewpoints, but
nevertheless we, too, believe that in the long run the energy sector has
to get to an equilibrium position where energy demand has to confine to
the renewable energy potential on the earth. In the short and medium run
we think that a smoothed but continuing growth of energy consumption is
probable.

Taking into account all uncertainties about the real long-term influencing factors on structure and development of the energy consumption
it seems to be necessary to start from the assumption of a relatively
close connection between energy consumption and economic development and
the increase in wealth that goes with it. But it is not valid - as was
often thought in the past - to assume a limitational relation between

these factors.(1) The functional relationships rather tend to be varying and variable with technological progress and changes in price relations, policy measures and tendencies towards a satisfaction of demand. Although this was already true before the oil crisis, these factors now have even got more weight.

But also the assessment of the future development of the exogenous variables - especially the rate of economic growth - differ considerably today from those of the past. One reason for the assumption that e.g. economic growth will be slower is the demographic development in the Federal Republic of Germany. After a high rate of growth during the post-war period (+ 20 per cent) a population decrease by 8-10 per cent now is expected till the year 2000.

Provided that this development is largely autonomous it can be assumed that other determinants of economic growth have at least partly been influenced by the oil price development and effects that were induced by it. The cut in investment activity, amplified by the recession, will reduce the production potential and increase the age structure of the capital stock. This leads to losses in productivity and the competitive position. As far as economic growth can be traced back to technological progress, further growth will be weakened because the implementation of technological progress is tied to new investments to a high degree. In addition, the danger of an unsatisfactory use of the production potential in the case of a diminishing population will negatively affect the propensity to invest once more. This again reduces demand. The development of energy reserves at increasing costs will absorb resources otherwise available for other purposes. On the other side, recycling of money and an increased economic development of energy exporting countries might result in incentives for the economic activity in developed nations. too. In any case, however, energy policy strategies without respect to economic and social costs will be developed more or less in all consuming countries and thus lead to a misallocation of resources and "necessary" protectionistic measures reducing international division of labour. A diminishing world economic growth potential must result in losses in national economic growth especially in an economy like the Federal Republic of Germany so heavily dependent on the world market.

---

1) A detailed analysis of the past and a cross comparison between countries shows remarkable deviations from the average. Therefore, the parallel development of GNP and PER (primary energy requirements) between 1960 and 1973 in the Federal Republic of Germany pointing to an average GNP/PER growth ratio of 1 is more or less the accidental result of a combination of various factors adding up or neutralising each other.

Beside a population decrease and a slower economic growth the future development of energy consumption may be influenced to a large degree, too, by saturation effects in some parts of the energy sector (especially those like transportation and room heating that promoted a considerable growth in the past), technical progress as well as a more economic use of energy and structural changes within the industry induced by the sharp increase of prices and supported by economic and technology policy measures to conserve energy.

On the basis of these considerations we expect only smoothed growth rates of energy consumption in the future and assume that the growth rates of the primary energy requirements (PER) that amounted to about 4.0 per cent/year between 1960 and 1974 may decline to 2.8 per cent/year between 1976 and 1985 and may further decline to 1.0 - 1.5 per cent/year during the time between 1985 and 2000; Case I, see Appendix I and Case II see Appendix II.(1) Basic assumptions for this forecast are:

- undisturbed economic development;
- reduced but remarkable economic growth: in average 3.5 - 4.0 per cent/year between 1976 and 1985 and 2.5 - 3.0 per cent/year between 1985 and 2000;
- energy prices increasing continuously but not abruptly in nominal terms, in real terms remaining constant till 1985 and between 1985 and 2000 increasing by 100 per cent;
- energy policy measures to reduce losses, to save energy and to increase production of indigenous resources being realised as far as still developed (Case I) and added by further measures but not neglecting the costs of these measures, not acting against the fundamental rules of a liberal economic system and not violating the scenario assumptions (Case II).

With the above-mentioned growth rates a decrease of the relation between the growth of energy consumption and the growth of GNP from 0.93 (for the time from 1960 to 1974) to about 0.8 (1976 - 1985) and to 0.4 - 0.5 (1985 - 2000) will ensue. According to this forecast the overall PER will increase from 366 Mill tce in 1974 and 370 Mill tce in 1976 to about

---

1) Case I corresponds to the reference case in the forecast of three research institutes for the Federal Government. Case II is slightly different than the "alternative case" analysed in this study. See: Liebrucks, M., Schmidt, H.W., Schmitt, D.: Die künftige Entwicklung der Energienachfrage in der Bundesrepublik Deutschland und deren Deckung bis zum Jahre 2000, Essen, 1978. (The updated Energy Program of the German Government from December 1977 was based on this forecast.)

475 Mill tce in 1985 and 550-600 Mill tce in the year 2000. Lower economic growth, and/or higher prices and/or further energy policy measures e.g. to conserve energy, assumptions that seem to many people in the Federal Republic of Germany by far more realistic, would result in a possibly far lower demand, increasing to less than 450 Mill tce in 1985(1) and 500-550 Mill tce in the year 2000. Even in this case PER would increase till the end of this century by more than 50 per cent.

Similar considerations hold for the development of the future electricity consumption in the Federal Republic of Germany. After decades of growth rates higher than those of the PER electricity consumption stagnated in 1974 and even decreased in 1975 in absolute terms. Here, too, the question has to be raised whether this development signifies a withdrawing from a so far esteemed energy source or whether this means only a short deviation from the trend. After all decreasing growth rates have to be stated in the electricity sector during the last years, they were only veiled by the expansion of a new sector of consumption, the electric storage heating.

On the basis of assumptions similar to those made above we estimate that in the electricity sector, too, the past growth rates (of more than 7 per cent/year) can no longer be attained.

However, with a sustaining or even reinforced use of clean, environment protecting, easy to handle, highly concentrated and reliable forms of energy that can be transformed into all final energy uses the growth of electricity consumption is estimated to be about 5.0 per cent/year during the time from 1976 - 1985 and 3.8 per cent/year until the year 2000 - by far higher than the growth rates of the total primary requirements.

Even given these growth rates far lower than former projections, electricity consumption will considerably increase on an absolute scale, namely from 335 TWh in 1976 to 520 TWh in 1985 and almost 830 TWh in 2000. This is a growth of nearly 250 per cent (about 4 per cent/year) until the year 2000 and means that by then 45 per cent of the overall primary energy consumption will fall to the share of electricity.

On the other side these results are only valid on the basis of the underlying assumptions. Less favourable conditions for producing and consuming electricity would also lead to lower growth rates in the electricity

1) For a set of scenarios about the future development of the energy sector in the Federal Republic of Germany, see: "Energy Demand Study for the FRG", by D. Schmitt and P.H. Suding, Energy Demand Studies: Major Consuming Countries (MIT Press, November 1976), and Energy Supply Demand Integrations to the Year 2000: Global and National Studies (MIT Press, June 1977).

sector. This development would be likely to come, if the competitive position of electricity in comparison to other energy sources did not improve because of unfavourable changes of relative costs and prices, if pollution control standards affected production and distribution of electricity heavily, if energy policy constrained the assumed economic development and electricity consumption etc. Under such circumstances growth rates of electricity consumption might also fall far below that of Case II (3.4 per cent/year until the year 2000). But if in this case demand grew only up to 650 TWh, this would still correspond to an almost 200 per cent increase compared to today. Electricity consumption might increase even to more than 1,000 TWh in the year 2000 if a development in favour of electricity production and consumption is assumed. This could take place if electricity on the basis of cheap, widely accepted, fast growing nuclear energy would substitute to a large extent fossil fuels with increasing prices.

Thus the future development of electricity demand can only be estimated on the basis of assumptions on important cost parameters of electricity production and the development of the environment for the electricity sector in our economy. This is valid, too, for energy consumption and the ways to meet this demand especially by nuclear energy.[1] If we assume that demand will increase as outlined above, the load curve will not totally change, energy policy measures and cost relations will remain the same as estimated today. An optimization shows that the bulk of the demand in the eighties and nineties especially in one of the fastest growing energy sectors will have to be met by nuclear energy and/or coal.[2] The decisive question, however, is to what extent our knowledge of today is sufficient to describe future developments. In this context, the following questions have to be raised:

- Which amount of various energy sources will be available for the Federal Republic of Germany in the long run?
- What will be the impact of an increasing demand for certain kinds of energy on its price and of the renunciation of other sources of energy on the energy price level?

1) Basically, development of demand and supply have to be estimated simultaneously. However, with respect to the high weight of factors like population, economic growth, income and industrial structure for the energy consumption in our country it seems to be possible to estimate first the demand (under certain assumptions) than to evaluate the possibilities to meet this demand and in case of failure to search iterative solutions.

2) See: Schmitt, D., Junk, H. et alii: Parameterstudie zur Ermittling der Stromerzeugungskosten, Oldenbourg, München, 1978.

- What will be the effects of a use of different energy sources with respect to safety and pollution standards?
- Will pollution control activities affect or delay the development and operation of various types of production, distribution and consumption of energy and reinforce the development of others?
- What will technological progress be like?
- What goals and measures of energy policy as well as research policy have to be considered in the Federal Republic of Germany as well as in other countries and in international organisations and what kind of impact will result on development and structure of energy consumption and supply?

As these questions cannot be answered clearly and unequivocally we have to put up with considerable technical, economic and political uncertainties concerning forecasts of the development of energy and electricity supply.

It seems to raise relatively few problems to assess to what extent we might draw upon domestic energy sources to meet the demand for energy and electricity in the Federal Republic of Germany:

- An extended economic development of domestic water-power is hardly possible because of the natural conditions in our country.
- A noteworthy increase of brown coal production (the cheapest energy resources in the Federal Republic of Germany) which will go beyond present plans might be impossible because of environmental restraints in surface mining; moreover it cannot be excluded that a growing amount of brown coal will be spared for gasification which may be fully developed by the year 2000.
- An increasing use of domestically produced natural gas cannot be expected, because in the nineties sources are likely to begin to deplete.
- Oil reserves in the Federal Republic of Germany amount to a six-month consumption.
- Only an intensified use of domestic hard coal seems to be possible because of huge, but under actual conditions, not economically exploitable reserves. As consumption in other sectors mainly because of negative consumer preferences is decreasing, additional amounts of hard coal might be available especially for electricity generation even without an increase of coal production. New technologies, as pebble bed combustion, will increase efficiency and improve the economic conditions of coal consumption. On the other hand, it has to be considered that, if the supply with coking coal remains scarce

all over the world as many believe still today, an increasing
share of the German hard coal production may be absorbed from the
European and world markets because of its good quality for coke
production. The same is true, if gasification of hard coal possibly on the basis of nuclear process heat, being developed now, will
be introduced on the market by 2000, as is planned. However, as
an increase of hard coal production of 20 per cent (i.e. from 90
to 110 Mill t SKE) is, because of lack of skilled labour, economic
problems and environmental restraints with respect to opening new
(even as substitute for old exhausted) mines, at best realistic,
this energy source, too, can only contribute to meeting a small
part of the huge additional demand. Nevertheless, it cannot be
doubted that the contribution of hard coal especially to electricity production could be raised in the future, its economy(1) and
early investment assumed. With respect to cost characteristics in
coal fired stations it will be particularly peak load and medium
load, where nuclear energy - because of its high fix costs - will
be competitive only under special conditions, if at all.
- Prospects concerning the contribution of "new" energy sources to
  supply our energy needs are even less optimistic. Our knowledge
  of the potential of these sources, the techniques to be employed,
  the costs and pollution problems is rather poor. Experts agree
  that nuclear fusion cannot be expected before 2000. A noteworthy
  use of tidal energy, of ocean thermal gradient and the waves will
  fail in the Federal Republic of Germany because of the topographic
  and climatic conditions. The use of geothermal energy will be confined to vulcanic areas, which are very scarce in the Federal
  Republic of Germany. Only the hot-rock-technology - not yet developed - might allow a wider use of geothermal energy. The prospects for solar electricity production (as well as wind power) are
  in spite of the huge potential also seen relatively pessimistic because of climatic conditions i.e. low concentration, diffuse radiation and lack of availability. As technologies for the use of solar
  energy are still in a nascent state it is quite uncertain, what
  technical progress will be made in this field. We do not yet know
  whether the development of polykrystaline silicon celles can be regarded as a technical breakthrough which might increase the market
  potential of solar power considerably. On the other side use of
  solar energy for heating purposes is expected to become economical

---

1) In case of lacking private profitability subsidies would work into the
   same direction.

with mean energy price rises and cost decreases. Nevertheless, the renewable energy resources are unlikely to contribute with significant quantities to supply the additional energy needs during this century. For the Federal Republic of Germany we estimate the share of "others" not to exceed 5 per cent of PER. These kinds of energy are likely to become increasingly important in the 21st century. They have to be developed now. But development and market penetration needs more time than often assumed in enthusiastic studies on the share of these kinds of energy in future energy supply.

Thus it follows that an increasing energy demand in the Federal Republic of Germany till the end of this century under realistic assumptions can only be met by an intensified use of traditionally imported energy sources and nuclear energy. Being dependent on imports from the world market creates in regard to the problem to supply this demand worldwide dimensions, as the future development of consumption and supply on the world markets will determine the supply conditions for a country like the Federal Republic of Germany, too. All uncertainties in this respect taken into account by far the most important factor in the future will be the growing process of political influence on the energy sector worldwide. Especially those kinds of energy like oil products that up till now supplied by far the biggest part of our energy consumption seem not to be possible to increase production steadily. Thus to believe that growing quantities might be held at the disposal of our country under favourable economic conditions is highly hypothetic beyond the year 2000, although it is not sound to judge future oil and gas supply by the reserves known today that will be depleted very soon. We must rather find out, how many reserves we can get, if more capital is used for the search and development of new deposits within the next decades. But model calculations that were carried out for oil by us in the Workshop on Alternative Energy Strategies(1) show that on the basis of by no means pessimistic assumptions oil production could culminate already in the early nineties and decrease rather quickly thereafter. Besides, (some) oil producing countries might limit production in order to optimise the returns from their resources. This would result in possibly considerable increases of oil prices. Other cheap energy sources being scarce, supply flexibility low, markets characterised by monopolistic structures and costs for the development and operation of other kinds of energy expected to rise, it is

1) WAES, Energy Global Prospects 1985-2000, McGraw Hill, 1977.

more and more doubted, whether oil will soon lose its function as a price leader on the energy market.

The use of huge tar sand and oil sand reserves as well as oil-shale as substitutes for traditional oil production, estimated realistically, is only possible at costs, which are twice the current oil price, and seem to create to some extent heavy environmental problems and can hardly be realised earlier than in the nineties.(1) Last but not least it has to be assumed that the availability of heavy fuel oil (the major oil product for electricity generation) will decrease with rising oil prices because of enforced cracking of residual.

Similar considerations are true for natural gas. Supply conditions for natural gas are estimated to be far more favourable than those of oil and increasing energy prices might as well as technical progress (heat pump) ease the development of so far uneconomic reserves and markets. But gas reserves are limited, too (production peak is estimated by WAES between 2010 and 2020), and the bulk of all reserves for export to Europe is either in the hands of the URSS or the OPEC countries which are believed to maximise profits for both kinds of energies. The firm linkages between oil and gas are shown by the fact that gas prices are closely tied to the oil price. In addition, the transportation of natural gas in pipelines and/or liquefaction systems from far distant areas to Europe and penetration into the market gets more and more costly and restricts the use of natural gas to preferred purposes and regions.

Therefore the main alternatives for satisfying the long-term energy demand for electricity generation are imported coal and nuclear energy. As far as the potential of reserves is concerned there are no restrictions at least for the next decades, as recently published studies show.(2) The huge deposits of coal as well as reactor fuels are distributed over a variety of countries. Though there is a high concentration in only a few countries (80 per cent of the technically producible coal reserves are held by four countries: URSS, China, Poland and the United States, and over 80 per cent of the assured and additionally assumed reserves - up till costs of 30 \$/lb U_3O_8 - of western world in five countries) it has to be stated that under a global view resource constraints do not exist

1) The reserves are concentrated in only a few countries (mainly United States, Canada and Venezuela) and the production in at least two of these countries will at first lower their import requirements and thus ease more than by exports the supply problems of the rest of the world.

2) See: Bundesanstalt für Geowissenschaften und Rohstoffe (Federal Institution for Geological Research and Resources): Das Angebot von Energie-Rohstoffen, Hannover 1976.

at least for the next decades. The bulk of reserves are to be found in countries that do not export oil, most of them belong to the industrialised nations. Much more relevant however is whether reserves will be developed quick enough and under what conditions these will be available for countries like the Federal Republic of Germany.

If coal and nuclear energy shall contribute to energy supply to the expected degree, high investments in exploring, developing and setting up production facilities as well as the necessary infrastructure especially in the coal sector with the necessity to move large quantities over long distances have to be made in the years to come. This is all the more urgent as these investments will need seven to ten years until production can be started.

The amount to which these two energy sources will contribute to the future supply in the Federal Republic has mainly to be decided on the basis of their cost and price relations (including) external effects concerning security of supply, balance of payments problems and contributions to economic growth). Another important question is that of the quantitative availability; this problem should be included in appropriate prices.

After considerable cost increases in the nuclear energy sector during the last years (uranium from 7 $/lb U_3O_8 up to more than 40 $/lb, enrichment from 30 $/kg SW to 60$-100$, plants from 1,000 to more than 2,000 DM/KW) the economy of nuclear power plants compared to power plants on the basis of imported coal has become questionable. In addition there are uncertainties concerning certain cost elements in the nuclear fuel cycle like reprocessing or waste-disposal. Additional safety standards (like bursting strength) would - similar to further delays - lead to additional cost increases. On the other hand advantages of improved standardization might be realised and real price decreases of uranium be possible. Cost and price increases can also not be excluded as far as imported coal is concerned. Costs for coal burning facilities have increased, too, especially because of stack gas desulphurisation which is compulsory by environmental legislation now. The most important factor is the development of long-term coal prices, especially if the demand on the world market, which is still narrow today, would accelerate.(1) In this connection we have to ask whether in the face of an increasing domestic demand (for electricity production but industrial purposes and gasification, too) in potential coal exporting countries like the United States coal production (in spite of huge reserves) can be extended fast enough at all,

---

1) The total coal trade runs up only to about 10 per cent of world coal consumption, coal trade, thus is by order of magnitudes lower than oil trade.

respectively, whether the impending environmental protecting restrictions and resistance will allow an increased production, especially for export purposes. As the Federal Republic of Germany will not be the only country with (an increasing) demand for coal, even steep price increases due to production shortages cannot be excluded especially in the case of a renunciation of nuclear energy. An intensified use of imported coal in the Federal Republic of Germany would of course presuppose that the current import restrictions would be changed (this presupposes a solution of the huge problems of our indigenous coal industry, too).

A satisfactory solution or acceptance of pollution problems (CO_2, NO_x) which go with the use of coal is another crucial point. Up till now there is a lack of studies which would include all economic and social costs of the use of nuclear energy versus coal and allow thus satisfying comparisons of the real economy of the use of certain kinds of energy. So the environmental impacts of an increased use of nuclear energy in the Federal Republic of Germany are intensively discussed at the moment whereas the long-term impacts of CO_2-, SO_2- and NO_x-emissions are nearly neglected, which could induce heavy problems especially in an energy economy without nuclear energy. If work on new technologies (fluidised bed combustion, e.g.) would come to a success the pollution problem of an increased use of coal especially for electricity production might lose weight, furthermore an improvement of the efficiency might counteract the rising cost trend, nevertheless problems of an increased use of coal will remain.

An evaluation of the relative price and cost developments between nuclear energy and coal including external effects therefore is extremely difficult, especially because of the long time to be regarded and our limited knowledge about the future development and impacts of main parameters. Assuming that the prices for nuclear fuels, the weight of which is relatively low with respect to the total costs, increase as those for coal this would influence overall costs for nuclear energy less than similar price increases for coal as long as inflation rates are lower than price increases for the fuel.

With respect to these tremendous uncertainties, forecasts can only describe possible paths of development and cannot represent more than subjective probabilities. But it is undoubtable that neither the growing electricity nor energy demand can be met without the use of nuclear energy and/or imported coal. Given the conditions of the Federal Republic, it seems, at least from a today's point of view, that the biggest part of the electricity demand in the year 2000 - the actual licencing problems being solved - will have to be met by nuclear energy. This is the result of a just finished analysis on costs of electricity in the Federal Republic of Germany;(1) together with domestic hard coal, imported coal, however, will also noteworthily contributed to electricity production, especially for medium and peak load, and possibly to gasification.

Supposed that 55 per cent of the electricity demand shall be met by nuclear power this represents for Case I described above an energy equivalent of more than 160 Mill tce the year 2000; this means that almost 30 per cent of the overall energy demand would be satisfied by nuclear energy. In this case the construction of about 60 power plants will be necessary. Each nuclear power plant that will not be set in operation, will require an additional import of hard coal of about 2.5 Mill tce per year.

Besides the problems discussed earlier supplying the electricity sector with imported coal raises questions regarding the additional balance-of-payments burden. Supposed the actual (real) prices for imported coal would remain constant (increase by 100 per cent) the coal import of 2.5 Mill t per year to fuel 1300 MW (2 x 650 MW) would require about 250 Mill DM (500 Mill DM) in foreign currency. For a capacity of 80,000 MW this amounts to 17 (34) billion DM. To fuel a similar nuclear capacity only an amount of about 2.0 (4.0) billion DM would be necessary. This is due to the fact, that to fuel nuclear power plants only imports of uranium seem to be necessary whilst a high share of the nuclear fuel cycle could be supplied indigenously.

Even though we do not believe that balance-of-payments arguments should play a very important role in the energy discussion in this country for the years to come the magnitude of foreign exchange necessary to pay for growing energy imports could impose serious problems in the long run. Energy imports in any case will be a major factor in the balance of payments, since other energy imports will certainly continue, which are difficult to substitute because of high consumer preferences and/or costs in the traffic, agriculture and non-energetic sectors or to sustain a clean gas supply using an existing infrastructure.

Demand for foreign currency could be additionally diminished by the development of fast breeder and converting technologies. Even assumed the open problems to be solved these technologies will probably not be introduced in a large scale in our energy system before the end of this century. Therefore, the above-mentioned impacts on the balance of payments should not be expected earlier, too. On the other hand there is no

---

1) For detailed information see the above-mentioned "Parameterstudie zur Ermittlung der Stromerzeugungskosten" by Schmitt, Junk et al.

doubt that only a satisfying implementation of nuclear power of the first generation into our economy will enable industry, public utilities as well as government authorities to finalise the ongoing development of advanced systems which include new uses of nuclear energy, too.

Finally, the price stabilizing effects connected with the use of nuclear energy, the impacts on economic development and employment and the positive effects on the security of supply must be taken into account. Thus only the considerations of all advantages and disadvantages of use or non-use of a certain kind of energy will enable rational evaluations and decisions.

In a world:

- where energy demand will further on increase,
- where the special energy needs of the less-developed countries with their increasing political pressure has to be supplied with high priority,
- where the supply conditions render more and more difficult,
- where the given options are limited and characterised partly by long lead times and high costs and are more or less available for the different countries

it cannot be justified in a country like Germany to give up even one of the few real energy policy options without sufficient justification: "Non use", coal and nuclear energy. This holds especially, the more capital-intensive and research-intensive these options are as it is the case with a strongly nuclear-based electricity and energy sector. Because of its high level of economic development and the structure of its energy consumption the Federal Republic of Germany faces in this area a comparatively favourable situation.

Table 1

PRIMARY ENERGY REQUIREMENTS BY SECTORS 1975 - 2000
IN THE FRG (CASE I)

- in 1,000,000 tce -

	1975	1985	1990	2000
- Final consumption	234.0	308.3	327.5	343.0
- Industry	84.0	111.2	120.5	137.0
- Traffic	46.2	56.5	60.0	62.0
- Residential and commercial (inc. military)	103.8	140.6	147.0	144.0
- Non-energy uses	23.5	39.4	43.1	45.0
- Final consumption+ non-energy uses	257.5	347.7	370.6	388.0
- Energy sector and transformation losses	90.2	134.8	159.4	212.0
- PER	347.7	482.5	530.0	600.0

Table 2

PRIMARY ENERGY REQUIREMENTS IN THE FRG 1975-2000
BY KINDS OF ENERGY - CASE I

- in 1,000,000 tce -

	1975	1985	1990	2000
PER	347.7	482.5	530.0	600.0
- Hard coal	66.5	74.6	80.1	102.0
- Lignite	34.4	35.3	35.4	38.0
- Oil	181.0	222.9	225.6	162.0
- Natural gas	48.7	7.8	89.5	97.0
- Nuclear	7.1	49.9	83.2	163.0
- Water / net imports of el.	7.8	9.6	11.2	13.0
- Other	2.2	2.4	5.0	25.0

Table 3

ELECTRICITY SUPPLY AND CONSUMPTION
IN THE FRG 1975 - 2000 (CASE I)

TWH	1975	1985	2000
- Final consumption	253.0	443.0	760.0
- Industry	125.0	215.0	390.0
- Traffic	8.9	14.5	28
- Residential and commercial	118.6	213.5	342
- Losses	56.6	91.0	140
Total	309.6	534	900
- Hard coal	73.3	111	170
- Lignite	85.3	96	100
- Fuel oil	30.0	96	100
- Natural gas	60.0	77	10
- Nuclear	21.4	156	510
- Water / net imports	24.9	30	40
- Others	14.4	22	35

APPENDIX II

Table 1

PRIMARY ENERGY REQUIREMENTS BY SECTORS 1975 - 2000
IN THE FRG (CASE II)
- in 1,000,000 tce -

	1975	1985	1990	2000
- Final consumption	234.0	302.3	320.7	319.0
- Industry	84.0	108.4	117.2	127.0
- Traffic	46.2	56.0	59.5	59.0
- Residential and commercial (incl. military)	103.8	137.9	144.0	133.0
- Non-energy uses	23.5	37.7	41.0	40.0
- Final consumption + non-energy uses	257.5	340.0	361.7	359.0
- Energy sector and transformation losses	90.2	130.0	153.3	191.0
PER	347.7	470.0	515.0	550.0

Table 2

PRIMARY ENERGY REQUIREMENTS IN THE FRG 1975-2000
BY KINDS OF ENERGY - (CASE II)
- in 1,000,000 tce -

	1975	1985	1990	2000
PER	347.7	470.0	515.0	550
- Hard coal	66.5	74.0	77.5	105
- Lignite	34.4	34.2	36.0	42
- Oil	181.0	217.2	221.5	145
- Natural gas	48.7	87.5	89.5	90
- Nuclear	7.1	45.5	74.5	126
- Water / net imports of el.	7.8	9.3	10.5	12
- Others	2.2	2.4	5.5	30

Table 3

ELECTRICITY SUPPLY AND CONSUMPTION IN THE FRG
1975 - 2000 (CASE II)

TWh	1975	1985	2000
- Final consumption	253.0	422.4	633
- Industry	125.5	205.0	333
- Traffic	8.9	13.8	25
- Residential and commercial	118.6	203.6	275
- Losses	56.6	86.6	117
Total	309.6	509	750
- Hard coal	73.3	111	180
- Lignite	85.3	92	110
- Fuel oil	30.0	40	10
- Natural gas	60.0	75	35
- Nuclear	21.4	142	345
- Water / net imports	24.9	29	40
- Others	14.4	20	30

AVERAGE GROWTH RATES / YEAR

	Case I		Case II	
	- Relatively high economic growth and electricity and nuclear consumption - Restrained policy		- Medium economic growth relatively low electricity and nuclear consumption - Vigorous policy	
	1985/1975	2000/1985	1985/1975	2000/1985
1) GDP	4%	3%	3.5%	2.5%
2) PER	3.3%	1.5%	3.06%	1.05%
3) $\frac{2}{1}$	0.85%	0.5%	0.886%	0.44%
4) Electri. consumption	5.6%	3.5%	5.1%	2.6%
5) $\frac{4}{1}$	1.4%	1.17%	1.46%	1.04%
6) Final energy consump.	2.8%	0.7%	2.6%	0.4%
ind.	2.8%	1.4%	2.6%	1.1%
Traffic	2.0%	0.2%	1.9%	0.3%
r. + c.	3.1%	0.2%	2.9%	0.2%
7) Non-energy uses	5.3%	0.9%	4.8%	0.4.
8) Energy sector losses	4.1%	3.1%	3.7%	2.6%

OECD SALES AGENTS
DÉPOSITAIRES DES PUBLICATIONS DE L'OCDE

OECD PUBLICATIONS, 2, rue André-Pascal, 75775 Paris Cedex 16 - No. 40.977 1979
PRINTED IN FRANCE